*Climate and Cultural Change
in Prehistoric Europe and the Near East*

THE INSTITUTE FOR EUROPEAN AND MEDITERRANEAN ARCHAEOLOGY
DISTINGUISHED MONOGRAPH SERIES

Peter F. Biehl, Sarunas Milisauskas, and Stephen L. Dyson, editors

The Magdalenian Household: Unraveling Domesticity
Ezra Zubrow, Françoise Audouze, and James G. Enloe, editors

Eventful Archaeologies: New Approaches to Social Transformation in the Archaeological Record
Douglas J. Bolender, editor

The Archaeology of Violence: Interdisciplinary Approaches
Sarah Ralph, editor

Approaching Monumentality in Archaeology
James. F. Osborne, editor

The Archaeology of Childhood: Interdisciplinary Perspectives on an Archaeological Enigma
Güner Coşkunsu, editor

Diversity of Sacrifice: Form and Function of Sacrificial Practices in the Ancient World and Beyond
Carrie Ann Murray, editor

Climate and Cultural Change in Prehistoric Europe and the Near East
Peter F. Biehl and Olivier P. Nieuwenhuyse, editors

CLIMATE AND CULTURAL CHANGE IN PREHISTORIC EUROPE AND THE NEAR EAST

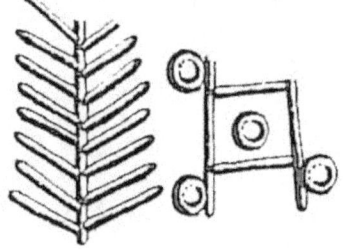

IEMA Proceedings, Volume 6

EDITED BY
Peter F. Biehl and
Olivier P. Nieuwenhuyse

STATE UNIVERSITY OF
NEW YORK PRESS

Logo and cover/interior art: A vessel with wagon motifs from Bronocice,
Poland, 3400 B.C. Courtesy of Sarunas Milisauskas and Janusz Kruk,
1982, Die Wagendarstellung auf einem Trichterbecher au Bronocice,
Polen, *Archäologisches Korrespondenzblatt* 12: 141–144

Published by
State University of New York Press, Albany

© 2016 State University of New York

All rights reserved

Printed in the United States of America

No part of this book may be used or reproduced in any manner whatsoever without written permission. No part of this book may be stored in a retrieval system or transmitted in any form or by any means including electronic, electrostatic, magnetic tape, mechanical, photocopying, recording, or otherwise without the prior permission in writing of the publisher.

For information, contact
State University of New York Press, Albany, NY
www.sunypress.edu

Production, Eileen Nizer
Marketing, Michael Campochiaro

Library of Congress Cataloging-in-Publication Data

Names: Biehl, Peter F., editor. | Nieuwenhuyse, Olivier, editor. | European Association of Archaeologists. Annual Meeting (16th : 2010 : Hague, Netherlands), sponsoring body. | 8.2 ka Climate Event and Archaeology in the Ancient Near East (Conference) (2010 : Leiden, Netherlands), sponsoring body
Title: Climate and cultural change in prehistoric Europe and the Near East / edited by Peter F. Biehl and Olivier P. Nieuwenhuyse.
Description: Albany : State University of New York Press, 2016. | Series: Institute for European and Mediterranean Archaeology distinguished monograph series | "The chapters of this book arose from two symposia on the archaeology of climate change: The 8.2 ka Climate Event and Archaeology in the Ancient Near East (Leiden, March 19, 2010), and Climate and Cultural Change in Prehistoric Europe and the Near East, (European Association of Archaeologists annual meeting in The Hague, 2010)."—Acknowledgements. | Includes bibliographical references and index.
Identifiers: LCCN 2015042619 (print) | LCCN 2016003288 (ebook) | ISBN 9781438461830 (hardcover : alk. paper) | ISBN 9781438461823 (pbk. : alk. paper) | ISBN 9781438461847 (e-book)
Subjects: LCSH: Paleoclimatology—Holocene—Congresses. | Paleoecology—Holocene—Congresses. | Prehistoric peoples. | Human beings—Effect of climate on.
Classification: LCC QC884 .C568 2016 (print) | LCC QC884 (ebook) | DDC 939.4/01—dc23
LC record available at http://lccn.loc.gov/2015042619

10 9 8 7 6 5 4 3 2 1

Contents

ILLUSTRATIONS ix

ACKNOWLEDGMENTS xix

INTRODUCTION *Olivier P. Nieuwenhuyse, Peter F. Biehl*
Climate and Culture Change in Archaeology 1

PART I
NEAR EAST

CHAPTER ONE *Mauro Cremaschi, Andrea Zerboni*
The Oasis of Palmyra in Prehistory: Late Pleistocene and
Early Holocene Paleoclimate and Human Occupation
in the Region of Palmyra/Tadmor (Central Syria) 13

CHAPTER TWO *Mandy Mottram*
When the Going Gets Tough: Risk Minimization Responses to the
8.2 ka Event in the Near East and Their Role in Emergence of the
Halaf Cultural Phenomenon 37

CHAPTER THREE *Olivier P. Nieuwenhuyse, Peter Akkermans,*
The 8.2 Event in Upper Mesopotamia: *Johannes van der Plicht, Anna Russell,*
Climate and Cultural Change *Akemi Kaneda*
67

CHAPTER FOUR *Patrick T. Willett, Ingmar Franz,*
The Aftermath of the 8.2 Event: Cultural and *Ceren Kabukcu, David Orton,*
Environmental Effects in the Anatolian Late *Jana Rogasch, Elizabeth Stroud,*
Neolithic and Early Chalcolithic *Eva Rosenstock, Peter F. Biehl*
95

CHAPTER FIVE — Philippa Ryan, Arlene Rosen
Managing Risk through Diversification in Plant Exploitation during the Seventh Millennium B.C.: The Phytolith Record at Çatalhöyük — 117

CHAPTER SIX — Bleda S. Düring
The 8.2 Event and the Neolithic Expansion in Western Anatolia — 135

PART II
EUROPE

CHAPTER SEVEN — Odile Daune-Le Brun, Alain Le Brun
"Singing in the Rain": Khirokitia (Cyprus) in the Second Half of the Seventh Millennium cal B.C. — 153

CHAPTER EIGHT — Catherine Perlès
Early Holocene Climatic Fluctuations and Human Responses in Greece — 169

CHAPTER NINE — Clive Bonsall, Mark Macklin, Adina Boroneanț, Catriona Pickard, László Bartosiewicz, Gordon Cook, Thomas Higham
Rapid Climate Change and Radiocarbon Discontinuities in the Mesolithic-Early Neolithic Settlement Record of the Iron Gates: Cause or Coincidence? — 195

CHAPTER TEN — Detlef Gronenborn
Climate Fluctuations, Human Migrations, and the Spread of Farming in Western Eurasia: Refining the Argument — 211

CHAPTER ELEVEN — Andrzej Pelisiak
Economic and Social Changes and Climate between 3200 and 2500 B.C.: Late Neolithic Transformations in Southeastern Poland — 237

CHAPTER TWELVE — Daniel Löwenborg, Thomas Eriksson
Climate and the Definition of Archaeological Periods in Sweden — 257

Part III
Commentary

CHAPTER THIRTEEN   *Ezra B. W. Zubrow*
Epilogue to a Prologue: The Changing Climate of the
Past, Present, and Future   279

CONTRIBUTORS   293

INDEX   295

Illustrations

Figures

Figure 1.1   Left, Landsat satellite imagery of the region surrounding Palmyra/Tadmor (central Syria) indicating the name of the main localities cited in the text; the inset illustrates the position of the study area in a regional context. Right, geomorphological sketch of the area described in the text. Key: 1, pre-Quaternary bedrock (limestone); 2, Quaternary deposits; 3, Pleistocene lake deposits; 4, delta system; 5, sabkhat; 6, erosional streams.   15

Figure 1.2   Ikonos satellite imagery indicating the distribution of Upper Pleistocene and Early Holocene lake (1) and aeolian (2) deposits in the sabkhat area, the archaeological sites (dots), and the main localities cited in the text (triangles: 1, tell Site 288; 2, wadi Aid section; 3, Site 250; 4, Site 389) (modified from Cremaschi and Zerboni 2012).   16

Figure 1.3   Stratigraphic section at wadi Aid, illustrating the results of sedimentological analyses (grain size, content of carbonate equivalent, and organic carbon). Black triangles indicate the locations of flint artifacts (a, lamellar débitage; b, Mousterian), dots indicate the position of sampling points; the dated layer and results are also indicated. Note in the cumulative grain size distributions from the upper part of the section (sampling points 2 and 3) a clear input of aeolian sand (for the interpretation of sedimentological data see: Krumbein and Sloss 1963; Kukal, 1971).   17

Figure 1.4   Geomorphological features at the southern margin of the Sabkhat al Mouh. (a) Late Pleistocene/Early Holocene lacustrine terraces at the southern margin of the sabkhat. (b) A gypsum-sand dune at the southern margin of the sabkhat. Cross-sections indicating the

x   ILLUSTRATIONS

	relationships between sand dunes (c) and lacustrine terraces (d) at the southern margin of the sabkhat (modified from Cremaschi and Zerboni 2012); the distribution of sites and the result of an AMS-^{14}C dating are also reported. 19
Figure 1.5	Some findings from the archaeological sites at the margin of the Sabkhat al Mouh dating to the Late Epipaleolithic and the Pre-Pottery Neolithic. 1: burin and backed point; 2: end scraper on blade; 3: rectangular geometric; 4: truncated and backed point; 5: partly backed point; 6: backed point; 7, 8: perforators; 9, 10: Khiam points; 11: Byblos point (recto/verso). 20
Figure 1.6	Geomorphological map of Abu Fawares area, diamonds indicated the position of Epipaleolithic and Neolithic archaeological sites; a geological cross section is also reported (modified from Cremaschi and Zerboni 2012). Key: 1, limestone outcrop; 2, slope deposits; 3, pediment; 4, windblown depression cut into Pleistocene alluvial sediments; 5, Holocene lake sediments; 6, archaeological mounds; 7, archaeological sites; 8, main wadis. 22
Figure 1.7	Flint artifacts from archaeological sites in the Abu Fawares area. 1: backed pointed bladelet; 2–4: backed blade and bladelet associated with truncations; 5: end scraper; 6: burin on a fracture, opposite to end scraper; 7: burins; 8: frontal end scraper on a flake; 9, 10: pedunculated backed points on bladelet; 11: perforator on a backed bladelet; 12, 13: lunates; 14: pedunculated flake; 15: perforator; 16: backed point on backed bladelet; 17: Khiam (?) point (recto/verso); 18–20: Byblos points (recto/verso). 23
Figure 1.8	(a) General view of the Neolithic tell at Site 288 (the present day oasis is in the background); the arrow indicates the position of the main stratigraphic section exposed in a quarry. (b) Stratigraphic section of the tell; and (c) a detail of the mud-bricks. 25
Figure 1.9	Some archaeological findings from the tell at Site 288. 1, 2: truncations; 3–6: Byblos points; 7: obsidian blade fragment; 8: White Ware fragment with basket impression; 9, 10: red slipped pottery fragments. 26
Figure 1.10	Cross-section of the study region (not to scale) reconstructing the main stratigraphic and landscape units described in the text and the chronological distribution of archaeological sites in each unit (modified from Cremaschi and Zerboni 2012). Key: 1, bedrock; 2, Pleistocene lake sediments; 3, Holocene lake sediments; 4, fluvial gravel; 5, lake terrace; 6, gypsum-sand dune; 7, sabkhat; 8, spring deposit; 9, oasis; 10, tell; GK, Geometric Kebaran; NT, Natufian; PPNA, Pre-Pottery Neolithic A; PPNB, Pre-Pottery Neolithic B; PPNBL, late Pre-Pottery Neolithic B; PN, Pottery Neolithic; CH, Chalcolithic. 29

ILLUSTRATIONS XI

Figure 2.1 Relative chronology for sites mentioned in the text showing current culture-historical phasing against the broader period of aridity associated with the 8.2 ka abrupt climate event (shown in darker grey). 39

Figure 2.2 Map of western Syria and Upper Mesopotamia showing mean annual rainfall and location of sites mentioned in the text. 42

Figure 2.3 Depictions on Early Halaf pottery of raised granaries/storage structures. Nos. 1–2, 4, 10: Sabi Abyad (after Akkermans 1989; Le Mière and Nieuwenhuyse 1996); 3: Fıstıklı Höyük (after Bernbeck et al. 2003); 5–6: Tell Halula (Mottram 2010); 7: Sakçe Gözü (after Garstang et al. 1937); 8–9: Arpachiyah (after Hijara 1978; Mallowan and Rose 1935); 11: Domuztepe (after Kansa et al. 2009). 49

Figure 2.4 Modern ethnographic examples of raised granaries and staple food stores analogous to those depicted on Early Halaf Pottery; 1: Toba Batak, Sumatra, rice granary (after Domenig 1980); 2: Trobriand *liku* yam house; 3: Baduy, Java, rice granary (after Schefold 2003); 4: Iron Age petroglyph of probable raised granary, Valcamonica, Italy (after Bradley 2005). 51

Figure 2.5 Depictions on Halaf pottery of probable habitation structures (Nos. 1–8) shown against analogous structures from southern Africa (Nos. 9–10) comprising both mudbrick and woven grass-and-branch huts; 1–2, 4, 8: Tell Halaf (after Oppenheim and Schmidt 1943); 5: Sakçe Gözü (after Garstang et al. 1937); 6: Yarim Tepe II (after Merpert et al. 1978); 3, 7: Arpachiyah (after Mallowan and Rose 1935); 9–10: Botswana (after Kent and Herbich 1989). 56

Figure 3.1 Map of Tell Sabi Abyad, showing the locations of areas of work (Operations). Solid line: the location of the section west-east shown in Figure 3.2. 71

Figure 3.2 Schematic section through the excavations at Operation III at Tell Sabi Abyad. Sequences A and B are discussed in this paper. 72

Figure 3.3 The stratigraphy and absolute date of the prehistoric sequence excavated at Tell Sabi Abyad, showing the relationships between Operations I to V. Grey shade indicates the duration of the 8.2 ka event as observed in the Greenland ice cores (after van der Plicht et al. 2011). 72

Figure 3.4 Distribution of Late Neolithic sites in the Balikh valley inhabited prior to, during and/or after the 8.2 ka climate event (University of Amsterdam survey). 75

Figure 3.5 Tell Sabi Abyad, Operation III. Proportions of the main livestock species by level groupings (animal exploitation phases) (after Astruc and Russell 2013). 77

Figure 3.6 Tell Sabi Abyad, Operation III. Stylized images of the bull, so-called bucrania, painted on Halaf Fine Ware serving vessels known as "cream bowls." 82

Figure 3.7 Tell Sabi Abyad, Operation III. Large storage vessels from level A1. 83

Figure 3.8 Tell Sabi Abyad, Operation III. The rising proportion of decorated ceramics from level A1 on. 84

Figure 3.9 Tell Sabi Abyad, Operation I. Transitional period painted serving vessel "Samarra style" (after Akkermans 1993; Nieuwenhuyse 2007). 86

Figure 4.1 Topographic map of Çatalhöyük—Konya, Turkey. 96

Figure 4.2 Pottery forms, decoration techniques, and firing temperatures at Çatalhöyük. 99

Figure 4.3 Excavation plan of Çatalhöyük West Mound Trench 5. 101

Figure 4.4 Diachronic change at Çatalhöyük from the late seventh to sixth millennium. 1Clockwise from top left: "classic" East Mount pottery, staple domesticates of naked barley (preferred), sheep/goat, and wheat. 2Clockwise from top center: typical West Mound painted pottery, staple domesticates of hulled barley (preferred), sheep/goat, cattle, and wheat. 109

Figure 5.1 Histogram showing phytoliths from storage bins (East Mound). 121

Figure 5.2 Histogram showing *Hordeum* sp. (barley) husk silica skeletons over time (East Mound). 123

Figure 5.3 Image of *Hordeum* sp. (barley) silica skeleton (East Mound), scale 100 microns. 123

Figure 5.4 Histogram showing *Phragmites* sp. (common reed) phytoliths over time (East Mound). 126

Figure 5.5 Image of *Phragmites* sp. (common reed) multicell phytolith, scale 100 microns (East Mound). 127

Figure 6.1 Neolithic sites of Anatolia. 1—Hoca Çesme; 2—Yarımburgaz and Yenikapı; 3—Fikirtepe; 4—Pendik; 5—Aktopraklik; 6—Ilipnar; 7—Mentese; 8—Barcin Höyük; 9—Ege Gübre; 10—Ulucak; 11—Dedecik-Heybelitepe; 12—Hacılar; 13—Kuruçay; 14—Bademağacı; 15—Höyücek; 16—Erbaba; 17—Çatalhöyük East and West; 18—Boncuklu Höyük; 19—Pınarbaşı; 20—Canhasan; 21—Aşıklı Höyük; 22—Kaletepe; 23—Mersin-Yumuktepe; 24—Knossos; 25—Mylouthkia; 26—Shillourakambos. 140

Figure 6.2 Cumulative radiocarbon plot for the Lake District, Aegean Anatolia, and the Marmara Region (n=135, data obtained from CONTEXT (http://context-database.uni-koeln.de/) database [Böhner and Schyle 2006] augmented with data from Özdoğan and Başgelen 2007). 141

ILLUSTRATIONS   XIII

Figure 7.1   Map of Cyprus showing the locations of the Late Aceramic Neolithic sites belonging to the "Khirokitia Culture" (7th–6th millenium B.C.).   154

Figure 7.2   Khirokitia. General view of the southern hillside (photo Th. Saggory) with the two successive walls enclosing the Neolithic village (Photos Mission Archéologique Française).   155

Figure 7.3   Khirokitia. The early village (levels J-B).   156

Figure 7.4   Khirokitia. The recent village after the shift (levels III and A).   157

Figure 7.5   Khirokitia. The Potamos sequence.   158

Figure 7.6   Khirokitia. View on the Potamos sequence. Levels P10–P4 (Photo Mission Archéologique Française).   159

Figure 7.7   Khirokitia. Temporal correlations between the excavated Reference sequence and the Potamos sequence.   160

Figure 7.8   Khirokitia. Level II. Abandonment of the northern slope.   163

Figure 7.9   Khirokitia. Level I. New constructions built over the remains of the enclosure wall 284.   165

Figure 8.1   Benthic foraminiferal Oxygen Index (OI) for eastern Mediterranean sediment cores along an NW-SE transect. The sapropel S1 is characterized by depleted OI values, interrupted by a sharp rise during the 8.2 event (after Schmiedl et al. 2010). Copyright Elsevier; used with permission.   172

Figure 8.2   Location of the cores discussed in the text (background map by G. Monthel).   173

Figure 8.3   Location of the sites and archaeological regions discussed in the text (background map by G. Monthel).   175

Figure 8.4   Dark line: relative abundance (%) of the cool-water dinoflagellate *Spiniferes elongatus* in core SL21 with respect to a gonyaulacoid-only dinocyst sum. Shaded area (with reverse scale): relative abundances (%) of warm-water planktonic foraminifera from core LC21. The light grey bands represent the 8.2 event. After Marino el al. 2009. Copyright Elsevier; used with permission.   181

Figure 8.5   Broad-leaved tree pollen percentage, (top) and index of evergreen *Quercus*/evergreen Quercus + deciduous *Quercus*) in core SL 152 (after Kotthoff et al. 2008). Copyright Elsevier, used with permission.   181

Figure 8.6   Simplified pollen diagram of Tenaghi Philippon (after Peyron et al. 2011). Copyright Peyron et al., used with permission.   182

Figure 8.7   Early Neolithic sites abandoned during the Middle Neolithic. The black oval circumscribes the old Voivi Lake. Figures for each site refers to Gallis 1992. CAD S. Ménard.   184

Figure 9.1 Mesolithic and Early Neolithic sites in the Iron Gates. Virtually all sites along the main trunk of the Danube were located immediately adjacent to the river and would have been vulnerable (in whole or part) to inundation at times of unusually high river flows (4 m+ floods). Named sites have ^{14}C dates that were used to generate the summed probability distributions in Figures 9.2–9.4.   196

Figure 9.2 Cumulative calibrated dating probability of radiocarbon data from Mesolithic and Early Neolithic sites along the Danube main channel in the Iron Gates. Datasets: A—all usable radiometric and AMS dates; B—AMS dates on terrestrial and human bone samples; C—AMS dates on terrestrial animal and plant samples. Calibrations performed with *CalPal* (29 May 2007) [http://www.calpal.de] and the IntCal04 dataset. ^{14}C data from Bonsall 2008; Bonsall et al. 1997, 2008, 2012, 2015, unpublished; Borić and Miracle 2004; Borić and Dimitrijević 2009; Borić 2011; Borić and Price 2013; Dinu et al. 2007. Prior to calibration human bone ^{14}C ages were corrected for the "freshwater reservoir effect" using Method 1 of Cook et al. (2002), assuming δ^{15}N endpoint values for purely terrestrial and purely aquatic diets of +8.3‰ and +17.0‰, respectively (cf. Cook et al. 2009).   197

Figure 9.3 Cumulative calibrated dating probability of radiocarbon data (terrestrial series) from Mesolithic and Early Neolithic sites along the Danube main channel in the Iron Gates, compared to climate proxy records from the North Atlantic and Europe 10–6 ka cal B.P. A—after Bond et al. (1997); B, D, F, G—smoothed records redrawn from Rohling and Pälike (2005); E—after Siani et al. (2013); C—horizontal bars represent cold phases recorded in δ^{18}O records from V11 Cave, NW Romania (Tămaş et al. 2005) and Katerloch Cave, Austria (Boch et al. 2009); vertical grey bars represent higher lake-level events in the Alps–Jura region (Magny 2004).   199

Figure 9.4 Summed calibrated probability distribution of radiocarbon dates from the Iron Gates for the period 12–7 ka B.P. (A, black, n = 143), superimposed on a summed probability distribution of radiocarbon dates simulated for calendar dates spaced every 25 years (B, grey, n = 213). Summed probability distributions produced with OxCal 4.2.4 (Bronk Ramsey 2009), using IntCal13 (Reimer et al. 2013) and ±25yr standard errors.   201

Figure 9.5 The Danube catchment showing the Iron Gates and key localities with climate proxy records for the Early to Middle Holocene: 1—Hölloch Cave (Wurth et al. 2004); 2—Katerloch Cave (Boch et al. 2009); 3—V11 Cave (Tămaş et al. 2007); 4—Lake Ammersee (von Grafenstein et al. 1998, 1999); 5—Lake Schleinsee (Tinner and Lotter 2001); 6—Lake Soppensee (Tinner and Lotter 2001); 7—Brunnboden and Krummgampen peat bogs (Kofler et al. 2005); 8—Preluca Tiganului

ILLUSTRATIONS   XV

and Steregoiu peat bogs (Feurdean 2005); 9—Alps-Jura lakes study region (Magny 2004); 10—Teleorman Valley (Macklin et al. 2011); 11—Durance Valley (Miramont et al. 2001); 12—Middle Rhône Valley (Berger et al. 2002); 13—South Adriatic Sea, core MD90-917 (Siani et al. 2013).   203

Figure 10.1   Comprehensive map of the spread of farming in western Eurasia; immediate study area is outlined by box (modified after Gronenborn 2010).   213

Figure 10.2   Paleoclimatic proxy data with global, North Atlantic, and Central European significance. Solar insolation: Berger and Loutre 1991; ^{14}C production rate: Kromer and Friedrich 2007; IRD events: Bond et al. 2001; NGRIP: Vinther et al. 2006; Labrador shelf freshwater forcing: Jennings et al. 2015; Ammersee: von Grafenstein et al. 1999; timberline eastern Alps: Nicolussi et al. 2005; Cold events Alps: Haas et al. 1998).   215

Figure 10.3   Simplified chronological table of the earliest appearance of domesticated cereals in selected regions across western Eurasia. PBO—Preboreal Oscillation; IRD—Ice Rafting detritus events, TBK—Funnel Beaker Culture; MK—Michelsberg Culture; LBK—Linear Pottery Culture; CHALC—Chalcolithic; PN—Pottery Neolithic; PPN—Pre-Pottery Neolithic (sources: NGRIP—Vinther et al. 2006; Rasmussen et al. 2007; ^{14}C production—Kromer and Friedrich 2007; Lang 1994; Stahm 2010; Böhner and Schyle 2006; Colledge et al. 2004; Willcox 2005; Willcox et al. 2009; Jacomet online; Reingruber 2011).   215

Figure 10.4   Adaptive cycles from Resilience Theory (modified after Holling and Gunderson [2002] and Bub [2011]): (a) basic build-up of adaptive cycles; (b) nested cycles in time with threshold values; (c) cycles and climate fluctuations.   217

Figure 10.5   Paleoclimatic proxy-data for period of IRD 5b and archaeological chronology of LBK. IRD events: Bond et al. 2001; Cold events Alps: Haas et al. 1998; Main oak deposition rate: Spurk et al. 2002; Main oak ring-width index: Spurk personal communication 2004; archaeological chronology: Gronenborn et al. 2013.   221

Figure 10.6   Paleoclimatic proxy-data for the end of IRD 5b and archaeologically dated events toward the termination of LBK in the Rhineland and Dutch Limburg. $\delta^{13}C$ Kückhoven: Helle and Schleser 1989; tree ring-width Kückhoven: Schmidt et al. 1998; ^{14}C production rate: Kromer and Friedrich 2007; house numbers Rheinland/Maas region (dASIS data bank RGZM).   222

Figure 10.7   Coupled climate and population dynamics model for early farming societies in western Eurasia. Shaded areas denote periods of adverse climate.   226

Figure 11.1 Map of southeastern Poland showing the area of research. 238

Figure 11.2 Map of southeastern Poland showing the locations of pollen diagrams discussed in the text. 239

Figure 11.3 The distribution of Funnel Beaker culture sites. Large dots: settlement sites; small dots: single finds of stone and chipped artifacts. 241

Figure 11.4 The distribution of Corded Ware culture sites. Vertical lines: barrows; small dots: single finds; larger dots: camp sites. 242

Figure 11.5 The distribution of Early Bronze Age Mierzanowice culture sites. 243

Figure 11.6 The distribution of Neolithic sites without cultural affiliation. 245

Figure 11.7 The distribution of Neolithic or Early Bronze Age sites without cultural affiliation. 245

Figure 11.8 Contemporary forest Southeast Poland in winter showing tree crowns covered by a thick, heavy icy coat. 247

Figure 11.9 Contemporary forest in Southeast Poland after a fierce winter when the tree crowns became too heavy, resulting in the collapse of wide areas of forest. Natural forces such as icing may result in the destruction and near deforestation of many hectares of woodland. 248

Figure 11.10 Contemporary park-type landscape in Southeast Poland. Late Neolithic peoples may have created very similar open space by subjecting broken and dried tree crowns and trunks to fire. 249

Figure 12.1 Map of northern Europe showing the location of the main research area in central Sweden, the Mälaren Basin (square) and the location of sites mentioned in the text. 259

Figure 12.2 Combined calibrations of the 14C dates from Uppland and Västmanland. Calibration made by IntCal04.14C (Reimer et al 2004). The result is compared to the number of sites with bronze hoards in the same area as well as climatic proxy-data from lakes in Northern Sweden (Grudd et al 2002; Gunnarsson 2008). There are clear overlaps with periods with cold and humid climate and low numbers of 14C-datings. 260

Figure 12.3 Number of uncalibrated time spans of 14C-datings from the northern part of Lake Mälaren Basin (staples). The diminishing numbers around 800 B.C. (ca 2650 B.P.), 400 B.C. (ca 2450 B.P.) and after A.D. 500 (ca 1500 B.P.) are clearly visible. The correspondence between the peak of sites with hoards, numbers of bronzes in hoards, and the end of the Bronze Age is also obvious. 261

Figure 12.4 The hoard from Spelvik, Spelvik parish in Södermanland (SHM inv. 813) The hoard consists of eleven intact neck-rings and fragments of more rings, two celts, two spears, and a hanging-bowl. It can be dated ca. 900–700 B.C. Photo Statens Historiska Museum © Historiska museet. 263

Figure 12.5 Some of the gold bracteates from the Söderby hoard, Danmark parish in Uppland. The hoard consists of ten bracteats, gold bullion for sword-beads, and other, small gold objects (SHM inv. 5802 & 33022). It was deposited south of Uppsala during the first half of the sixth century A.D. The décor on the bracteates consists of geometrical symbols and a male figure that can be interpreted as the god Odin and his two ravens. (Lamm et al 1999). Photo Ulf Bruxe; Statens Historiska Museum © Historiska museet. 264

Figure 12.6 Illustration of the chronology of sites from a recent large archaeological project in central Sweden. Lines represent burial grounds (black) and nearby settlements (grey) (after Wikborg 2007:179). 269

Figure 13.1 The general locations of ICAP field sites (Kamchatka Russia, Old Factory Lake Canada, Yli-ii Finland). 283

Figure 13.2 Simulated Temperatures in Centigrade from 7000 to 3900 BP from Finland, Canada, and Kamchatka at the locations of ICAP field sites. 284

Figure 13.3 The distribution of Betula, Pincea, and Picea over time from Finnish sites north of 64 degrees. 284

Figure 13.4 Trend analyses of Betula (birch), Pincea (pine), and Picea (spruce) pollen from northern Finland sites between 64 degrees to 67 degrees North. 285

Figure 13.5 General North American Air Masses Patterns (mPMaritime Polar air masses, cPContinental Polar air masses). Linking Holocene climate records to the AMO index (top right). The AMO index (black) and the instrumental precipitation record from the Yucatan peninsula (green) is shown together with the coral based δ18O record from Puerto Rico14 (red) and the δ18O record from lake Chichancanab29 (dashed line). (Bottom left) Location of the climate proxy records (dots) and schematic overview of the major atmospheric systems (bottom right). Modern atmospheric systems of the North Atlantic region. This scenario represents average conditions for the last 3,000 years BP, where an overall 'neoglacial' regime with more frequent meridional atmospheric circulation patterns and an ITCZ located close to the Cariaco site was prevalent during Northern Hemisphere summer. The arrows indicate the dominant wind directions. 287

Figure 13.6 The Mediterranean seesaw. 291

Tables

Table 3.1 Tell Sabi Abyad. The results of the Bayesian analysis: date ranges for Operation III, Sequence A and B. 73

Table 3.2 University of Amsterdam survey in the Balikh Valley. Sites listed with occupation in the Early Pottery Neolithic (pre-event), Pre-Halaf (8.2

	ka. event) and Transitional to Early Halaf period (post-event), showing estimated maximum site extent. 76
Table 8.1	Composition of the seed assemblages at Franchthi Cave, from the Bölling to the Preboreal (data after Hansen 1991). 176
Table 8.2	Composition of the marine molluscan assemblages from the Lower Mesolithic of trench FAS (data after Shackleton 1988). 178
Table 8.3	Relative proportion of the domesticated faunal species in Early and Middle Neolithic settlements. 186
Table 13.1	Predicted values for the percentage of pollen by taxa with and without noise for 50 years, 100 years, and 150 years in the future. 286
Table 13.2	Each chapter of this volume classified by title, author, issue, time, area, culture and methodology. 289

Acknowledgments

The chapters of this books arose from two symposia on the archaeology of climate change: The 8.2 ka Climate Event and Archaeology in the Ancient Near East (Leiden, March 19, 2010), and Climate and Cultural Change in Prehistoric Europe and the Near East, (European Association of Archaeologists annual meeting in The Hague, 2010). We would like to thank the organizers of the European Association of Archaeologists (EAA) for accepting our session, and especially Willem Willems who recently unexpectedly passed, and the University at Leiden for hosting our workshop. The Leiden workshop was made possible by the Faculty of Archaeology of Leiden University, the Netherlands Organization for Scientific Research (NWO; dossier 360-62-040), and Archaeological Research and Consultancy (ARC; many thanks, Jan Schoneveld!). We would also like to thank Thomas Harper, Hannah Quaintance, and Heather Rosch for their superb editorial assistance, copyediting, and production. Furthermore, we would like to express our gratitude to The Institute for European and Mediterranean Archaeology and the publisher of the Monograph Series at SUNY Press for the smooth production process. And finally, we would like to thank all contributors for their contributions, patience, and support in publishing this book.

Introduction

Climate and Culture Change in Archaeology

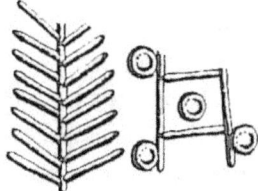

Olivier P. Nieuwenhuyse, Peter F. Biehl

This book examines how humans responded to climate change that occurred during the Holocene in Europe and the Near East. It grew out of two workshops on the archaeology of climate change organized in the Netherlands in 2010. One of these was organized with a specific past climate event and geographic focus in mind, investigating the socioeconomic repercussions of the so-called 8.2k abrupt climate event in the ancient Near East, while the other aimed at a broader range of topics. What both workshops held in common was an interest not merely in disseminating new data, insights, and tentative conclusions but rather in the ways we as archaeologists proceed to investigate the role of climate change as one of many causal factors in culture change.

The chapters scrutinize new archaeological and paleoenvironmental data in order to contextualize key climatic events such as the *8.2k cal B.P.*, *4.2k cal B.P.* and other events with cultural changes and transitions. One of the main threats for the authors is to discuss when, how, and if changes in climate and environment caused people to adapt, move, or perish. Contrary to perceptions of threatening global warming in our popular media, and in contrast to grim images of collapse presented in some archaeological discussions of past climate change, the papers in this book rather unanimously reject outright societal collapse as a likely outcome (cf. McAnany and Yoffee 2010, Schwartz 2006, Tainter 1988). Yet this does not keep them from considering climate change as a potential factor in explaining culture change. The authors in this book started from the view that when climate changes, societies may change by either adapting or transforming. Yet, shying away from simplistic, mono-causal explanations, they adopt a critical stance with regard to the long-standing practice of equating synchronicity with causality, and explicitly consider alternative explanations. The chapters are illustrated with case studies to analyze human responses to climatic events on a micro scale as well as on a macro scale.

Archaeology and Climate Change

Archaeologists have always underscored that how humans respond to changes in their natural environment plays a crucial part in the formation of society. In an era of unprecedented, threatening global warming and massive species extinctions, this message is clear even to a broader contemporary audience. From world leaders expressing their concerns to popular blockbusters such as *The Day After Tomorrow* to even the leader of the Roman Catholic church (Schiermeier 2015), there is a general sense that we are constrained by our environment even today, despite our tremendous capacity to adapt and bend nature to our will. Environmental reconstructions, including climate, have always formed part of larger narratives of long-term human development, and of who we are as a species. As many authors have observed, archaeology as a discipline is uniquely situated to contribute to current climate-versus-culture debates (Dann 2015:24; Danti 2010; Dawdy 2009; Roberts 2011). By arriving at closely contextualized understandings of the heterogeneous ways in which humans interacted with climate changes in the past, archaeologists may offer valuable alternative strategies for today's challenges.

With the rapid development of paleo-climatic sciences over the past few decades, archaeologists have become confronted with a wealth of new data unprecedented both quantitatively and qualitatively (Alley 2000; Birks 2008; Mayewski and White 2002). This poses challenges of a methodological character and of a more conceptual kind. In terms of archaeological method, it may be suggested that up until now much theorizing of climate change versus culture change in the past was based upon poor data sets. Just as climatologists have long felt the need to collect improved records, so archaeologists interested in these relationships feel compelled to do the same. As stressed in several chapters in this book, this brings a much more critical stance toward those earlier data sets: Do they *really* synchronize well with climate change? And, have we *really* been using the most relevant cultural proxies?

Conceptually, as Ur (2015:69–70) and many others have pointed out, models for human-environment interaction tend to fall along a continuum. One extreme position holds that the environment is seen as the primary determining factor causing social change or even collapse. At the other end are those explanations that see human societies as entirely independent from their environments, with social change emerging only from shifting social relationships between individuals and groups. As Ur notes (2015:70), such interpretations have their actors "exist in an abstract world without physical environment." Human societies are seen as existing entirely independent of their environments (Willet et al., this vol.).

As with current global warming itself, neither extreme would appear to be sustainable in the long run. Most scholars today would rather opt for some intermediate position, arguing for a nonexclusive causal role for culturally mediated environmental factors (Oldfield 2008; Rosen and Rosen 2001; Rosen 2007). Climatic factors may constrain human decision making, occasionally in quite extreme ways, but they do not fully determine specific responses and adaptations (McIntosh et al. 2000).

Archaeological scholarship as reflected in this volume is diverse in its paradigmatic leanings. Some authors are more inclined toward an explicitly processual perspective (e.g., Gronenborn, this vol.) or they are more outspoken in their preference for climate as primary causal factor instigating cultural innovations (e.g., Mottram, this vol.). But they would all subscribe to the view that climate change is one among many factors, and that archaeologists should strive to evaluate multiple, alternative explanations. For instance, Perlès (this vol.) is explicit in searching for nonclimate alternatives: "This is not to deny, evidently, that the climatic changes I have considered had no effect whatsoever on human populations in Greece. Before such a conclusion could be reached, however, a number of theoretical and methodological problems would first have to be solved."

As Zubrow points out (this vol.), climate anomalies may also yield effects that do not so much disturb a system as exacerbate already ongoing trends (see also Nieuwenhuyse et al., this vol.). The classic systems view, as elaborated by Gronenborn (this vol.), portrays a dynamic equilibrium as starting and ending points. In reality, systems and cultures are always changing. The assumption that stability was a natural condition seems to be misleading, as well as that some external factor—climate—is needed to bring about culture change and reorganization, and new stability. Taking the systems approach very literally would suggest that human societies were passive and climate active—in short, a very one-sided relationship. Today, we see societies changing climate, and archaeology offers good examples from the past. Thus, the relationship is mutual, dialectic. What this means is that climate narratives based on a too-strict reading of the systems perspective neglect the importance of cultural *context*, the insight that existing social structures and worldviews affect precisely *how* human groups respond to climate change.

Finally, archaeologists sometimes stress that adapting to climate changes does not only need to be a dramatic tale of struggling for survival. Even abrupt climate change does not need to have negative impacts, and severe events such as the 8.2 event may even have had positive impacts. For Upper Mesopotamia, Nieuwenhuyse et al. (this vol.), and perhaps also Mottram (this vol.), argue that the event may have pushed people toward taking the final steps to a very successful farming-herding strategy, which would continue to dominate the cultural landscape for centuries after the event. For Greece, Perlès (this vol.) asks us to consider that, against the background of a typically dry Mediterranean climate, the cooler and wetter summers brought by the 8.2 event provided better conditions for farming. In her view, they were a treat that all too briefly benefited human beings, animals, and cultivars alike.

The Issue of Synchronicity

The issue of synchronicity is key in each of the chapters in this book, as it is in climate archaeology in general. At first sight this seems pretty straightforward. As Neil Roberts observes (2015:30–31), accurate dating plays a key role in equating environmental proxies with cultural sequences, as causal relations are often inferred from the timing between events. Simply, because societal consequences cannot have preceded a purported environmental cause, this logic is generally used to falsify or verify a climate hypothesis (Sandweiss and Quilter 2012).

On closer consideration, the relationship is far less easy than is often acknowledged. Methodologically, formidable obstacles are still to be overcome to synchronize archaeological cultural sequences with paleoclimate records. Typical for the culture-historical paradigm that dominated archaeology until very recently, and in many parts of the world still reigns superior, long-term cultural change was often conceived of in the form of a static succession of cultural entities. The climatologist's job then may simply seem to be to seek synchronization between some climate event and some wholesale culture-historical shift. However, as absolute dating frameworks are often still rather poor for many prehistoric archaeological contexts globally, it remains all too easy to slide important cultural boundaries up and down the chronological ladder so as to allow perfect synchronization to emerge (Coombes and Barber 2005; Maher et al. 2011:19).

Settlement data, for instance, are key to understanding regional responses, yet they are often problematic as to their synchronicity with climate change. To some degree this is a general methodological issue found in many parts of the world. The issue is that the establishment of synchronicity in settlement data depends on the nature of ceramic style change: if this is very slow, before/during/after some climate event becomes invisible in settlement patterns. If it is very quick and very conspicuous, before/during/after the event can be isolated. Case studies for the latter scenario include the Upper Mesopotamian steppes and Central Anatolia (Düring, this vol.; Nieuwenhuyse et al., this vol.). Which scenario applies to a particular climate-culture discussion should be assessed carefully for each specific case.

Then there's the issue of scale, both diachronically and spatially. Paleoclimate proxies of global extent may be mute when it comes to understanding concrete human affairs in the past. As Daune-Le Brun and Le Brun (this vol.) emphasize, climate changes should be investigated at a local level as well, not only at a supralocal or even global level. Local effects may be very diverse and different from global patterns, but it is these localized effects that impacted prehistoric communities. Thus, assessing the effects of the 8.2 climate event on patterns of plant exploitation in Central Anatolia, Ryan and Rosen (this vol.) argue that the lack of independent information about how the Konya region was *locally* affected makes it difficult to assess the potential impact of climate as distinct from other social and ecological factors. Similarly, cultural adaptations to the same climate event should not a priori be expected to play out similarly across larger regions.

Conceptually, the processual assumption that societal innovations are primarily the *consequence* of environmental *causes* is seen by many archaeologists as problematic, or at least as potentially too simplistic. Archaeologists today move beyond earlier deterministic approaches: societies do not simply roll and flow with the environmental tide. Our own situation today provides a valuable ethnographic example testifying to the importance of cultural factors mediating responses to climate change: will it be "abandoning fossil fuels" or "business as usual"? Whether our own societies of today will choose to fail or survive (Diamond 2005) in the near future will ultimately depend on contemporary cultural perceptions of what climate is, on extant technologies, and, above all, on asymmetrical relations of power.

Apart from the lurking determinism, of course, the relationship may equally work the other way. Again looking at our modern world, a growing majority of people now accept that socioeconomic innovations *preceded* environmental changes, even *caused* them. In this regard, we may hardly be unique. In the past, as Zubrow reminds us (this vol.), there may have been several instances of humans instigating climate changes already in the Early Holocene (Oldfield 2008). The Near East and the Mediterranean, regions that yielded most of the contributions in this book, are often seen as environmentally marginal, hence vulnerable to even minor climatic perturbations. The other way round, however, anthropogenic factors such as deforestation, overgrazing, or intense exploitation of natural resources would have had noticeable effects on local environmental conditions, especially in these regions (Glantz 1994; Redman 1999).

Cultural ecology in its strictest sense should therefore not be the only paradigm adopted. Or rather, our reconstructions of social responses should consider how individual members and subgroups of past societies would have perceived environmental changes and how they would have adapted to them. A famous case in point is the cultural response of Norse groups in Greenland when encroaching arctic conditions gradually starved them out. Rather than adapting their subsistence, for instance by emulating the successfully advancing Inuit, the Norse attributed their ill fortune to divine wrath and erected more churches (Diamond 2005:248–76). Such social representations of climate, and ecology in general, are not very explicit in the various contributions to this book but implicitly surface in some of the papers.

As Roberts observes (2015:30), the danger of determinism, of simply equating synchronicity with causality, can only be avoided when looking at the evidence from a broad variety of perspectives. Bonsall et al. (this vol.) make this very clear: "Establishing a causal link between climatic events and archaeological phenomena requires a convincing mechanism for transmitting cause to effect." Apart from establishing synchronicity, as with all well-excavated and carefully documented cases discussed in this book, a direct cause-and-affect scenario remains difficult. We argue that this is characteristic of climate archaeology as a whole, but has been generally overlooked by people seeking perfect synchronization. Several authors in this book discuss situations where they believe the evidence was messy and confusing and inviting to alternative, nonclimatic explanations. It is only through multidisciplinary, contextualized analysis of social dynamics over multiple scales that we may start to identify the differential impact of climate change.

Climate Archaeology in European and Near Eastern Archaeology

As readers will have observed, our collection is not a homogeneous series discussing a single climatic event, a circumscribed region, or a specific cultural context. Nor do authors subscribe to a similar methodology or academic worldview. The papers differ vastly in geographic and temporal scale. This choice was deliberate. We wished to bring together contributors that collectively reflect the methodological and paradigmatic heterogeneity that is today characteristic of climate archaeology in Europe and the ancient Near East. As a result, conclusions drawn in individual papers are sometimes in opposition; this, too, is typical of climate archaeology. Rather than imposing our editorial verdict on any conclusions reached, we hope that papers will speak for themselves and stimulate much further debate.

Cremaschi and Zerboni provide in their opening chapter *The Oasis of Palmyra in Prehistory* a paradigmatic link to the "Oasis Archaeology" of earlier generations of archaeologists, influencing among others Childe's famous "Oasis Theory" for the origins of agriculture. Using the oasis of Palmyra in the Syrian Desert as a case study, they show how long-term geomorphological changes affected human settlement and subsistence in this challenging landscape.

The following seven chapters all focus explicitly on the cultural repercussions of the 8.2 ka abrupt climate event in the ancient Near East and Mediterranean. Mottram's chapter *When the Going Gets Tough* steps away from monocausal, simplistic cause-effect relationships, but argues for a strong causal role of climate change in the emergence of the Halaf culture. The argument is fairly similar to the one made by the Tell Sabi Abyad team since 2006, but both chapters differ in their emphasis on the role of climate change as prime mover. Whereas for Mottram climate change was key, for Nieuwenhuyse, Akkermans, van der Plicht, Russell, and Kaneda in their chapter *The 8.2 ka Event in Upper Mesopotamia*, climate remains but one potential factor among many. Moreover, the Sabi Abyad team argues that its role is not particularly easy to isolate: climate may have accelerated changes that already had begun.

Both chapters stress continuity and adaptation, not collapse as had been proposed by earlier scholars. This "change-and-continuity-not-collapse" approach to climate change is also evidenced in *The Aftermath of the 8.2 Event* by Willett, Franz, Kabukcu, Orton, Rogasch, Stroud, Rosenstock, and Biehl. They use Çatalhöyük as another example showing the importance of collecting meticulously detailed local data on specific episodes of culture change that happen to coincide with climate change. The examples of Tell Sabi Abyad and Çatalhöyük show that once you get down to the ground level of messy archaeological data neat theories no longer look so neat. Interpretations become confused, as multiple causal agents offer themselves, and a causal role for, specifically, climate change becomes much harder to demonstrate in practice (see also Düring's discussion of Western Turkey and Perlès's discussion of data from Greece).

Ryan and Rosen's chapter *Managing Risk through Diversification in Plant Exploitation* provide a nuanced discussion of plant exploitation evidence during the 8.2 event. The authors suggest the possibility of climate change as a prime causal factor but in the end they leave this as an open question, emphasizing the difficulty of disentangling climate effects from other causal agents, for instance, anthropogenic factors. In *The 8.2 Event and the Neolithic Expansion in Western Anatolia* Düring provides a critical reassessment of issues of chronology and synchronicity with regard to the 8.2 event in Anatolia. Pointing out several difficulties with the synchronicity perceived between the climate event and Neolithic expansion in western Anatolia, Düring cautions forcefully for what he perceives to be climate determinism (see also Gronenborn's chapter on the spread of farming in Europe).

Moving on toward Europe, in *"Singing in the Rain,"* Daune-Le Brun and Le Brun document another case for change-and-continuity and adaptation through the 8.2 ka event in the eastern Mediterranean, not collapse. The authors highlight the complexities of correlating culture change with climate change methodologically. Apart from establishing synchronicity, as with all well-excavated and carefully documented cases discussed in this book the cause and affect scenario remains difficult. The authors emphasize the "insularity" of the cultural setting they describe for Khirokitia, cautioning against super-generalizing climate or cultural modeling.

Perlès continues in *Early Holocene Climatic Fluctuations and Human Responses in Greece* with a discussion of settlement patterns in Greece before, during, and after two climate events, one of which is the 8.2 ka event. Pottery styles in the seventh millennium did not change so quickly that synchronicities with climate change can so easily be observed in settlement data. With many other authors in this book, Perlès explicitly engages in scrutinizing alternative, nonclimate explanations for explaining culture change (also Nieuwenhuyse et al.; Bonsall et al.; Pelisiak, Löwenborg and Eriksson, this vol.). As in the preceding chapter, the author emphasizes the importance of *localized* ecological impacts of global climate events such as the 8.2; in Greece, this event may not have instigated drought but instead may have locally led to more severe flooding.

Moving north into the Danube, Bonsall, Macklin, Boroneant, Pickard, Bartosiewicz, Cook, and Higham contextualize the question of cause or coincidence in *Rapid Climate Change and Radiocarbon Discontinuities in the Mesolithic-Early Neolithic Settlement Record of the Iron Gates*. They stress that establishing synchronicity alone is not enough and explicitly discuss alternatives including social changes and taphonomic effects. If this and previous chapters mostly adopted a microregional, or even site-based perspective, this contrasts with the macro-scale, almost global approach of Gronenborn in his *Climate Fluctuations, Human Migrations, and the Spread of Farming in Western Eurasia*. Together with, to some degree, Mottram (this vol.), Gronenborn is among the few authors in the book to explicitly adopt a systems or processual approach. Thus, Culture as extrasomatic adaptation rests in dynamic equilibrium with the ecological environment, until factors external to the system—i.e., climate change—push the system off balance leading to adaptation and renewed balance. Gronenborn's contribution is valuable in making explicit what remains implicit in many other chapters, and by extrapolation climate change discussions elsewhere in the archaeological literature.

Pelisiak's *Climate Change in the Polish Upland Bronze Age* sketches in detail the environment/climate background to culture change and shifts in settlement patterns. However, the author describes that the potential impacts of climate were intimately connected with internal social, economic, and political processes, and hence climate change may have been only one of many factors involved in prehistoric culture change.

Finally, Löwenborg and Eriksson's *Climate and the Definition of Archaeological Periods in Sweden* offers an explicit, theoretically informed discussion of climate archaeology and its methods used so far. They provide a historical summary of the role of climate in Swedish archaeology, starting with the rise of scientific archaeology and processual archaeology when climate was elevated to the role of prime causal agent for culture change. In contrast, notwithstanding increasing methodological sophistication in recovering environmental data and vibrant theoretical debate, in much post-processual archaeology in Sweden today climate is categorically rejected as a causal agent. The authors call for a more balanced, theoretically informed consideration of both cultural and environmental factors. We hope the various contributions in this book may offer a means toward this end.

Epilogue

In December 2001, several weeks of intermittent northerly winter outbreaks caused serious disruptions around the Black Sea–Aegean Sea region. The severe conditions included sustained periods of subzero temperatures, snowstorms and blizzards, heavy rains, and strong winds. Athens and Istanbul received about thirty cm of snow, and city governor Erol Cakir even declared conditions in Istanbul a "national disaster." In Larissa, Greece, night temperatures plummeted to a minimum of $-20.2°$ C. More than 300 villages in northern and central Greece were snowed in, while airports and schools were closed in the North. In Bulgaria, heavy snowfall cut power lines, while frosts cut off water supplies.

Of course, what happened here was simply *weather*. What climatologists formally term climate is weather averaged over a period of thirty years. But these horrifying conditions to some degree match the scenario often reconstructed for the infamous 8.2 ka abrupt climate event, estimated to have lasted for about two centuries or more, the equivalent of perhaps ten human generations. Based on the above, we may start to *imagine* the climatic impacts experienced around Europe and the Near East in Prehistory. Winter conditions would have been characterized by extremes much more pronounced than today, and very extensive rainfalls and snowfalls would have given rise to serious problems with crops, grazing, flooding, and the attendant destabilization of hillsides and mud-brick dwellings. During other winters conditions may have remained very dry, again with considerably detrimental effects on crops and grazing. Overall, we can imagine a considerable amount of pressure on resources, and general environmental stress during climatic events.

Yet, as the examples in this book show, prehistoric communities in Europe and the ancient Near East did not perish and in some cases may even have flourished. Nor are contributors to this book unanimous in ascribing all culture change they observe in the archaeological record to adaptation to environmental adversities. Climate archaeologists in Europe and the Near East are certainly aware of the importance of climate in explaining cultural innovation, but they remain unimpressed by simplistic tales of collapse caused by the "serial killer of civilization" (Linden 2009) called climate.

References Cited

Alley, R. B. 2000 *The Two-Mile Time Machine. Ice Cores, Abrupt Climate Change, and Our Future*. Princeton University Press, Princeton.

Birks, H. J. B. 2008 Holocene Climate Research—Progress, Paradigms, and Problems. In *Natural Climate Variability and Global Warming. A Holocene Perspective*, edited by R. W. Battarbee and H. A. Binney, pp. 7–57. Wiley-Blackwell, West Sussex.

Coombes, P., and K. Barber 2005 Environmental Determinism in Holocene Research: Causality or Coincidence? *Area* 37(3):303–311.

Cooper, J., and P. Sheets (eds.) 2012 *Surviving Sudden Environmental Change. Answers from Archaeology*. University of Colorado Press, Boulder.

Dann, R. 2015 Introduction: Can Archaeology Save the World? In *Climate and Ancient Societies*, edited by S. Kerner, R. J. Dann, and P. Bangsgaard, pp. 19–25. Tusculanum, Copenhagen.

Danti, M. D. 2010 Late Middle Holocene Climate and Northern Mesopotamia: Varying Cultural Responses to the 5.2 and 4.2 ka Aridification Events. In *Climate Crises in Human History*, edited by B. A. Mainwaring, R. Giegengack, and C. Vita-Finzi, pp. 139–172. American Philosophical Society, Philadelphia.

Dawdy, S. L. 2009 Millennial Archaeology: Locating the Discipline in the Age of Uncertainty. *Archaeological Dialogues* 169(2):131–142.

Diamond, J. 2005 *Collapse. How Societies Choose to Fail or Survive*. Allan Lane, New York.

Glantz, M. H. 1994 Drought, Desertification, and Food Production. In *Drought Follows the Plow*, edited by M. H. Glantz, pp. 7–32. Cambridge University Press, Cambridge.

Linden, E. 2009 *The Winds of Change: Climate, Weather, and the Destruction of Civilizations*. Simon and Schuster, New York.

Maher, L. A., T. Banning, and M. Chazan 2011 Oasis or Mirage? Assessing the Role of Abrupt Climate Change in the Prehistory of the Southern Levant. *Cambridge Archaeological Journal* 21(1):1–29.

Mayewski, P. A., and F. White 2002 *The Ice Chronicles. The Quest to Understand Global Climate Change*. University Press of New England, London.

McAnany, P. A., and N. Yoffee 2010 Why We Question Collapse and Study Human Resilience, Ecological Vulnerability, and the Aftermath of Empire. In *Questioning Collapse. Human Resilience, Ecological Vulnerability, and the Aftermath of Empire*, edited by P. A. McAnany and N. Yoffee, pp. 2–20. Cambridge University Press, Cambridge.

McIntosh, R. J., J. A. Tainter, and S. K. McIntosh 2000 Climate, History, and Human Action. In *The Way the Wind Blows: Climate, History, and Human Action*, edited by R. J. McIntosh, J. A. Tainter, and S. K. McIntosh, pp. 1–42. Columbia University Press, New York.

Oldfield, F. 2008 The Role of People in the Holocene. In *Natural Climate Variability and Global Warming. A Holocene Perspective*, edited by R. W. Battarbee and H. A. Binney, pp. 58–97. Wiley-Blackwell, West Sussex.

Redman, C. L. 1999 *Human Impact on Ancient Environments*. University of Arizona Press, Tucson.

Roberts, N. 2015 Holocene Climate Changes and Archaeological Implications, with Particular Reference to the East Mediterranean Region. In *Climate and Ancient Societies*, edited by S. Kerner, R. J. Dann, and P. Bangsgaard, pp. 27–39. Tusculanum, Copenhagen.

Rosen, A., and S. A. Rosen 2001 Determinist or not Determinist? Climate, Environment, and Archaeological Explanation in the Levant. In *Studies in the Archaeology of Israel and Neighboring Lands in Memory of Douglas L. Esse*, edited by S. R. Wolff, pp. 535–549. Oriental Institute, Chicago.

Rosen, A. 2007 *Civilizing Climate. Social Responses to Climate Change in the Ancient Near East*. Altamira, Lanham.

Sandweiss, D. H., and J. Quilter 2012 Collation, Correlation, and Causation in the Prehistory of Coastal Peru. In *Surviving Sudden Environmental Change. Answers from Archaeology*, edited by J. Cooper and P. Sheets, pp. 117–139. University of Colorado Press, Boulder.

Schiermeier, Q. 2015 Why the Pope's Letter on Climate Change Matters. *Nature News 18 june 2015* (doi:10.1038/nature.2015.17800).

Schwartz, G. M. 2006 From Collapse to Regeneration. In *After Collapse. The Regeneration of Complex Societies*, edited by G. M. Schwartz and J. J. Nichols, pp. 3–17. University of Arizona Press, Tucson.

Tainter, J. A. 1988 *The Collapse of Complex Societies*. Cambridge University Press, Cambridge.

Ur, J. 2015 Urban Adaptations to Climate Change in Northern Mesopotamia. In *Climate and Ancient Societies*, edited by S. Kerner, R. J. Dann, and P. Bangsgaard, pp. 69–95. Tusculanum, Copenhagen.

PART I

Near East

CHAPTER ONE

The Oasis of Palmyra in Prehistory

Late Pleistocene and Early Holocene Paleoclimate and Human Occupation in the Region of Palmyra/Tadmor (Central Syria)

Mauro Cremaschi, Andrea Zerboni

Abstract *In this contribution we discuss the evidence, collected during a geoarchaeological survey, for Upper Pleistocene to Early Holocene environmental modifications and human occupation in the oasis of Palmyra/Tadmor (central Syria) and its surroundings. In the area, archaeological evidence consists of Geometric Kebaran, Natufian, and Early Neolithic (Pre-Pottery) archaeological sites, distributed within the limits of the present-day oasis, along the margin of the Sabkhat al Mouh, and in the Abu Fawares area. Most of the archaeological findings are in connection with lacustrine and spring deposits, suggesting a Late Pleistocene/Early Holocene period with a humid climate, sustaining human exploitation of natural resources. Wet environmental conditions presumably turned toward aridity in correspondence with the Early Neolithic, and later archaeological sites were concentrated in the area of the present-day oasis and marginally within the Abu Fawares basin. In this phase, archaeological communities presumably changed their subsistence strategies toward food production, exploiting resources in the vicinity of the oasis. The integrated study of the environmental changes that occurred in the region and the distribution of archaeological features permits us to delineate the reduction in size of the oasis of Palmyra, which dates back to the Early Holocene, and compare it to equivalent processes that occurred in other circum-Mediterranean arid regions.*

INTRODUCTION

The evolution of the oases in the arid and semi-arid regions of the Old World is one of the most evident consequences of climate changes that occurred in the last millennia of the Holocene. Oases originate and develop after major changes in the precipitation regime that lead to the decrease in water availability, the reduction of natural resources and the contraction of environments favorable to life to limited ecological niches (Cremaschi and Zerboni 2011, 2013; Mori et al. 2013). In this sense, we may interpret the environmental history of an oasis in the complex mosaic of the desert as a geomorphological process, which requires the existence of underground water reservoirs feeding few springs and therefore sustaining the residual vegetation. In the Sahara Desert, along the Nile Valley, and in many parts of the Arabian Peninsula, the main environmental changes occurred after the withdrawal of the monsoon systems and the nucleation of the oases are related to the Mid-Holocene transition (e.g., Brooks 2006; Cremaschi 2001; Cremaschi and Zerboni 2011, 2013; Hassan 2002; Kuper and Kröpelin 2006; Kühn et al. 2010; Rognon 1980). Additionally, in many cases, the deep changes in the territory that contributed to the formation of the oases also influenced the human groups living therein; the necessity to survive aridity promoted a profound population relocation over the long term and the adjustment of settlement strategies and modalities of land exploitation by archaeological communities (e.g., Anderson et al. 2007; Kuper and Kröpelin 2006).

The timing and steps of the formation of the oases systems in the Sahara and in the Arabian Peninsula is relatively well known and described by several authors; most of them focused their attention on the climatic crisis (interruption of water supply) that drove the geomorphological process. The archaeological and cultural responses to new environments have also been investigated, postulating, for instance, a positive contribution to the spreading of agriculture within the green limits of the oases, its development and the introduction of advanced technologies for water management and irrigation. On the contrary, the change in size of the oases in the continental, arid Near East is rather poorly known, and it seems to be older than in other circum-Mediterranean regions. Also, the possible cultural repercussions remain to be investigated much further.

In this contribution, we shall introduce the first results of geoarchaeological research carried out in the region surrounding Palmyra (Tadmor, central Syria; Figure 1.1), mostly dedicated to the study of climate changes and human responses between the end of the Pleistocene and the first millennia of the Holocene (Al-Maqdissi et al. 2009; Cremaschi and Zerboni 2012). In the region, this phase is coincident with a significant change in environmental settings and human adaptation to the contraction of the oasis. The data discussed here should be considered as preliminary, as recent uprising in Syria sadly determined the sudden interruption of the geoarchaeological research in the region.

The Environmental Context and the Open Questions

The cuvette of Palmyra (Figure 1.1) is located at the northern margin of the Arabian Desert and the present-day climate is semi-arid, alternating rainy and dry seasons, with a mean annual precipitation less than 140 mm. The investigated region includes the margins of the green oasis of Palmyra/Tadmor, the lacustrine terraces lining the present-day Sabkhat al Mouh (a seasonally flooded, saline marsh), and the lake formations at Abu Fawares (Cremaschi and Zerboni 2012; Soulidi-Kondratiev 1966). These areas, connected by the narrow saddle separating Jebel Qayad from Jebel el Madjur, belong to the same endorheic depression centered on the present-day oasis (Figure 1.1).

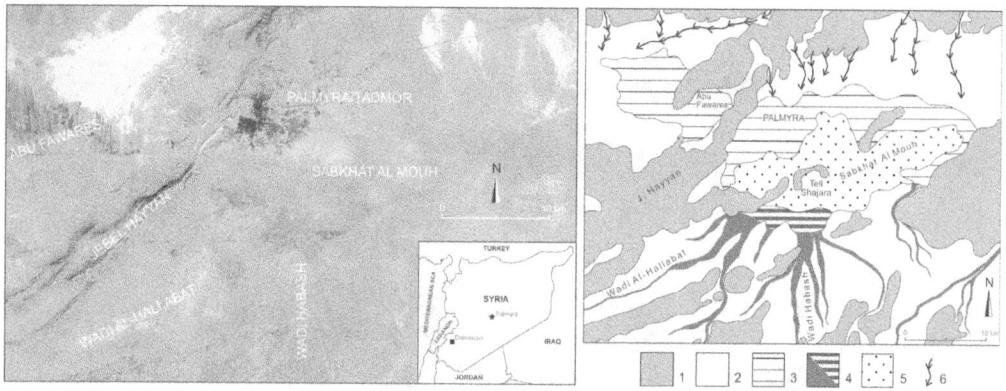

FIGURE 1.1. Left, Landsat satellite imagery of the region surrounding Palmyra/Tadmor (central Syria) indicating the name of the main localities cited in the text; the inset illustrates the position of the study area in a regional context. Right, geomorphological sketch of the area described in the text. Key: 1, pre-Quaternary bedrock (limestone); 2, Quaternary deposits; 3, Pleistocene lake deposits; 4, delta system; 5, sabkhat; 6, erosional streams.

The region was shaped as of the Miocene and then filled with fluvial and lake sediments during the Late Quaternary (Soulidi-Kondratiev 1966). The lower part of the Palmyra basin is occupied by the Sabkhat al Mouh (Figure 1.2), a salty marsh seasonally fed by rainfall and runoff from the wadis located at its southern margin; in the wet seasons the sabkhat turns into an ephemeral hyperhaline lake. The region surrounding the sabkhat is rather open and scarcely covered by vegetation; therefore, the eastern winds can easily erode the surface of the lake and transport sand to the slopes at the margin of the oasis.

Notwithstanding that, various geological and archaeological indicators confirm that in the recent geological past the region experienced wetter environmental conditions. For instance, evidence for formerly high-standing lake levels dates back to the Pleistocene; a large freshwater basin during the Middle Paleolithic phase attracted human groups, which established their base camps at the margin of the lake and in the rockshelters surrounding it (Akazawa 1979; Hanihara and Sakaguchi 1978; Julig et al. 1999). A further wet episode, marked by the high stand of the lake level and a possible freshwater environment, was attributed by several authors (Akazawa 1979; Hanihara and Sakaguchi 1978) to the period encompassing the last millennia of the Upper Pleistocene and the beginning of the Holocene. Besançon et al. (1997) refused any hypotheses on persistent wet environmental conditions after Marine Isotope Stage (MIS) 4, but the identification of late Epipaleolithic sites close to the present oasis, and of deposits connected to a post-Pleistocene high lake stand, imply a revision of Besançon's conclusion.

METHODS

Each physiographic unit of the study region was assessed during a geomorphological survey, mainly based upon high-resolution satellite imagery (Ikonos and Google Earth™)

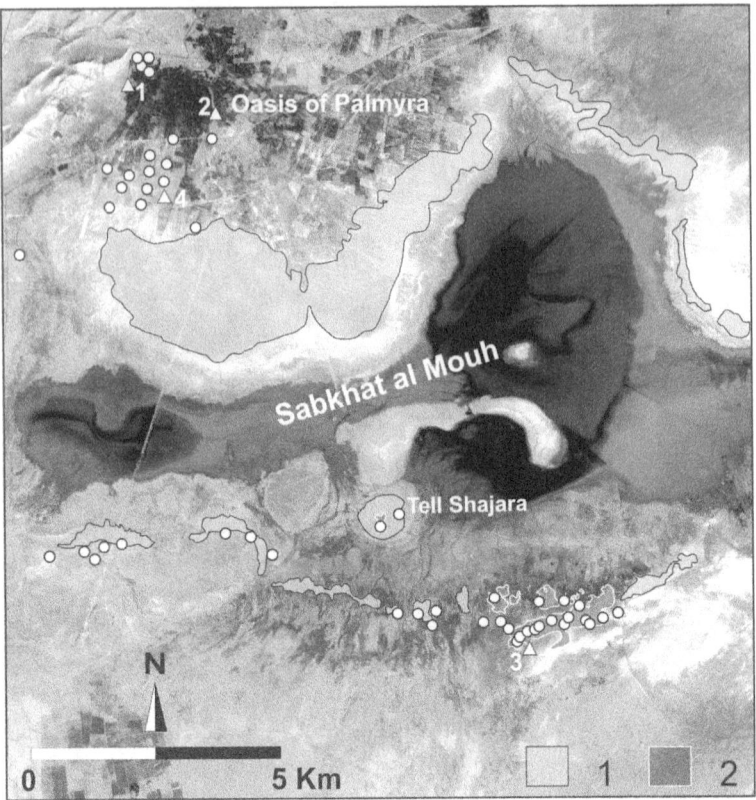

FIGURE 1.2. Ikonos satellite imagery indicating the distribution of Upper Pleistocene and Early Holocene lake (1) and aeolian (2) deposits in the sabkhat area, the archaeological sites (dots), and the main localities cited in the text (triangles: 1, tell Site 288; 2, wadi Aid section; 3, Site 250; 4, Site 389) (modified from Cremaschi and Zerboni 2012).

and field control, and sampled from an archaeological point of view. The identified sites were positioned through GPS, described, and then integrated in a geographic information system (GIS). A limited set of diagnostic pottery fragments and lithics were collected from each site, together with samples for radiometric dating and laboratory paleoenvironmental analyses. Additionally, a sedimentary section along the wadi Aid, in part investigated in the past (Sakaguchi 1978, 1987), was described and bulk samples were collected for sedimentological analyses. They furnished supplementary paleoenvironmental information on surface processes; applied laboratory methods are summarized in Note 1. The chronological framework discussed here is based upon the archaeological contexts and several new and published radiocarbon dates obtained from organic sediments and charcoal from hearths of archaeological sites. AMS-^{14}C dates were calibrated, with a precision of 2σ, using the IntCal13 calibration curve (Reimer et al. 2013).

Paleoenvironmental Evidence and Distribution of Archaeological Sites

Paleoenvironmental Evidence at the Margin of the Sabkhat al Mouh

Within the limit of the present-day oasis, the Upper Pleistocene and Early Holocene geological formations of the terrace are deeply cut by the wadi Aid, where a significant stratigraphic section, already described by Sakaguchi (1978, 1987), is exposed (Figure 1.3). Due to its high interest for the understanding of Late Quaternary environmental changes, we decided to investigate the section again to collect samples for laboratories analyses. The base of the section, the lower boundary of which is unknown, consists of silty clay deposits cemented by calcite and gypsum crystals and quite rich in organic

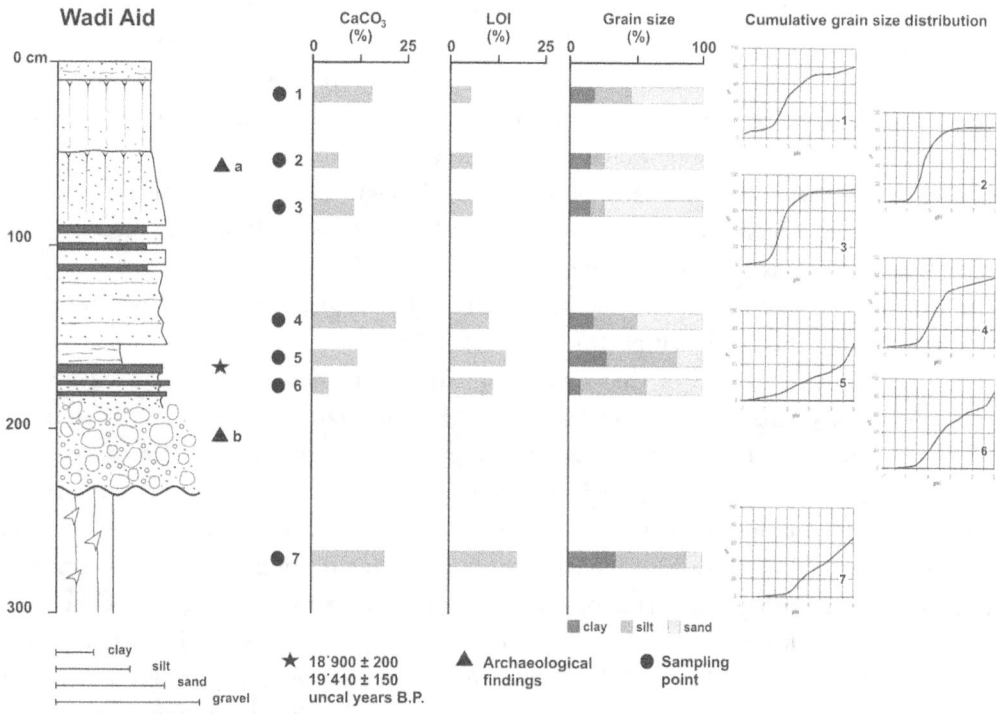

FIGURE 1.3. Stratigraphic section at wadi Aid, illustrating the results of sedimentological analyses (grain size, content of carbonate equivalent, and organic carbon). Black triangles indicate the locations of flint artifacts (a, lamellar débitage; b, Mousterian), dots indicate the position of sampling points; the dated layer and results are also indicated. Note in the cumulative grain size distributions from the upper part of the section (sampling points 2 and 3) a clear input of aeolian sand (for the interpretation of sedimentological data see: Krumbein and Sloss 1963; Kukal, 1971).

matter. These are followed by a matrix-supported gravel with discontinuous concave bedding, including rolled Mousterian artifacts. This layer is overlaid by a massive, greenish lacustrine silt rich in carbonates, intercalated by thin planar layers of organic matter–rich (TOC up to 15%) sand and silt. This layer was radiocarbon dated (Sakaguchi 1978, 1987) to 18,900 ± 200 and 19,410 ± 150 uncal years B.P., corresponding to 21,360–20,438 and 21,826–21,012 cal years B.C. (23,331–22,388 and 23,776–22,962 cal years B.P.) respectively. A plane-laminated to massive hydromorphic sandy unit follows; silt and clay fractions and carbonate content increase toward the top as an effect of pedogenesis, while the organic fraction is lower.

The sequence was interpreted by Sakaguchi (1987) as originating in a lacustrine sedimentary environment, but this attribution was later refused by Besançon et al. (1997). However, our data support the older interpretation: the silty clay at the base of the sequence is interpreted to be deposited in a lacustrine environment and represents the sedimentary bedrock of the oasis; it is attributed to a high stand of the lake dated to the Upper Pleistocene. The gravel that follows was accumulated probably inside erosive scars cut into the Upper Pleistocene lacustrine formation, indicating the aggradation of fluvial deposits toward the sabkhat, while the erosive surface at its base points to a drop (or desiccation) of the lake level as a consequence of a dry phase during the Last Glacial Maximum. On the contrary, the greenish laminated silts are indicative of standing water inside the valley of wadi Aid, whose occurrence was determined by a high stand of the lake level (transgressive phase). Radiocarbon ages (Sakaguchi 1987) put this phase at the end of the Last Glacial Maximum, but radiocarbon dating on lacustrine organic mud may overestimate the true age of the deposit (Walker 2005; Lee et al. 2010). Finally, the cumulative grain size curve from the upper part of the sequence indicates the prolonged input of aeolian dust, originating from the deflation of the sabkhat.

The southern margin of the Sabkhat al Mouh is marked by the delta formed by the confluence of the wadi al-Hallabat, al-Annan, and Habash (Figure 1.1). Remnants of fossil dunes have been observed at the eastern side of the delta (Figure 1.2); main outcrops are shaped in ridges up to 3 m high, oriented east to west (Figure 1.4). Furthermore, smaller outcrops are present, covering the whole delta front, in the form of small yardangs dissected by the extant fluvial net. Dunes there are composed of cross-stratified cemented sand, with grains made of reworked gypsum crystals. At the top of the sand ridges, a hard calcareous crust up to 30 cm thick is often present; it represents the remnant of a pedogenetic caliche (a carbonatic crust produced by pedogenesis under climatic conditions of alternating arid and wet phases). The dune slopes facing the sabkhat are lined with lacustrine deposits, consisting of loose to weakly cemented sand, composed of finely subdivided gypsum crystals; they are laminated close to the dunes, becoming massive in an outer position. Also, on these deposits, a discontinuous caliche has been observed (Figure 1.4). On the basis of stratigraphic and archaeological contexts we can provide a minimum dating for the formation of dunes: they may be regarded as having been deposited during a dry phase in the Upper Pleistocene. Field investigation highlighted to the existence of two distinct facies within the lacustrine deposits: (1) a shore facies with laminated sand at the margin of the dunes, possibly formed after the

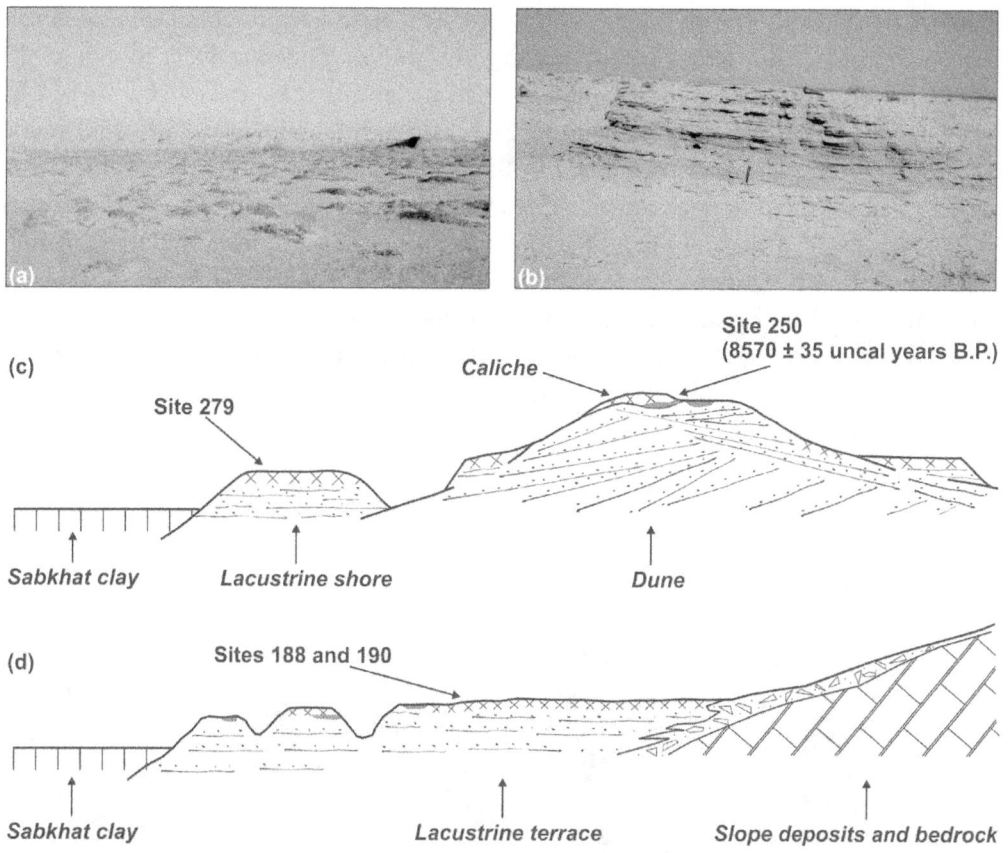

FIGURE 1.4. Geomorphological features at the southern margin of the Sabkhat al Mouh. (a) Late Pleistocene/Early Holocene lacustrine terraces at the southern margin of the sabkhat. (b) A gypsum-sand dune at the southern margin of the sabkhat. Cross-sections indicating the relationships between sand dunes (c) and lacustrine terraces (d) at the southern margin of the sabkhat (modified from Cremaschi and Zerboni 2012); the distribution of sites and the result of an AMS-^{14}C dating are also reported.

reworking of dune, and (2) a distal facies, composed of massive sediments due to alkalis precipitation. The stratigraphic relationships and the archaeological contexts suggest a Late Glacial or Early Holocene age. The lacustrine deposits give way to a discontinuous flat terrace, lying at ca. 1.5 m above the salty mud of the sabkhat and extending to the west of the delta system, at the margin of the rock outcrops delimiting the sabkhat (Figure 1.2).

ARCHAEOLOGICAL SITES AT THE SOUTHERN MARGIN OF THE SABKHAT AL MOUH

On the sand ridges and on the terraces of lacustrine sediments several archaeological sites have been recorded, consisting of clusters of lithic débitage (Figure 1.2). Most of the flint artifacts have a whitish patina, suggesting their persistence in an alkaline sedimentary environment; this also suggests that they are to be regarded as lying in situ at the top of the sand deposits, possibly inside the caliche, and later exposed at the surface by wind erosion (Cremaschi and Zerboni 2012). Charcoal from a fireplace (Site 250) found within the caliche crust has been dated to 8570 ± 35 uncal years B.P. (UGAMS-10474, $\delta^{13}C = -18.3‰$), corresponding to 7632–7538 cal years B.C. or 9582–9488 cal years B.P.; this result indicates the time of attendance of the site, and, from the paleoenvironmental point of view, it represents a the limit *ante quem* for the deposition of dunes at the margin of the sabkhat and a period of intense pedogenesis in the area.

Some clusters of lithics include several formal tools (Figure 1.5): end-scrapers on retouched blades, burins of different types, geometrics, and backed and truncated bladelets. Tool assemblages allow us to attribute the sites to the Geometric Kebaran.

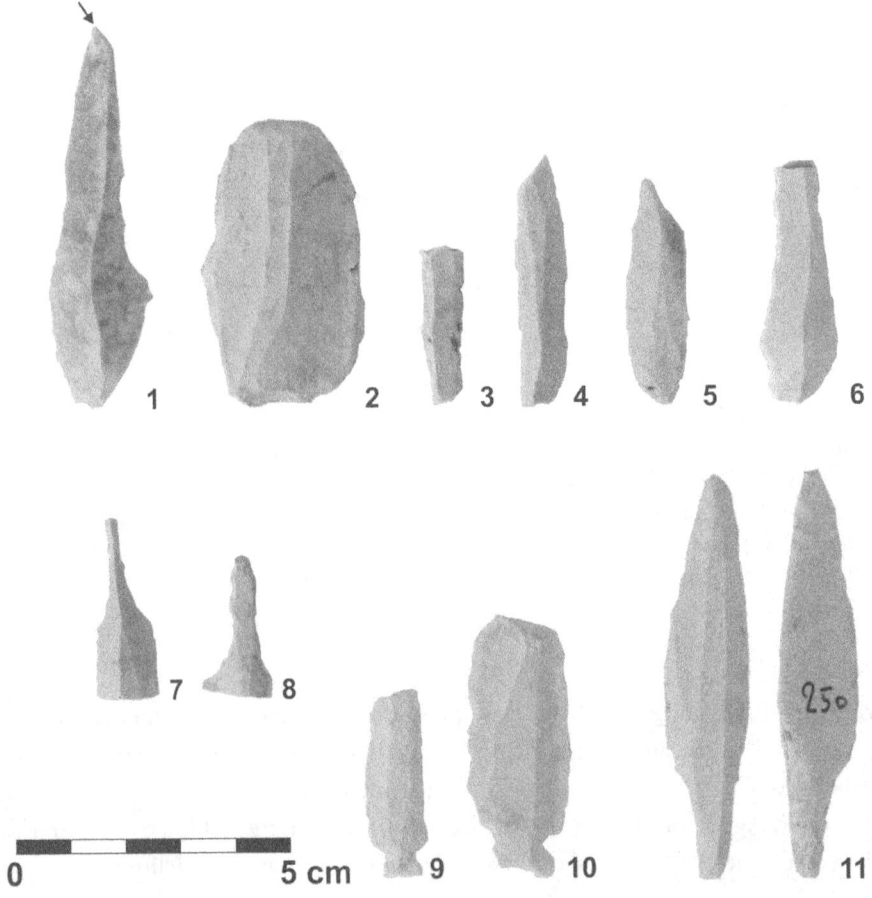

FIGURE 1.5. Some findings from the archaeological sites at the margin of the Sabkhat al Mouh dating to the Late Epipaleolithic and the Pre-Pottery Neolithic. 1: burin and backed point; 2: end scraper on blade; 3: rectangular geometric; 4: truncated and backed point; 5: partly backed point; 6: backed point; 7, 8: perforators; 9, 10: Khiam points; 11: Byblos point (recto/verso).

Elsewhere, lithic clusters consist of Natufian microliths (lunates and rectangles), and are apparently associated with artifacts dating to the PPNA (Khiam arrowheads and perforators) and PPNB (Jericho/Byblos arrowheads) phases. Some sites preserve their original configuration, including dense concentrations of débitage and stone hammers and anvils in connection with sub-circular spots of burned sediments, interpreted as fireplaces. Although a commixture of artifacts of different age is expected, since we are dealing with residual surfaces, the stratigraphic relationships between the different litho-complexes must be checked by archaeological excavation. The occurrence of lithics dating from the Late Epipaleolithic to the Pre-Pottery Neolithic testify for a long-lasting presence (between the end of the Pleistocene and the beginning of the Holocene) of human groups along the margins of the sabkhat.

Tell Shajara and the Central Part of the Sabkhat al Mouh

Tell Shajara is a small, isolated terrace in the middle of the Sabkhat al Mouh (Figure 1.2). It is elevated a few meters above the level of the sabkhat, whose bedrock consists of gravel dating to the Pliocene buried by the remnants of fossil dunes. Around the Pliocene bedrock there are remains of lake deposits dating back to the Upper Pleistocene, which may be correlated with lacustrine sediments located at the northern margin of the sabkhat (Sakaguchi 1978). Mousterian artifacts of Levallois technique, lacking any evidence of post-depositional transportation, were found in situ in the Pleistocene lacustrine marls. In the southern part of Tell Shajara we found a recently excavated well; in the shaft, black, organic mud belonging to a lacustrine facies has been observed. Unfortunately, the stratigraphic position of these deposits is uncertain, but they were originally located at shallow depth below the ground level and are testimony to lacustrine sedimentation in a waterlogged environment. Geometric Kebaran tools (backed and truncated bladelets) and large flint blades and scrapers (possibly Neolithic in age), together with faunal remains and some undecorated sherds, have been also recovered inside the organic mud; the sherds present a bad state of preservation, but some of them display similarities to those described at some desert sites by Borrell et al. (2013). We suggest that the original position of the artifacts was within the organic mud and they were deposited during a high stand of the lake level; in fact, flint artifacts and the animal bones present a blackish patina, while pottery fragments are corroded, confirming a long persistence under anoxic conditions. Furthermore, few clusters of Epipaleolithic and Neolithic flint artifacts were found lying on the surface of the lacustrine deposit in the vicinity of Tell Shajara.

Evidence between the Northern Limit of the Oasis and the Abu Fawares Basin

The slopes surrounding the sabkhat and the saddle separating the oasis of Palmyra from the Abu Fawares depression have been surveyed and most of the lithic dispersions identified may be attributed to the Paleolithic (Mousterian). A large Late Pleistocene site, represented by a dense cluster of lithics dating back to the Geometric Kebaran, was discovered on the western slope of the saddle linking the oasis with the valley of Ad Daww and Abu Fawares regions. The concentration of artifacts is very high; unfortunately, the archaeological deposit is not preserved and lithics lie on the exposed bedrock, which is covered by only a few centimeters of aeolian sand. The spatial distribution of artifacts within the limits of the site is worth mentioning, as distinct areas with geometric tools, end-scrapers, burins, and débitage associated with naviform cores were described; they indicate the preservation of functional clusters.

The locality of Mazraet Abu Fawares is located at the eastern end of the Ad Daww depression; in the lower part of the basin we found lacustrine deposits dating to the Pleistocene (Figure 1.6). Subsequently, sediments were incised by fluvial activity, and the resulting fluvial net was later enlarged by wind erosion. Thus, the smooth bottom of the basin valley is dissected by elongated interconnected shallow basins, displaying a flat bottom.

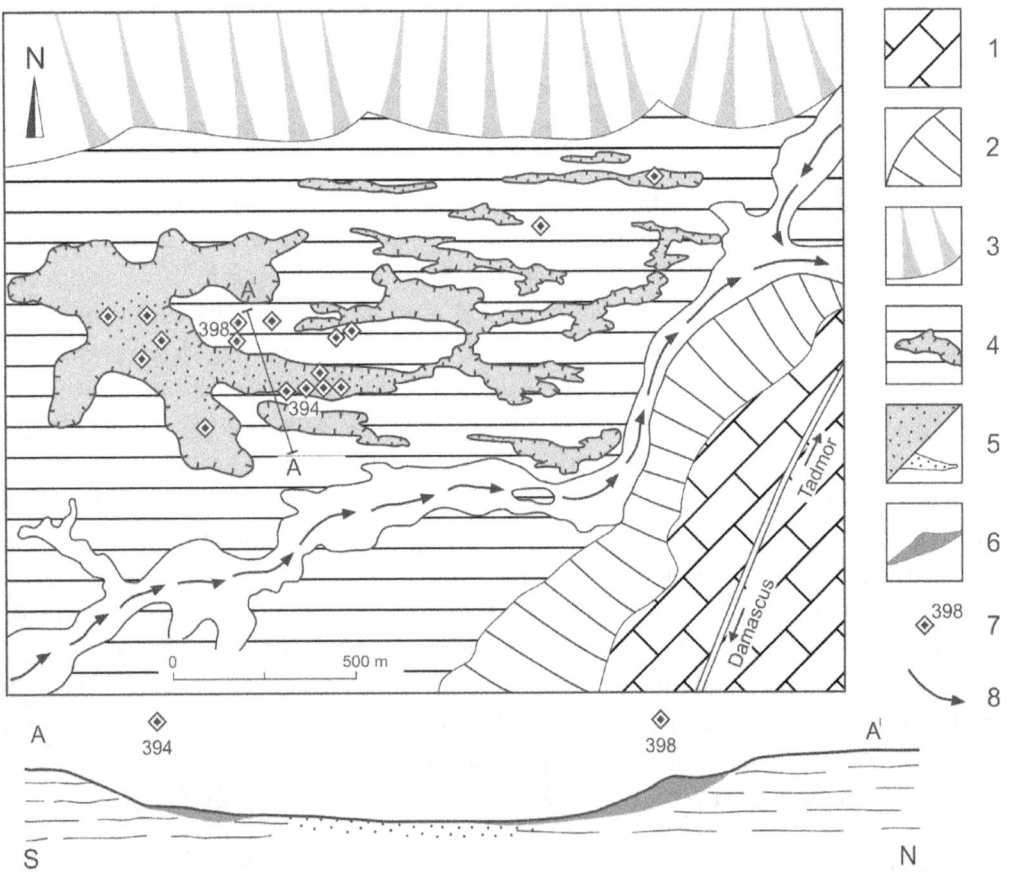

FIGURE 1.6. Geomorphological map of Abu Fawares area, diamonds indicated the position of Epipaleolithic and Neolithic archaeological sites; a geological cross section is also reported (modified from Cremaschi and Zerboni 2012). Key: 1, limestone outcrop; 2, slope deposits; 3, pediment; 4, windblown depression cut into Pleistocene alluvial sediments; 5, Holocene lake sediments; 6, archaeological mounds; 7, archaeological sites; 8, main wadis.

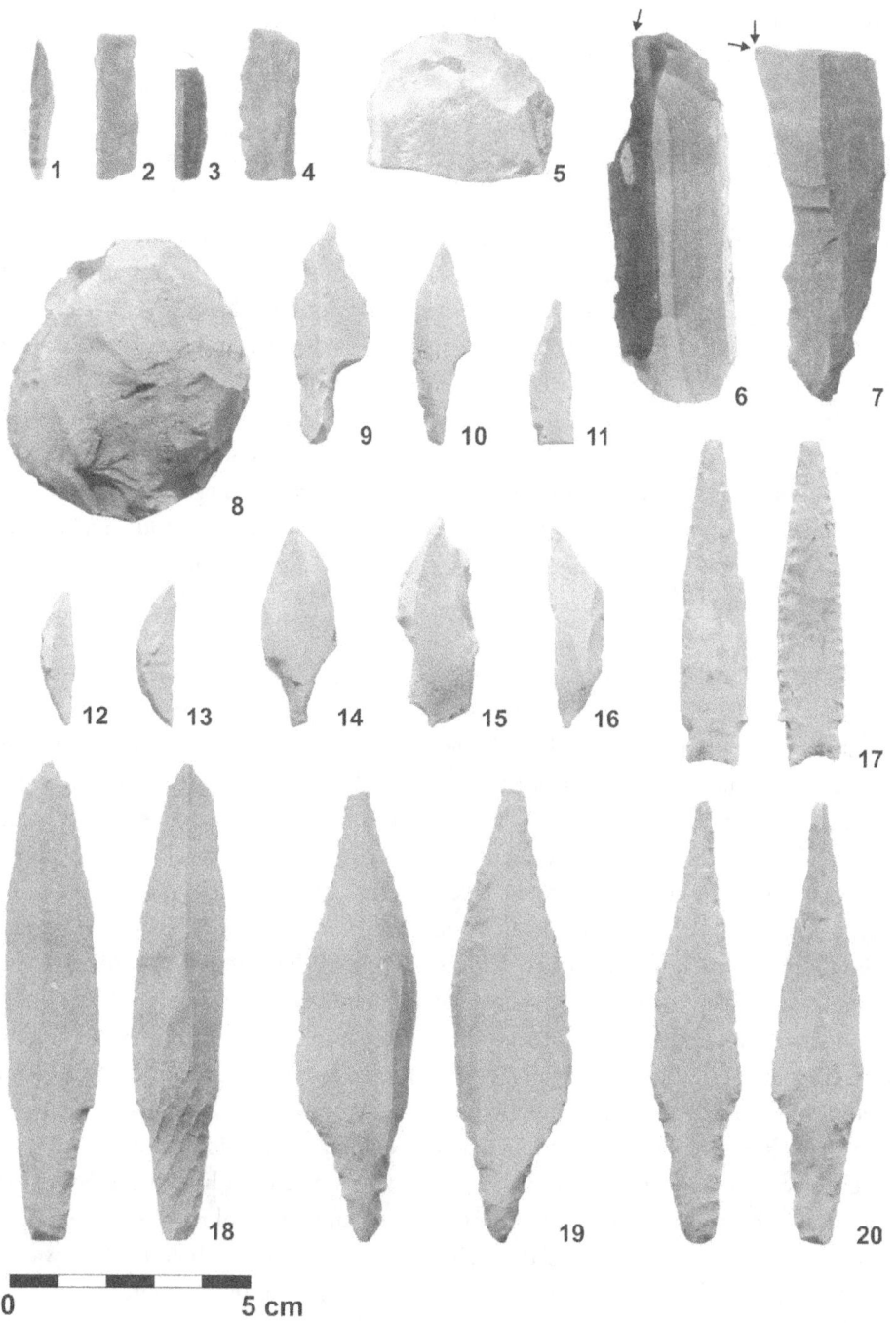

Figure 1.7. Flint artifacts from archaeological sites in the Abu Fawares area. 1: backed pointed bladelet; 2–4: backed blade and bladelet associated with truncations; 5: end scraper; 6: burin on a fracture, opposite to end scraper; 7: burins; 8: frontal end scraper on a flake; 9, 10: pedunculated backed points on bladelet; 11: perforator on a backed bladelet; 12, 13: lunates; 14: pedunculated flake; 15: perforator; 16: backed point on backed bladelet; 17: Khiam (?) point (recto/verso); 18–20: Byblos points (recto/verso).

Inside, thin discontinuous lacustrine marls have been observed, consisting of planar layers of dark organic matter–rich sand alternating whitish sandy-silty strata. Therein, a cluster of very rich Neolithic sites was discovered; they consist mostly of concentrations of lithics on deflated surfaces (Figure 1.7). However, many sites occur at the base of the shallow scarps bordering the interconnected basins, which are well preserved in the form of shallow mounds of dark, organic sediments. Neolithic sites in the area of Abu Fawares are very rich in archaeological finds: clusters of lamellar débitage, associated with cores and formal tools, have been observed; a few, well-preserved primary archaeological structures represented by stone circles delimiting fireplaces were also found. In one case, within the limit of the shallow mound representing the extension of a site rich in lithics, several limestone blocks, which may have been used as hammers or anvils in flint processing, are present. Formal tools include Khiam and Jericho arrowheads that are attributed to the PPNA and PPNB periods, but lithics dating to the Natufian period were also found. The latter include crescents, long backed-points on thin blades, microlaminar débitage, and naviform cores.

The distribution of archaeological sites in this part of the study region suggests a direct connection between the Sabkhat al Mouh basin and the area of Abu Fawares during the late Epipaleolithic.

The Northern Margin of the Sabkhat al Mouh

The northern margin of the sabkhat is delimited by a terrace consisting of Pleistocene alluvial gravel and lacustrine marls (Besançon et al. 1997; Sakaguchi 1978, 1987; Soulidi-Kondratiev 1966), representing the substrate of the palm gardens and the ruins of Palmyra. The margin of the terrace of Palmyra is dissected in narrow valleys separated by branching ridges, merging into recent lake deposits. These are generally made up of pale brown to pale yellow carbonates and gypsum-rich fine sand. Such deposits give rise to a shallow lacustrine terrace up to two kilometers wide. The external part of the lacustrine terrace is completely deprived of archaeological material, while several clusters of lithics were discovered along the margins of the valleys dissected into the terrace (Cremaschi and Zerboni 2012). These sites consist of scattered lithics dating to the late Geometric Kebaran and mostly to the Natufian phases (geometrics, backed points, end-scrapers, burins, and lunates, with common microblade débitage and microlaminar cores). A high concentration of artifacts was recorded in correspondence with well-preserved primary structures such as stone circles delimiting hearths, and archaeological sediments forming a flat and large mound. At one of these sites, a tethering stone (Pachur 1992) was found, its surface weathered and corroded by wind erosion, rendering its contextual link to the prehistoric material reliable. The occurrence of tethering stones in the arid Syrian steppe may be related to the exploitation of wild fauna (Cremaschi and Zerboni 2012), which has up to now been witnessed, outside the margins of the oasis, by desert kites (Morandi Bonacossi and Iamoni 2012). Archaeological sites found in this part of the sabkhat are connected to dark organic matter–rich silt and sand (with lithics laying upon and in some cases are entombed within), often cemented by gypsum, superposed on the lacustrine carbonates, and formed in a marsh environment. A sample of the organic matter–rich sediment (Site 389) was submitted for radiocarbon dating and the result of 5730 ± 30 uncal years B.P. (UGAMS-10475, $\delta^{13}C = -24.5‰$), corresponding to 4683–4496 cal years B.C. (6633–6446 cal years B.P.), indicates a later phase of activity of springs at the base of the terrace that occurred during the fall of the lake level.

WITHIN THE OASIS

The perfect conservation of the traditional gardens and the intense cultivation make the archaeological visibility in the palm grove very low. Some concentrations of Late Epipaleolithic and Neolithic flint artifacts have been found (Figure 1.2). The most relevant site consists of an archaeological tell (Site 288; Cremaschi and Zerboni 2012), more than 25,000 m² in area, located at the western fringe of the palm garden along the main road to Damascus, at the boundary of the limestone slope (Figure 1.8).

FIGURE 1.8. (a) General view of the Neolithic tell at Site 288 (the present day oasis is in the background); the arrow indicates the position of the main stratigraphic section exposed in a quarry. (b) Stratigraphic section of the tell; and (c) a detail of the mud-bricks.

The survey of the area revealed a dense concentration of pottery and lithics (Figure 1.9), including fragments of White Ware with basketry impressions. We found some lithics dating to the late PPNB (and possibly they are associated with the White Ware fragments), and sherds dating to the Early Pottery Neolithic (red painted pottery fragments, possibly of the Pre-Halaf red-slipped Standard Ware type); to the latter period fragments of White Ware can also be attributed (see Borrell et al. 2013). Some material dating to the Late Chalcolithic (Chaff-faced ware, in some cases red slipped) and the Classical period (Palmyrene, Byzantine, Islamic) were also found. A rich set of lithics was also collected, including Jericho arrowheads and obsidian blades related to the Neolithic cultures; the occurrence of obsidian artifacts suggests a long-distance provenience of the raw material. A stratigraphic sequence, exposed at the western fringe of the mound by a quarry (Figure 1.8), shows some sand lenses discontinuously dipping toward the northeast, including charcoal, structural stones, and faunal remains, covered by recently reworked massive organic sands, rich in ash and including pottery dating to the Classical age. At the base of the sequence, a large portion of a fallen wall was identified, composed of mud-bricks made up of silty-clay rich in organic

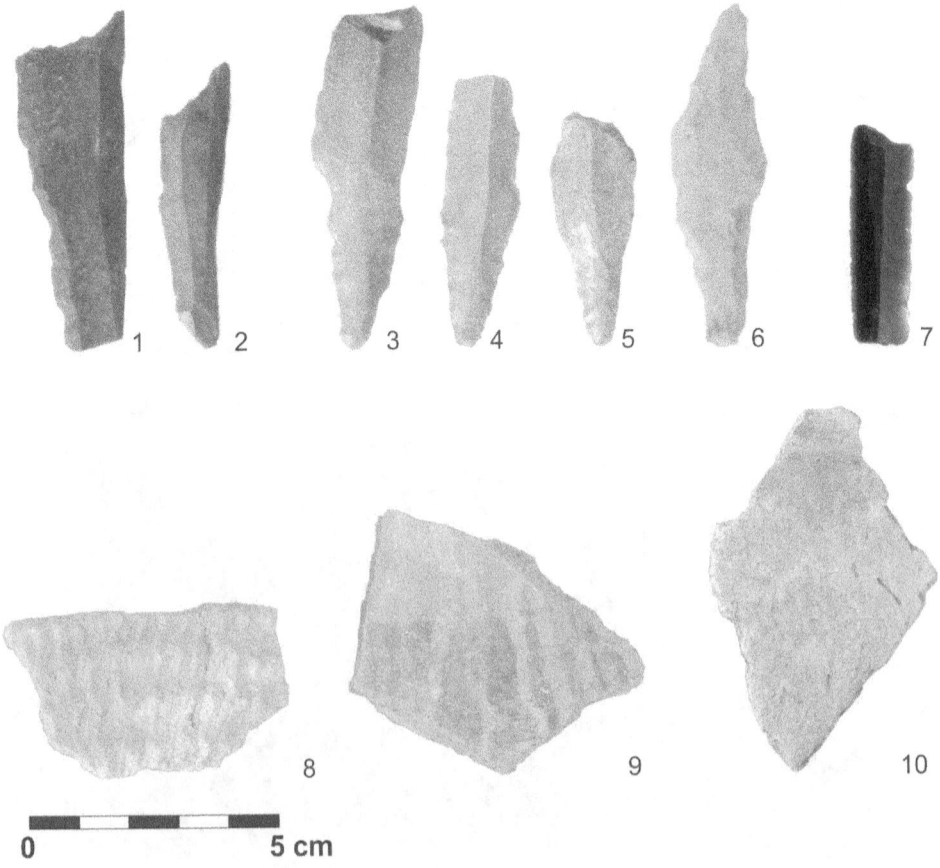

FIGURE 1.9. Some archaeological findings from the tell at Site 288. 1, 2: truncations; 3–6: Byblos points; 7: obsidian blade fragment; 8: White Ware fragment with basket impression; 9, 10: red slipped pottery fragments.

matter (similar to that observed along the sequence of wadi Aid and associated with dry springs) mixed with vegetal inclusions (Figure 1.8).

Discussion

On the basis of our data and some observations published by Sakaguchi (1978), we can trace the paleoenvironmental evolution of the region of Palmyra since the Upper Pleistocene. The territory of Palmyra since MIS 5 has been characterized by two opposite environmental settings. The first corresponds to the high stand of the lake occurring contemporary to the early Mousterian settlement in the area, while the second is represented by the present-day semi-arid condition, with a salty marsh occupying the lower part of the basin.

During MIS 5 the cuvette of Palmyra was occupied by a large and deep freshwater lake, which contributed to the formation of lacustrine terraces at the margin of the Sabkhat al Mouh. This is confirmed by paleoenvironmental information from the Douara and Jerf el-Ajla caves (Julig et al. 1999; Sakaguchi 1978) that points for that period to the same climatic settings, marked by high precipitation. A further confirmation of wet environmental conditions during the Upper Pleistocene comes from the new U/Th age of the calcareous tufa sealing the stratigraphic sequence at Jerf el-Ajla cave. We dated this speleothem to ca. 42,000 years B.P., suggesting that substantial wet conditions persisted also during MIS 4, allowing spring activity in the mountain areas and enabling the Mousterian groups to inhabit the region. After that, the former lake basin was replaced by a sabkhat that was less extensive than the freshwater lake. The margins of the basin were affected by strong wind erosion; the windblown sands were deposited as dunes along the shores of the former lake and were pushed by the eastern winds to wrap the leeward slopes of mountains surrounding the oasis of Palmyra. The environmental transition from a freshwater into an evaporitic environment required a major climatic change. The transition from wet to dry conditions had consistent geomorphologic consequences on the landscape; it enhanced the surface processes of erosion and sedimentation connected to the drop of the lake level. The erosional surface and the gravel occurring at the base of the wadi Aid sequence is interpreted as an effect of the drop of the lake level, thus indicating a dry phase. The latter can be dated to after the Mousterian period, thanks to the artifacts abraded by fluvial transportation entombed in the gravel unit, and attributed to the Last Glacial Maximum (MIS 2). To the same period, we date the aggradation of the fossil dunes occurring at the front of the delta of wadis al-Hallabat, al-Annan, and Habash (Cremaschi and Zerboni 2012).

Our research suggests that a new rise of the water level in the sabkhat occurred after increased precipitation during the period between the Late Pleistocene and Early Holocene, leading to the formation of a new freshwater lake. The rise of the lake level is indicated by lacustrine sand and silty-clay, quite rich in organics, found in the wadi Aid section, and by the terraces composed of reworked gypsum sand distributed all around the present-day sabkhat. The radiocarbon age of the organic sediments of wadi Aid (Sakaguchi 1987), suggesting a Pleniglacial age, is to be regarded as too old (Cremaschi and Zerboni 2012), but the archaeological context offers the opportunity to date environmental changes. The concentration of open-air sites, interpreted as hunter-gatherers' temporary camps ranging from the Geometric Kebaran to the PPNB or, at least, to the Early PN, indicates conditions suitable for human and animal life between at the Pleistocene/early Holocene transition. This attribution is further confirmed by the radiocarbon dating of charcoal from a fireplace at the top of a dune on the southern fringe of the sabkhat.

The aforementioned tell at the western margin of the sabkhat was established since the PPNA and PPNB and indicates the first stable exploitation of the oasis environment. The portion of the collapsed wall at the base of the sequence suggests the presence of mud-brick buildings entombed within the tell. Moreover, the composition of the mud-bricks (organic matter–rich silt and sand) confirms the occurrence of a swamp with thick organic sediments close to the site and therefore increased water availability. On the contrary, the sand that forms most of the exposed stratigraphic sequence is aeolian, originating from the deflation of the freshly exposed sabkhat surface in consequence to a drop of its level. From the paleoenvironmental point of view, this indicates that since the late PPNB (and into the later seventh millennium, according to the age of the White Ware fragments included in the sequence) the area suffered substantial aridification. Moreover, the occurrence of a mud-brick structure further suggests a permanent settlement, and we can hypothesize the occurrence of productive activities in the surrounding area. If we compare this evidence with the tell site at Al Kowm (some 150 km East of Palmyra), which represents a key site for understanding the cultural transition from the Epipaleolithic to the Neolithic (Stordeur 1993, 2000; Borrell et al. 2013), we can note a substantial difference. At Al Kowm, its investigators highlight a substantial continuity between the environmental settings during the Neolithic, the present-day conditions, and the adaptation of prehistoric societies to a dry steppe environment. On the contrary, in the area of Palmyra we found evidence for a climatic discontinuity occurring between a wet phase dating to the Lateglacial and Early Holocene and a later dry period, which presumably started at the end of the ninth millennium B.P. The same climatic discontinuity at that time has also been described elsewhere in the Levant (e.g., Blockley and Pinhasi 2011; Robinson et al. 2006). In the study area, evidence from the wadi Aid section and the archaeological context suggests that the aggradation of aeolian sand occurred in the late PPNB period, possibly coinciding with the cooling/drying episode during the worldwide interruption of the Holocene climatic optimum at ca. 8,200 years B.P. (Alley et al. 1997; Wiersma and Renssen 2006). This phase led to the markedly cold and arid conditions that were observed at 8200 cal years B.P. in a number of high-resolution climate proxies in the eastern circum-Mediterranean region (e.g., Berger and Guilaine 2009; Weninger et al. 2006). In the Palmyrene, climatic changes may have favored a major shift in settlement patterns and subsistence strategies, which also includes a population relocation (Barker 2006; Sherratt 1997) within the limits of the oasis. Before this time the hunter-gatherer groups exploited all physiographic units of the area, as documented by archaeological sites distributed from the Abu Fawares region to the southern shores of the Sabkhat al Mouh. Later, permanent settlements were concentrated within the oasis, where water resources survived despite reduced rainfall, while the surrounding dry steppe was exploited for seasonal hunting (Morandi Bonacossi and Iamoni 2012). In any case, the oasis did not represent a stable geomorphic system and underwent a progressive reduction of water reservoirs, testified by the later extinction of springs at the foot of the terrace.

In a regional perspective, the paleoclimatic reconstruction inferred from environmental changes occurred in the oasis of Palmyra since the Last Glacial Maximum fit well with information recorded in independent paleohydrological records described in the Levant. The sequence at wadi Aid and geomorphological features at the southern fringe of the sabkhat testify to an arid phase dated to the Last Glacial Maximum (Bar-Yosef 2011). During the same period in the Levant a cold and dry phase occurred, confirmed, for instance, by the reduction of the water level at Lake Lisan (Bartov et al. 2002). The latest Pleistocene rise of the lake level in the Sabkhat al Mouh basin corresponded to a humid period recorded in speleothems and in lake cores in the Hula and Ghab valleys (van Zeist et al. 2009) after the Last Glacial Maximum. After that, several archives for proxy data described an early Holocene phase marked by intense precipitation; for instance, a significant increase in precipitation is documented in the speleothems from the Soreq Cave in central Israel (Bar-Matthews et al. 1999). Additionally, the abrupt climate switch at the time of the 8.2 B.P. event is reported in various paleoarchives from the Levant and western Mediterranean region (e.g., Berger and Guilaine 2009; Mayewski et al. 2004; Spötl et al. 2010; Weninger et al. 2006), where it contributed to some rapid and irreversible environmental changes.

The change of the settlement pattern in the area of Palmyra underlines the process of nucleation of the oasis (Figure 1.10). From the Geometric Kebaran up to the PPNB,

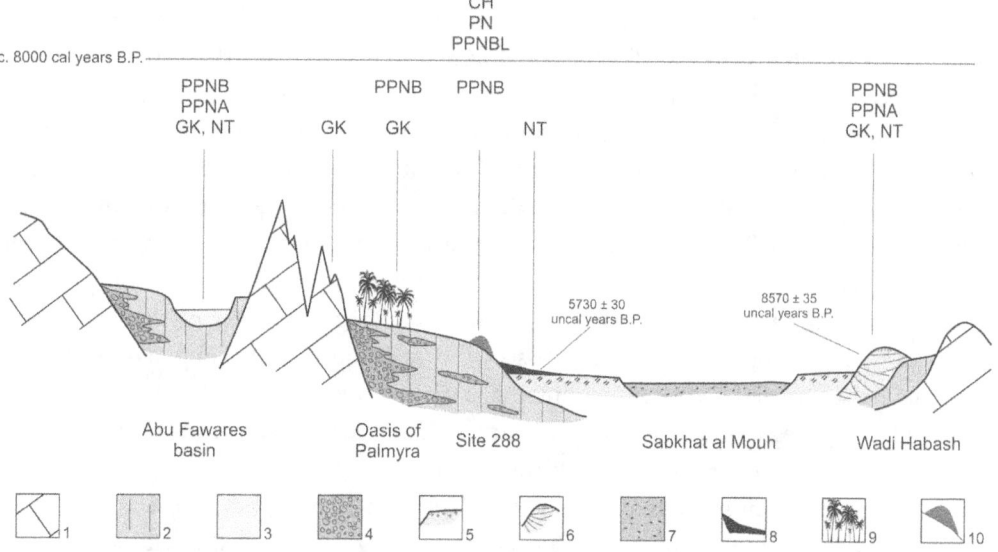

FIGURE 1.10. Cross-section of the study region (not to scale) reconstructing the main stratigraphic and landscape units described in the text and the chronological distribution of archaeological sites in each unit (modified from Cremaschi and Zerboni 2012). Key: 1, bedrock; 2, Pleistocene lake sediments; 3, Holocene lake sediments; 4, fluvial gravel; 5, lake terrace; 6, gypsum-sand dune; 7, sabkhat; 8, spring deposit; 9, oasis; 10, tell; GK, Geometric Kebaran; NT, Natufian; PPNA, Pre-Pottery Neolithic A; PPNB, Pre-Pottery Neolithic B; PPNBL, late Pre-Pottery Neolithic B; PN, Pottery Neolithic; CH, Chalcolithic.

archaeological vestiges are not homogeneously distributed in the region, but concentrated in some ecological niches sustained by environmental humidity (at Abu Fawares and in the eastern part of the Sabkhat al Mouh), while the western part of the sabkhat proved to be empty of settlements. The distribution of sites seems to be unrelated to the modifications of the margins of the sabkhat, but centered on the corridor connecting the basins of Palmyra and Abu Fawares, where a permanent settlement grew from the PPNB to the ceramic Neolithic and the Chalcolithic. Furthermore, no other sites of this period have been found in the surrounding of the oasis; consequently, a main gap in information exists up to the beginning of the Early Bronze Age, when inside the oasis and in the surrounding steppe a different settlement pattern developed. The growth of sedentary settlements within the oasis, which possibly developed toward a complex social system in the face of a changing landscape, represents the interconnection between historical socioeconomic and climatic factors in the social organization. Even though it is not easy to quantify the relative contribution of climate and socioeconomic factors, our case study can be assumed as a model to interpret the cultural trajectory of poorly known proto-urban systems developed in arid and semi-arid environmental contexts of the Levant.

Conclusions

A multidisciplinary approach, integrating the distribution of archaeological evidence, geomorphological observations, and laboratory analyses allows the reconstruction of the environmental history of the region surrounding Palmyra since the Upper Pleistocene, with special emphasis on the Pleistocene/Holocene transition and the first millennia of the Holocene. This specific phase saw the systematic occupation of the region, from the southern margin of the Sabkhat al Mouh to the Abu Fawares basin, by hunter-gatherer groups under wetter environmental conditions. Greater than present-day water availability persisted for several millennia, even sustaining a freshwater lake, but from the eighth/seventh millennium B.P. onward it was progressively reduced, leading to the contraction of the vegetal cover and the nucleation of the oasis. During the same period, human groups moved toward the present-day Palmyra, occupying the shores of the lake (which was slowly shrinking) and the green oasis. The relocation of the population and the exploitation of natural resources within the limits of the oasis possibly followed, as in other adjoining regions, the 8.2 B.P. event; the latter may have contributed to push a change in subsistence strategies started some centuries before.

This study demonstrates that a major change in rainfall led to the reduction of water reservoirs in the region; the main effects were the drop of the lake level in the Sabkhat al Mouh, which turned into a saline marsh, the desiccation of the Abu Fawares basin, and the increase of wind activity. Moreover, the same geomorphological process contributed to the contraction of the areas suitable for life, and the consequent reduction of the size of the oasis of Palmyra. This natural phenomenon was almost synchronous with the development of a sedentary settlement within the oasis, and likely also with a change in subsistence strategies. Even though it is not easy to quantify the relative contributions of climate and socioeconomic factors in this transition, on the basis of our preliminary data we can suggest a third way. The stable occupation of the oasis started under relatively humid conditions (as confirmed by mud-bricks made of lacustrine organic mud), but its exploitation continued in a drying environment; incoming aridity and the necessity of exploiting the natural resources found in the oasis may have accelerated preexisting cultural processes. Regardless, Late Neolithic and later communities in reality never completely abandoned the semi-arid territories surrounding the oasis, which were continually exploited by herders and especially by hunters. In the region, the seasonal hunting of gazelles is confirmed by the extraordinary number of desert-kites (Morandi Bonacossi and Iamoni 2012) distributed on the slopes of the mountains surrounding Palmyra and located along the north-south migration route of Persian Gazelle between Syria and Jordan (Akkermans and Schwartz 2003; Bar-Oz et al. 2011).

From the archeological point of view, our preliminary data on the pottery and lithic materials found in the region suggest that they are remarkably similar to those found in other desert sites (e.g., El Kowm, Umm el Tlel). As reported by Borrell et al. (2013), significant differences exist in the material culture between the El Kowm region and the Palmyra (and Douara) basin. Notwithstanding this, our research permitted us to highlight the occurrence of White Ware in the Palmyrene and, most importantly, to discover the evidence of a permanent settlement within the limits of the oasis. The occurrence of a stratified archaeological sequence and buried architectural remains at site 288 permits to propose a first comparison with the archaeological record in the area of El Kowm, and should be carefully considered in the elaboration of models describing the complex pattern of colonization of the central Syrian steppe.

To broaden our perspective, we may compare this case study with the evolution of the oasis in the central Sahara some millennia later (Cremaschi 2001, 2005; Cremaschi and Zerboni 2011; Zerboni 2013). In the Fezzan region (SW Libya) the formation of the oases system began with the Mid-Holocene transition, which corresponded to a dramatic reduction in the intensity of monsoon precipitation and the progressive size-reduction of the so-called Green Sahara (deMenocal et al. 2000; Gatto and Kerboni 2015). In this phase, the adaptive strategies of the late Pastoral groups also changed toward a more sedentary model based upon the agricultural exploitation of the oasis and growing social complexity (Cremaschi and di Lernia 2001). Evidently, also in this case the climatic factor did not play the only role in causing the relocation of population. Socioeconomic factors were surely relevant but it is also meaningful to note that this phase corresponds, in the paleohydrological record, to a strong decrease of water availability, as consequence of a prolonged drought, and to the reduction of natural resources. However, even after aridification, dry lands were never completely abandoned, but were subject to reduced human exploitation. As occurred in the Syrian Desert, these areas, apparently remote and abandoned, were frequented by nomads, hunters, and shepherds, who exploited their scarce resources (Cremaschi and Zerboni 2011). Finally, we suggest that in these fragile environments, the seemingly marginal but protracted human pressure may have ultimately contributed to accelerating landscape degradation. Large-scale soil erosion and the spread of desertification may have resulted from the human-induced intensification of such interconnected processes as aeolian erosion, soil stripping, and the extreme reduction of vegetal cover. The development and intensification of farming in this fragile landscape aggravated the effects of aridity.

Acknowledgments

The Syrian-Italian Geo-Archaeological Mission in the area surrounding Palmyra was established in the year 2008 and carried out in cooperation between the universities of Milano and Udine (Italy) and the Directorate General of Antiquities and Museums of Damascus (Syria). The Mission is directed by Prof. M. Cremaschi (University of Milano) and co-directed by Prof. D. Morandi Bonacossi (University of Udine) and Dr. Michel Al-Maqdissi (DGM-Damascus). The research concerning the paleoenvironmental and prehistoric research was financially supported by the Italian Ministry for Foreign Affairs (MAE) and the Università degli Studi di Milano (Contributo Rettorale). Warm thanks to M. Hammoudi, Y. Kanhoush, E. Sam'an, S. Shabo (DGAMS-Damascus), F. Garbasi (University of Milano), and A. Savioli and M. Iamoni for helping in the fieldwork; we are also indebted to the late Khaled Al-Assad (Museum of Palmyra) for indicating the existence of the tell at Site 288. We are grateful to A. Perego for remote sensing study and C. Zanardi for performing sedimentological analyses. Finally, this work is dedicated to the people of Palmyra/Tadmor. Our warmest thanks go to P. Biehl and O. Nieuwenhuyse for inviting us to join this editorial project.

Note

1. Total organic carbon was estimated by loss on ignition (LOI; Heiri et al. 2001); samples were air-dried and organic matter was oxidized at 500–550°C to carbon dioxide and ash, then the weight lost during the reaction was measured by weighing the samples before and after heating. Calcium carbonate equivalents were chemically performed using a Dietrich–Frühling calcimeter; the method is based on the measurement of the volume of CO_2 developed by acid reacting with the bulk sample, which is proportional to the carbonate concentration. Grain-size analyses (Gale and Hoare 1991) were performed after removing organics by hydrogen peroxide (130 vol) treatment; sediments were wet sieved (diameter from 1000 to 63 mm), then the finer fraction (<63 mm) was determined by hydrometer on the basis of Stokes's law; results are expressed as cumulative grain size distributions.

References Cited

Akazawa, T. 1979 Prehistoric Occurrences and Chronology in Palmyra Basin, Syria. In *Paleolithic Site of Duara Cave and Paleogeography of Palmyra Basin. Part II: Prehistoric Occurrences and Chronology in Palmyra Basin*, edited by K. Hanihara and T. Akazawa, pp. 201–220. The University of Tokyo Bulletin 16. The University Museum, Tokyo.

Akkermans, P. M. M. G., and G. M. Schwartz 2003 *The Archaeology of Syria: From Complex Hunter-Gatherers to Early Urban Societies (c. 16,000–300 BC)*. Cambridge University Press, Cambridge.

Alley, R. B., P. A. Mayewski, T. Sowers, M. Stuiver, K. C. Taylor, and P. U. Clark 1997 Holocene Climatic Instability: A Prominent, Widespread Event 8200 years ago. *Geology* 25:483–486.

Al-Maqdissi, M., Y. Kanhouche, M. Cremaschi, and D. Morandi Bonacossi 2009 Présentation sommaire des travaux archéologiques de la mission syro-italienne: Les fouilles archéologiques à Mishirfeh–Qatna et la prospection de la Palmyrène occidentale. *Studia Orontica* I:5–20.

Anderson, D. G., K. A. Maasch, and D. H. Sandweiss (editors) 2007 *Climate Change and Cultural Dynamics*. Academic Press, London.

Barker, G. 2006 *The Agricultural Revolution in Prehistory*. Oxford University Press, Oxford.

Bar-Matthews, M., A. Ayalon, A. Kaufman, and G. Wasserburg 1999 The Eastern Mediterranean Paleoclimate as a Reflection of Regional Events: Soreq Cave, Israel. *Earth and Planetary Science Letters* 166:85–95.

Bar-Oz, G., M. Zeder, and F. Hole 2011 Role of Mass-Kill Hunting Strategies in the Extirpation of Persian Gazelle (*Gazella subgutturosa*) in the Northern Levant. *PNAS* 108:7345–7350.

Bar-Yosef, O. 2011 Climatic Fluctuations and Early Farming in West and East Asia. *Current Anthropology* 52:175–193.

Bartov, Y., M. Stein, Y. Enzel, A. Agnon, and Z. Reches 2002 Lake Levels and Sequence Stratigraphy of Lake Lisan, the Late Pleistocene Precursor of the Dead Sea. *Quaternary Research* 57:9–21.

Berger, J.-F., and J. Guilaine 2009 The 8200 cal BP Abrupt Environmental Change and the Neolithic Transition: A Mediterranean Perspective. *Quaternary International* 200:31–49.

Besançon, J., A. Delgiovine, M. Fontugne, C. Lalou, P. Sanlaville, and J. Vaudour 1997 Mise en évidence et datation de phases humides du Pléistocène supérieur dans la région de Palmyre (Syrie). *Paléorient* 23:5–23.

Blockley, S. P. E., and R. Pinhasi 2011 A Revised Chronology for the Adoption of Agriculture in the Southern Levant and the Role of Lateglacial Climatic Change. *Quaternary Science Reviews* 30:98–108.

Borrell, F., E. Boëda, M. Molist, and H. Al-shakel 2013 The First Half of the 7th Millennium cal. B.C. in the El Kowm basin in Central Syria: Umm el-Tlel Revisited. In *Interpreting the Late Neolithic of Upper Mesopotamia*, edited by O. P. Nieuwenhuyse, R. Bernbeck, P. M. M. G. Akkermans, and J. Rogasch, pp. 277–287. Papers on Archaeology of the Leiden Museum of Antiquities, Brepols, Turnhout, Belgium.

Brooks, N. 2006 Cultural Responses to Aridity in the Middle Holocene and Increased Social Complexity. *Quaternary International* 151:29–49.

Cremaschi, M. 2001 Holocene Climatic Changes in an Archaeological Landscape: The Case Study of Wadi Tanezzuft and its Drainage Basin (SW Fezzan, Libyan Sahara). *Libyan Studies* 32:3–27.

Cremaschi, M. 2005 The Barkat Oasis in the Changing Landscape of the Wadi Tanezzuft during the Holocene. In *Aghram Nadharif. The Barkat Oasis (Sha'abiya of Ghat, Libyan Sahara) in Garamantian Times*, edited by M. Liverani, pp. 13–20. Arid Zone Archaeology, Vol. 5. Edizioni All'Insegna del Giglio, Firenze.

Cremaschi, M., and S. di Lernia 2001 Environment and Settlements in the Mid-Holocene Palaeo-Oasis of Wadi Tanezzuft (Libyan Sahara). *Antiquity* 75:815–825.

Cremaschi, M., and A. Zerboni 2011 Human Communities in a Drying Landscape. Holocene Climate Change and Cultural Response in the Central Sahara. In *Landscape and Societies, Selected Cases*, edited by P. I. Martini and W. Chesworth, pp. 67–89. Springer Science, Dordrecht.

Cremaschi, M., and A. Zerboni 2012 Adapting to Increasing Aridity: The Cuvette of Palmyra (central Syria) cur from Late Pleistocene to Early Holocene. In *Variabilités environnementales, mutations sociales: natures, intensités, échelles et temporalités des changements. XXXIIes* rencontres internationales d'archéologie et d'histoire d'Antibes, edited by F. Bertoncello, and F. Braemer, pp. 37–52. Editions APDCA, Antibes.

Cremaschi, M., and A. Zerboni 2013 Fewet: An Oasis at the Margin of Wadi Tanezzuft (central Sahara). In *Life and Death of a Rural Village in Garamantian Times. Archaeological Investigations in the Fewet Oasis (Libyan Sahara)*, edited by L. Mori, pp. 7–15. Arid Zone Archaeology Monographs, 6, Edizioni all'Insegna del Giglio, Firenze.

deMenocal, P., J. Ortiz, T. Guilderson, J. Adkins, M. Sarnthein, L. Baker, and M. Yarusinsky 2000 Abrupt Onset and Termination of the African Humid Period: Rapid Climate Responses to Gradual Insolation Forcing. *Quaternary Science Reviews* 19:347–361.

Gale, S. J., and P. G. Hoare 1991 *Quaternary Sediments*. Belhaven Press, New York.

Gatto, M. C., and A. Zerboni 2015 Holocene supra-regional environmental changes as trigger for major socio-cultural processes in Northeastern Africa and the Sahara. *African Archaeological Review* 32:301–333.

Hanihara, K., and Y. Sakaguchi (editors) 1978 Palaeolithic Site of Douara Cave and Paleogeography of Palmyra Basin in Syria. Part I: Stratigraphy and Paleogeography in the Late Quaternary. The University of Tokyo Bulletin 14. The University Museum, Tokyo.

Hassan, F. A. (editor) 2002 *Droughts, Food, and Culture*. Kluver Academic/Plenum, New York.

Heiri, O., A. F. Lotter, and G. Lemcke 2001 Loss on Ignition as a Method for Estimating Organic and Carbonate Content in Sediments: Reproducibility and Comparability of Results. *Journal of Paleolimnology* 25:101–110.

Julig, P. J., D. G. E. Long, H. B. Schroeder, W. J. Rink, D. Richter, and H. P. Schwarcz 1999 Geoarchaeology and New Research at Jerf al-Ajla Cave, Syria. *Geoarchaeology* 14:821–848.

Krumbein, W. C., and L. L. Sloss 1963 *Stratigraphy and Sedimentation*. Freeman, San Francisco.

Kühn, P., D. Pietsch, and I. Gerlach 2010 Archaeopedological Analyses Around a Neolithic Hearth and the Beginning of Sabaean Irrigation in the Oasis of Ma'rib (Ramlat as-Sab'atayn, Yemen). *Journal of Archaeological Science* 37:1305–1310.

Kukal, Z. 1971 *Geology of Recent Sediments*. Central Geological Survey, Prague.

Kuper, R., and S. Kröpelin 2006 Climate-Controlled Holocene Occupation in the Sahara: Motor of Africa's Evolution. *Science* 313:803–807.

Lee, M. K., Y. I. Lee, H. S. Lim, J. I. Lee, J. H. Choi, and H. I. Yoon 2010 Comparison of Radiocarbon and OSL Dating Methods for a Late Quaternary Sediment Core from Lake Ulaan, Mongolia. *Journal of Paleolimnology* 45:127–135.

Mayewski, P. A., E. J. Rohling, C. J. Stager, W. Karlen, K. A. Maasch, D. L. Meeker, E. A. Meyerson, F. Gasse, S. Van Kreveld, K. Holmgren, J. Lee-Thorp, G. Rosqvist, F. Rack, M. Staubwasser, R. R. Schneider, and E. J. Steig 2004 Holocene Climate Variability. *Quaternary Research* 62:243–255.

Morandi Bonacossi, D., and M. Iamoni 2012 The Early History of the Western Palmyra Desert Region. The Change in the Settlement Patterns and the Adaptation of Subsistence Strategies to Encroaching Aridity: A First Assessment of the Desert-Kite and Tumulus Cultural Horizons. *Syria* 89:31–58.

Mori, L., M. C. Gatto, F. Ricci, and A. Zerboni 2013 Life and Death at Fewet. In *Life and Death of a Rural Village in Garamantian Times. Archaeological Investigations in the Fewet Oasis (Libyan Sahara)*, edited by L. Mori, pp. 373–387. Arid Zone Archaeology Monographs, 6, Edizioni all'Insegna del Giglio, Firenze.

Pachur, H.-J. 1992 Tethering Stones as Palaeoenvironmental Indicators. *Sahara* 4:13–32.

Reimer P. J., E. Bard, A. Bayliss, J. W. Beck, P. G. Blackwell, C. Bronk Ramsey, C. E. Buck, H. Cheng, R. L. Edwards, M. Friedrich, P. M. Grootes, T. P. Guilderson, H. Haflidason,

I. Hajdas, C. Hatté, T. J. Heaton, A. G. Hogg, K. A. Hughen, K. F. Kaiser, B. Kromer, S. W. Manning, M. Niu, R. W. Reimer, D. A. Richards, E. M. Scott, J. R. Southon, C. S. M. Turney, and J. van der Plicht 2013 IntCal13 and MARINE13 radiocarbon age calibration curves 0–50000 years cal BP. *Radiocarbon* 55:1869–1887.

Robinson S. A., S. Black, B. W. Sellwood, and P. J. Valdes 2006 A Review of Palaeoclimates and Palaeoenvironments in the Levant and Eastern Mediterranean from 25,000 to 5000 Years B.P.: Setting the Environmental Background for the Evolution of Human Civilization. *Quaternary Science Reviews* 25:1517–1541.

Rognon, P. 1980 Pluvial and Arid Phases in the Sahara: The Role of Non-Climatic Factors. *Palaeoecology of Africa* 27:45–62.

Sakaguchi, Y. 1978 Palmyra Pluvial Lake-Paleogeography around the Paleolithic Site of Douara. In *Palaeolithic Site of Douara Cave and Paleogeography of Palmyra Basin in Syria. Part I: Stratigraphy and Paleogeography in the Late Quaternary*, edited by K. Hanihara and Y. Sakaguchi, pp. 5–28. The University Museum, The University of Tokyo Bulletin 14, Tokyo.

Sakaguchi, Y. 1987 Paleoenvironments in Palmyra District during the Late Quaternary. In *Paleolithic Site of Duara Cave and Paleogeography of Palmyra Basin in Syria. Part IV: 1984 Excavations*, edited by T. Akazawa and Y. Sakaguchi, pp. 271–287. The University of Tokyo Bulletin 14. The University Museum, Tokyo.

Sherratt, A. 1997 Climatic Cycles and Behavioural Revolutions: The Emergence of Modern Humans and the Beginning of Farming. *Antiquity* 71:271–287.

Soulidi-Kondratiev, E. D. 1966 The Geological Map of Syria, Scale 1:200,000, Sheet I-37-XV (Tudmor). Explanatory Notes. Ministry of Industry, Danascus, Syria.

Spötl, C., Kurt N., G. Patzelt, R. Boch, and Daphne Team 2010 Humid Climate During Deposition of Sapropel 1 in the Mediterranean Sea: Assessing the Influence on the Alps. *Global and Planetary Change* 71:242–248.

Stordeur, D. 1993 Sedentaire et Nomades du PPNB final dans le desert de Palmyre (Syrie). *Paleorient* 19:187–204.

Stordeur, D. 2000 *El Kowm 2, une ile dans le desert*. CNRS editions, Paris.

van Zeist, W., U. Baruch, and S. Bottema 2009 Holocene Palaeoecology of the Hula Area, Northeastern Israel. In *A Timeless Vale: Archaeological and Related Essays on the Jordan Valley in Honour of Gerrit van der Kooij on the Occasion of His Sixty-Fifth Birthday*, edited by E. Kaptijn and L. P. Petit, pp. 29–64. Leiden University Archaeological Studies 19. Leiden University Press, Leiden.

Walker M. 2005 *Quaternary Dating Methods*. John Wiley and Sons, Chichester, UK.

Weninger, B., E. Alram-Stern, E. Bauer, L. Clare, U. Danzeglocke, O. Jöris, C. Kubatzki, G. Rollefson, H. Todorova, and T. van Andel 2006 Climate Forcing Due to the 8200 cal BP Event Observed at Early Neolithic Sites in the Eastern Mediterranean. *Quaternary Research* 6:401–420.

Wiersma, A. P., and H. Renssen 2006 Model-Data Comparison for the 8.2 ka BP Event: Confirmation of a Forcing Mechanism by Catastrophic Drainage of Laurentide Lakes. *Quaternary Science Reviews* 25:63–88.

Zerboni, A. 2013 Early Holocene Palaeoclimates in Northern Africa: An Overview. In *Neolithisation of Northeastern Africa*, edited by N. Shirai N., pp 65–82. Studies in Early Near Eastern Production, Subsistence, and Environment, 16. ex Oriente, Berlin.

CHAPTER TWO

When the Going Gets Tough

Risk Minimization Responses to the 8.2 Event in the Near East and Their Role in Emergence of the Halaf Cultural Phenomenon

Mandy Mottram

Abstract *Archaeologists have long sought explanations for the apparently sudden and widespread appearance across northern Mesopotamia during the Late Neolithic of sites characterized by a vibrantly painted and stylistically homogeneous pottery known as Halaf, after the site where it was first identified. Throughout much of the twentieth century, explanations centered around either large-scale human migrations or the rapid expansionary diffusion of cultural traits via such means as intensified trading, simple low-level exchange, or even itinerant potters. Discoveries of the last two decades, as well as a growing rejection by archaeologists of the culture-historical models on which these explanations are based, have led to a reappraisal of Halaf origins, with scholars now recognizing that the emergence of the Halaf ceramic style was a gradual process that occurred simultaneously across a wide area. Few studies have attempted to resolve what triggered these events, although their coincidence with the 8.2 event is now well recognized.*

In this article I contend that the processes that culminated in the emergence of the distinctive Halaf painted pottery style and associated developments were climate-driven. Evidence of changes in subsistence practices, settlement patterns and organization, architecture, and other aspects of material culture from sites across the Mesopotamian north and adjacent regions is presented to support the contention that many of the expressions of cultural and social change observed in the archaeological record during the period ca. 6300–5950 B.C. correspond to progressively more drastic strategies employed by Late Neolithic societies for coping with resource scarcity and minimizing future risk during an extended period of environmental unpredictability. The Early Halaf represents the point at which the more costly of these measures—i.e., those involving the mobilization of social relationships both within and outside the immediate social group—were becoming institutionalized within social and cultural practice. Alliance creation and contractual obligations aimed at ensuring against future food scarcity were central to the structuring of Halaf societies and were established and maintained through communal activities in which the decorated pottery played an important symbolic role.

INTRODUCTION

Archaeologists have long sought explanations for what was once perceived as the sudden and widespread appearance across Upper Mesopotamia, and adjacent regions of the Late Neolithic, of sites characterized by a vibrantly painted and stylistically homogeneous pottery known as Halaf, after the site where it was first identified. Throughout much of the twentieth century, explanations for the widespread dispersal of Halaf sites and the painted pottery centered around either large scale migrations or the diffusion of cultural traits via such means as intensified trading, simple low-level exchange, and even itinerant potters (e.g., Davidson 1977; Garstang 1953; Mellaart 1975; Perkins 1949; Watson 1983). Whatever its origins, there was general concurrence that such a widespread style was evidence of a greater degree of social interaction than had occurred previously.

More recently, however, archaeologists have come to recognize that well before the start of the Halaf era there was already widespread sharing of cultural traits and interregional interaction, with the result that the general consensus now seems to be that what was once regarded as a single, bounded "culture" originating from some "'homeland" in Upper Mesopotamia, in fact constituted the aggregate of a number of simultaneous local developments (Akkermans 1993; Akkermans and Schwartz 2003; Campbell 1992; Campbell et al. 1999; Nieuwenhuyse 2000). The main catalyst for this paradigm shift was the discovery at Tell Sabi Abyad, located in the Balikh Valley of northern Syria, of occupation levels that documented for the first time the local evolution of a Halaf cultural assemblage out of earlier Neolithic foundations.

Starting around 6300 B.C., during what is now known as the Pre-Halaf (Figure 2.1), ceramic assemblages began to undergo a number of significant changes, including a diminution in the role of plain, coarse wares and a commensurate rise in importance of fine wares, the latter occurring in conjunction with the appearance of more complex vessel shapes and increasingly intricate forms of painted decoration (Cruells and Nieuwenhuyse 2004). At the same time, important changes occurred in architecture and settlement organization, including the appearance of circular *tholos* buildings identical to those usually associated with Halaf sites. Also documented for the first time are various stamp seals, figurines and other small objects, all of which were, again, usually associated with Halaf sites. The foundations for many of the key characteristics of the Halaf cultural assemblage were therefore laid well before it is possible to observe the conjunction of traits usually identified as Halaf, which should consequently be viewed as the culmination of earlier developments.

At the same time, archaeologists began to abandon the culture-historical models that had dominated theoretical approaches to the study of the Late Neolithic in Mesopota-

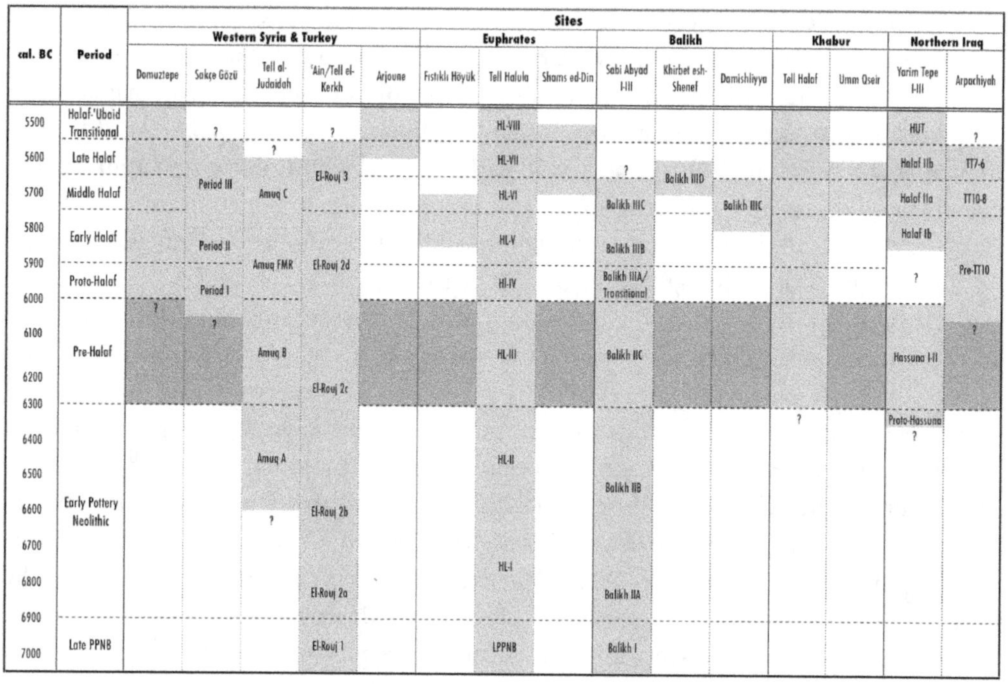

cal. BC	Period	Western Syria & Turkey					Euphrates			Balikh			Khabur		Northern Iraq	
		Domuztepe	Sakçe Gözü	Tell al-Judaidah	'Ain/Tell el-Kerkh	Arjoune	Fıstıklı Höyük	Tell Halula	Shams ed-Din	Sabi Abyad I-III	Khirbet esh-Shenef	Damishliyya	Tell Halaf	Umm Qseir	Yarim Tepe I-III	Arpachiyah
5500	Halaf-'Ubaid Transitional		?		?			HL-VIII							HUT	?
5600	Late Halaf			?				HL-VII		?	Balikh IIID				Halaf IIb	TT7-6
5700	Middle Halaf		Period III	Amuq C	El-Rouj 3			HL-VI		Balikh IIIC		Balikh IIIC			Halaf IIa	TT10-8
5800	Early Halaf		Period II					HL-V		Balikh IIIB					Halaf Ib	
5900	Proto-Halaf			Amuq FMR	El-Rouj 2d			HL-IV		Balikh IIIA/Transitional					?	Pre-TT10
6000			Period I													
6100	Pre-Halaf	?	?		Amuq B			HL-III		Balikh IIC					Hassuna I-II	?
6200																
6300					El-Rouj 2c									?	Proto-Hassuna	
6400								HL-II							?	
6500					Amuq A					Balikh IIB						
6600	Early Pottery Neolithic				El-Rouj 2b											
6700					?											
6800								HL-I								
6900					El-Rouj 2a					Balikh IIA						
7000	Late PPNB				El-Rouj 1			LPPNB		Balikh I						

FIGURE 2.1. Relative chronology for sites mentioned in the text showing current culture-historical phasing against the broader period of aridity associated with the 8.2 ka abrupt climate event (shown in darker grey).

mia, in which different ceramic traditions were seen as the product of different peoples. Campbell (1992:208–214) has long asserted the deceptiveness of the Halaf "culture" model, and as an alternative has suggested that assessment of stylistic variation might be a more effective means of identifying and understanding the spatial patterning of cultural traits and the changes in the use of style evident in the advent of painted pottery. Certainly, when comparing Halaf assemblages, it is clear that there was no single, coherent Halaf culture group (Akkermans 1993; Campbell 2007; Mottram 2010; Nieuwenhuyse 2007). Nevertheless, there has been a tendency to exclude from our considerations of Halaf development sites outside the old Halaf "core area" despite the many commonalities that exist between them. Consequently, we may be ignoring potential clues as to how and why the phenomenon—of which the Halaf pottery is only the most visible expression—arose and functioned. If, over a matter of a few centuries at the end of the seventh millennium B.C., the ceramics produced by Neolithic societies across northern Syria, Mesopotamia, and southeastern Anatolia transformed from a rather limited array of plain, simply shaped vessels to the finely made, symbolically charged painted ceramics we associate with the Halaf, then it is almost inevitable that the societies that produced them were undergoing, or had undergone, some changes of their own that were articulated through these artifacts (Nieuwenhuyse 2007:29).

In this article, I contend that the processes that culminated in the emergence of the distinctive Halaf painted pottery and associated sociocultural developments were climate-driven, in the sense that they were an outcome of both local and regional adaptive responses to a period of global climate instability climaxing around 8,200 cal. B.P., now known as the 8.2 event (Alley et al. 1997; Barber et al. 1999; Rohling and Pälike 2005; Weninger et al. 2006). This event apparently brought about a reduction in rainfall over northern Africa and the Near East, resulting in cold, dry conditions and enhanced aridity of already arid areas (Rohling and Pälike 2005; Brooks 2006).

This is not to say that the possibility that climate change had a role to play in these developments has been completely overlooked (e.g., Akkermans et al. 2006; Campbell 1992; Nieuwenhuyse 2007). In 2006, the University of Leiden embarked on a new program of excavations at Tell Sabi Abyad designed specifically to address this issue, and since then attention has been drawn to a number of changes to subsistence patterns and material culture which the excavators attribute, at least in part, to human adaptations to cool, arid conditions resulting from the 8.2 event (Akkermans et al. 2010; van der Plicht et al. 2011; Nieuwenhuyse et al. this volume). Even so, the possible linkages between this period of climate deterioration and developments that culminated in the Halaf have yet to be adequately explored in terms of documented responses by human groups to long-term drought. For example, while noting that the Halaf ceramic tradition was the outcome of processes set in motion during the Pre-Halaf, Nieuwenhuyse (2007:213–223) does not attempt to resolve *why* these changes occurred, instead offering the explanation that the painted Halaf pottery, as a distinct ceramic style, arose through the course of emulation among competing Late Neolithic groups. I would suggest this situation occurred as a consequence of feasting and ritual activities aimed at mitigating food stress and ensuring both food security and group survival during an extended period of resource instability (Bollig 2005; Halstead and O'Shea 1989; Rautman 1996). From various climate proxy records it is evident that the 8.2 event occurred as a sharp anomaly within a longer period of aridity commencing ca. 6300 B.C., at the start of the Pre-Halaf, and lasting until ca. 6000 B.C., shortly before the start of the Halaf proper (Alley et al. 1997; Barber et al. 1999; Rohling and Pälike 2005; Mayewski et al. 2004; Wiersma and Renssen 2006). Viewed in this light, the developmental continuum from the Pre-Halaf to the Early Halaf provides a rich context for interpretation.

Risk Management and Social Change

Resource scarcity, or even the threat of it, is a condition understood to exert a strong influence on all facets of culture, shaping societal organization and providing the crucial conditions that give rise to social change and transformation (Halstead and O'Shea 1989:4–5). Change is propelled not just through modifications to economic activities and competition for resources, but through the institution of progressively more drastic and resource-intensive buffering mechanisms aimed at ensuring an equal, or at least equitable, distribution of resources, and by the deployment of rituals that attempt to influence the course of events (Bollig 2005; Halstead and O'Shea 1989). For the more costly or high-level mechanisms—i.e., those involving the mobilization of social relationships—to be effective in the longer term, there is a strong selective pressure for them to become embedded within more regular cultural practices, thus facilitating the development of social institutions that ensure that they are regularly reinforced. Over a prolonged period, the institution of more inclusive responses to productive risk is likely to result in significant social changes (Braun and Plog 1982; Halstead and O'Shea 1989).

Several researchers have investigated the behavioral responses or risk-reduction strategies employed by subsistence societies when faced with variability in the food supply caused by prolonged drought or other calamities. Although examined from several different theoretical perspectives, most scholars are of the view that the magnitude or social cost of the response will correspond directly to the severity of the problem, resulting in a hierarchical or ordered sequence of responses (Colson 1980; Halstead and O'Shea 1989; Minnis 1996). Consequently, there is also likely to be a time lag between the onset of drought conditions and the implementation of more drastic coping strategies—i.e., over the longer term we might expect to observe several iterations of adjustment.

In considering the likely responses by Late Neolithic groups to a drought of such apparent magnitude and duration, it is important to bear in mind that no matter how rapid the onset of arid conditions associated with the 8.2 event, the changes in the weather record would be neither monotonic nor unidirectional, but over time there would be fewer good years and more bad years. Thus, one should not necessarily expect to find in the archaeological record evidence for either sudden or dramatic responses to climate events, and certainly not of the catastrophic nature suggested by some scholars (Staubwasser and Weiss 2006; Weninger et al. 2006). Instead, what might be expected is a gradual pattern of change that at first may appear to be no more than an extension, or intensification, of previous practices. Indeed, given that the regional arid phase probably commenced some time prior to the sharp anomaly of the 8.2 event, communities occupying the more arid areas are likely to have been familiar with periods of drought, and already possess measures for reducing the impact of future droughts, and for maximizing the benefits in good years.

It is also important to note some marked differences in the proxy climate records for the region. For example, sediment cores from the Dead Sea record a rapid drop in lake levels ~8.1 cal kyr B.P., indicating arid conditions (Migowski et al. 2006), whereas geochemical and pollen data from Lake Van in Turkey point to an increase in relative humidity for the same period (Wick et al. 2003). Certainly, in a region as ecologically diverse as that in which Halaf sites are found, and with the pronounced variations in rainfall (Figure 2.2), not all areas would have been affected to the same degree. Human responses to climate variability are therefore unlikely to have been consistent across the entire region. Indeed, Rautman (1996) has suggested that during periods of climatic instability, exchange networks are most likely to emerge between regions experiencing differences in environmental stress.

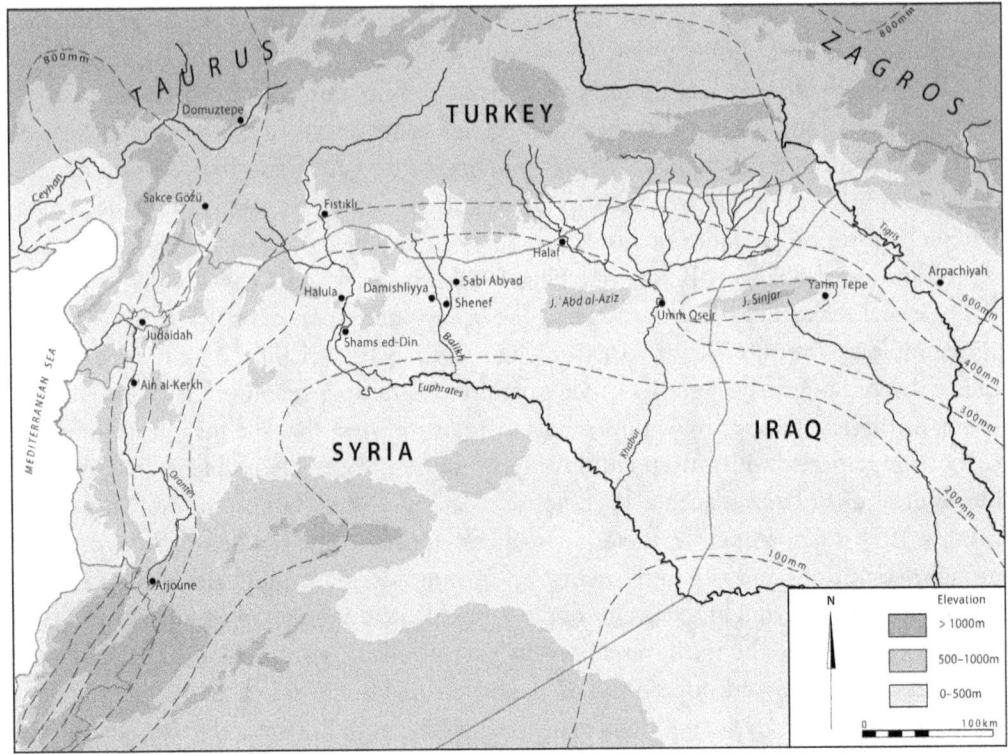

FIGURE 2.2. Map of western Syria and Upper Mesopotamia showing mean annual rainfall and location of sites mentioned in the text.

DIVERSIFICATION OF SUBSISTENCE ACTIVITIES

Probably the most common means by which people seek to lessen their vulnerability to food shortages caused by adverse conditions is through the diversification of subsistence activities (Bollig 2005; Colson 1979; Halstead and O'Shea 1989; Minnis 1996). Although in the Near Eastern context it is often assumed that threats to the agricultural base would lead to a greater reliance on pastoral nomadism (Weiss and Bradley 2001), such subsistence specialization represents risky behavior that is unlikely to have been sustainable over the longer term, especially given the probable reduction in pasture caused by increased aridity (Colson 1979). Moreover, it is inconsistent with the archaeological evidence for the period in question, which provides no indication of the sorts of de facto territorial markers that might be expected of pastoral nomadic groups, such as stone cairns, rock art stations, and seasonal camps (Wilkinson 2003). Nor is it supported by site distributions, which steadily contract to the areas normally most suited to rain-fed agriculture or to the main river valleys, rather than extending into areas of low rainfall or along seasonal watercourses, as attested for the earlier and climatically more favorable PPNB era or even for the later Halaf period (Akkermans 1993; Nieuwenhuyse 2007). However, economic specialization, in the form of craft specialization, may occur among communities that are poorly situated for adequate food production, in order to supply goods for exchange (Minnis 1996). Greater mobility is also to be expected, to take advantage of variability in the distribution of resources (Bollig 2005; Halstead and O'Shea 1989; Minnis 1996).

To date, few detailed studies have been undertaken of Late Neolithic subsistence practices; nevertheless, they point to the implementation by different communities of increasingly diversified subsistence strategies that could be seen as reflecting responses to actual and potential food shortages and regional variability in the food supply. In the case of Tell Halula, located in the Syrian Euphrates valley, the late Pre-Halaf and Proto-Halaf phases are characterized by an intensification of hunting practices, with a far greater amount of the biomass being supplied by equids and gazelles (Saña Seguí 1999). Pigs also become an increasingly important meat source, in contrast to cattle and sheep, which were being kept for longer, presumably for their milk and other byproducts and, quite likely, as contingency food storage.

A shift in strategies indicative of a period of economic uncertainty is also attested for Sabi Abyad. The Proto-Halaf faunal assemblage, for example, indicates a much more opportunistic hunting strategy than previously and, in contrast to Tell Halula, a lesser reliance on large herbivores (Cavallo 1996). Exploitation of the riverine biotope, especially, became more intensive, with a broadening of the faunal spectrum to include birds, fish, mollusks, and tortoises. At the same time, there was a significant shift in herding strategies involving an increase in the exploitation of sheep and goats, more focused and intensive breeding of cattle for meat production, and the virtual abandonment of pig husbandry (Cavallo 1996; Russell 2010). Since pigs require daily watering and adequate shade and wallow (Zeder 1996:301), this implies that drier conditions prevailed along the course of the Balikh than along the Euphrates. Also indicative of a diversification of subsistence strategies is the occurrence in Proto-Halaf contexts of a large deposit of two-seeded einkorn wheat together with a quantity of wild barley. According to Van Zeist and Waterbolk-Van Rooijen (1996:534–535), it is extremely unusual for this subspecies of wheat to be cultivated as a crop in its own right, and, evidently, the practice was abandoned by the Early Halaf. Another sign of economic diversification, evidenced by changes in the mortality profiles for ovicaprids, and corroborated by a significant increase in spindle whorls, is a growing emphasis on fiber production (Russell 2010; Van der Plicht et al. 2011). As hardier animals, sheep and goats would have been better adapted to the marginal environment of the Balikh, and thus more productive during droughts. Woolen textiles and felt products could have provided "stored wealth" to be subsequently traded for needed items.

Culling strategies for the period also appear to have been geared for the first time toward maximizing both meat and milk production. At Halula, culling of ovicaprids for secondary products such as milk and fleece is already attested by the start of the Pottery Neolithic; however, it becomes increasingly important at both Halula and Sabi Abyad from the start of the Pre-Halaf (Helmer et al. 2007; Russell 2010). By the start of the Halaf era, cattle were also being managed for their milk products and were often only slaughtered on special occasions (Becker et al. 2007; Cavallo 1996; Grigson 2003; Kansa and Campbell 2004). The production of milk would no doubt have contributed to the animal's symbolic value, although it is likely that the ideological significance and value of cattle as a feasting food was already well established by the time of its domestication (Twiss 2008). The importance of milk products in the Pre- and Proto-Halaf diet has been confirmed by residue analysis of pottery from ʿAin el-Kerkh in western Syria and from Sabi Abyad (Evershed et al. 2008; Shimoyama and Ichikawa 2000). Significantly, the ʿAin el-Kerkh sherds include two strainers from a type of vessel that was probably a forerunner to the Halaf bow-rim jar (e.g., Tsuneki et al. 1997:Figure 19.2, 2000:Figure 6.18), as well as a carinated "cream bowl." Mallowan and Rose (1935:131) were probably not so wrong in their choice of epithet for the latter type, the implication being that both vessel types were used in the service and consumption of dairy products.

All these developments are consistent with ethnographically attested responses to drought conditions—the focused slaughter of small stock, especially those that produce fewer secondary products; the consumption of substitute or "famine" foods that do not usually form part of the diet; a greater than usual reliance on hunting; and foraging for wild plant foods (Bollig 2005; Colson 1979; Halstead and O'Shea 1989; Minnis 1996).

Settlement Dispersal

Another common response to food shortage is intensified territorial mobility, or settlement dispersal, which enables groups to take advantage of spatial variability in the distribution of resources (Bollig 2005; Colson 1979; Halstead and O'Shea 1989; Minnis 1996). Although even coarse-grain survey data for many parts of northern Syria and Mesopotamia are limited, the available data tend to suggest a greater number and dispersal of settlements than previously, but only in certain regions. The data for the Balikh Valley, for example, rather than indicating an increase in settlement numbers, point to a shift of settlement toward the northern part of the valley, presumably to take advantage of typically higher precipitation (Akkermans 1993; but see Nieuwenhuyse et al., this vol., for a different perspective). This commences during the Pre-Halaf and it is only late in the Early Halaf that the number of sites increases, thus casting doubt on the "nomad" scenario that has been put forward to explain Proto-Halaf and Halaf social developments at Sabi Abyad (Akkermans and Duistermaat 1997; Verhoeven 1999). For the wetter Khabur region, however, the survey data suggest not only a shift to the north, but also increased settlement density (Nieuwenhuyse 2000), while for the Iraqi Jazireh there is evidence for a dramatic increase in the number of sites during the late Pre- and Proto-Halaf followed by a reduction in numbers during the Early Halaf (Campbell 1992). Already Campbell (1992:220–221) has noted the probable link between this sudden proliferation of sites and the growing use of painted symbolism on pottery, a development he suggests was most likely triggered by increasing contacts between groups as they sought new and more viable areas in which to farm and graze their livestock. Territorial organization would have come under pressure as people competed for scarce resources. The best "common good" resolution of this competition is therefore likely to have involved the development of institutions that enabled the integration of different groups and communities into a wider system (Campbell 1992:223).

This does not mean that negative responses were entirely absent (Colson 1979). It is almost certain that raiding would have occurred, targeting crops, granaries, and cattle (Halstead and O'Shea 1989; Keeley 1996; Minnis 1996). The increasing incidence of such occurrences is implied by the stockpiles of clay sling bullets which appear for the first time in Pre-Halaf villages and become increasingly common throughout the later Neolithic (e.g., Akkermans and Wittmann 1993; Merpert and Munchaev 1993b; Munchaev and Merpert 1973; Spoor and Collet 1996; Tsuneki 1998b). The arsenals of these projectiles found at both Proto-Halaf and Halaf sites suggest that their inhabitants were not just intending to use them for hunting or discouraging predators (e.g., Akkermans and Wittman 1993:159), but also expecting to have to discourage raiders.

Resource Sharing and Exchange

The principal mechanism by which the integration of different social groups is likely to have been achieved is via the cultivation of social relationships and the creation of co-operative alliances, probably through marriage, which, in the context of food security and the exchange of durable valuables, crop foods, and livestock, enabled communities to tap into the food resources of other regions. In time, the institutions developed to manage these transactions are likely to have become normalized within the broader framework of culture, leaving the way open for the sorts of competitive responses Nieuwenhuyse (2007:219–223) suggests were responsible for the numerous and rapid stylistic developments attested from the Pre-Halaf on.

Social links with those outside one's immediate terrain are the ultimate insurance against food scarcity (Colson 1979:23); however, before drawing on these becomes necessary it is usual for people to appeal to other members of their group or village for assistance, which may be given freely, but often involves some measure of obligation and anticipated reciprocity (Halstead and O'Shea 1989; Rautman 1996). Alliance creation and reciprocal obligation, or "social storage," are established and manipulated through commensal hospitality, including the sharing of economic surpluses, giving of loans and gifts to relatives and neighbors, and donations of livestock to community rituals. Among many societies it is common for food sharing to intensify during periods of hardship and drought in order to reinforce existing alliances (Bollig 2005; Clarke 2001; Hayden 2001; Rautman 1996).

The context for establishing these social and economic bonds, and through which transactions of various kinds typically take place, is the feast. Already, Nieuwenhuyse (2007, 2008, 2009) has suggested that feasting was the venue for much ceramic innovation during the later Neolithic, as individuals and social groups engaged in a process of competitive emulation. Some of the stylistic and technological innovations he describes could perhaps be explained in terms of generational change and shifting social alliances. However, one possibility he does not address is that feasting and food sharing for Late

Neolithic societies were an important means of ensuring economic security through the construction of alliances and contractual obligations. This probably occurred on several levels, including within the clan, lineage, or village, and also between communities (e.g., Clarke 2001; Hayden 2001). Although essential to promoting group solidarity and social cohesion, feasts would have also entailed competition between groups, involving displays of items indicative of a group's success such as food and livestock, craft products, and trade items (Dietler 2001; Hayden 1996, 2001).

An important facet of the social obligations established through marriage alliances, feasting and gift exchange is the need for specialized witnesses and record keepers who ensure that debts are not forgotten or neglected (cf. Akkermans and Duistermaat 1997). Because at times the amounts involved in these transactions may be sizeable, simple recording devices or counters are employed to record debt (Hayden 2001), while obligations tend to be denoted by tokens, which become imbued with the latent equivalent value of the gifted food item or service (Halstead and O'Shea 1989). The many chipped sherd and stone calculi and *jetons* found on Proto-Halaf and Halaf sites are self-evident, as are the clay tokens (e.g., Bernbeck et al. 2003; Mallowan and Rose 1935). So far the best examples of the storage and management of such items come from the Sabi Abyad Proto- and Early Halaf levels, where they were concentrated in several large storage buildings (Akkermans and Verhoeven 1995; Verhoeven 1999). However, in contrast to the site's excavators who consider these items and the associated clay sealings to signify transactions conducted between resident and nomadic sectors of the communities (Akkermans and Duistermaat 1997; Verhoeven 1999), I would suggest that they recorded contractual debts established through feasting and other social obligations in order to ameliorate nutritional stress. Similarly, the broken female and animal figurines found in the same buildings, while no doubt representing services and goods—the female figures probably productive wives and the animals either meat or livestock—rather than "giving life" to society in the purely symbolic sense (Verhoeven 1999:230–231), were, along with the broken seals and tokens, records of debts repaid.

As regards feasts, there is probably no clearer evidence of their growing relevance to Late Neolithic societies than the increasing prevalence of painted pottery. Various ethnographic studies, as well as some detailed analyses of archaeological assemblages, indicate that elaborate decoration is a common characteristic of serving vessels and utensils used in communal feasting contexts, owing to the capacity of such items to convey visually notions of solidarity and unity, identity, and the nature of social alliances, or to assert the ambitions and success of particular individuals and groups (Hayden 1996, 2001; Mills 2007; Schiffer and Skibo 1997). As others have remarked, this degree of stylistic-symbolic referencing implies a need to ensure the active co-operation and sociocultural integration of different groups of people, as interactions became more frequent (Akkermans 1993:320–321; Campbell 1992:220–221).

It is around this time also that important advances in food technology take place, with the development of efficient, locally manufactured cooking pottery. The increasing importance to late Pre-Halaf and Proto-Halaf communities of ceramics that were suitable for cooking is evidenced first by the widespread occurrence of Syro-Cilican Dark-Faced

Burnished Ware (DFBW) and the regularity with which vessels of this pottery were remodeled to serve as cooking pots, and by its subsequent replacement for this purpose by the so-called Mineral Coarse Ware, which prevailed across Upper Mesopotamia, and its western correlate, the Dark-Faced Unburnished Ware (Cruells and Nieuwenhuyse 2004; Le Mière and Picon 1994; Nieuwenhuyse 2007; Mottram 2010). Presumably, some pottery was used for cooking before this, using stone boiling; however, cooking using pots rather than roasting or baking provides a major broadening of culinary options, especially, if animals were being kept for longer and their meat required lengthy stewing or braising to make it palatable.

Another ceramic development specific to the Proto- and Early Halaf periods that appears to have been central to the structuring of consumption and feasting is that of the cream and other carinated bowls. While the appearance of such complex vessel shapes may be seen as a purely technological achievement, its real significance is likely to have been symbolic and tied to the capacity of such vessels to convey feelings of solidarity and ideological unity, particularly early on when many examples are uniformly decorated (Nieuwenhuyse 2007). The capacity for symbolism of these open drinking and service vessels is also embodied in their profiles which replicate the flaring "S" shape of the bucranium as it came to be depicted on Halaf Painted Ware (e.g., Le Mière and Nieuwenhuyse 1996:Figure 3.41.3, 7, 9; Mallowan and Rose 1935:Figure 74.3–4, 9). The economic and symbolic importance of cattle to Halaf groups is demonstrated most dramatically in the contents of the Late Halaf Death Pit at Domuztepe, in southeastern Turkey, which indicate that prime-age female cattle were selected preferentially for the feasts that accompanied the funerary rites (Kansa and Campbell 2004; Kansa et al. 2009). However, even before this it is clear that cattle were only consumed on rare occasions and probably represented a major source of prestige for their owners (Kansa et al. 2009). Both the shape of the vessels and the later use of the bucranium motif were evidently critical to displays of public identity and unity amongst Proto- and Early Halaf groups. This is best illustrated by the occurrence at ʿAin el-Kerkh and Tell Judaidah, in western Syria, of cream bowls made from pattern-burnished, rather than painted, DFBW (Braidwood and Braidwood 1960; Tsuneki et al. 1997).

The continuing importance of feasts to the consolidation of Halaf social relationships is further evidenced by the occurrence at Tell Halula, Khirbet esh-Shenef, Arpachiyah, Shams ed-Din and elsewhere, of areas set aside for communal cooking and baking (Akkermans and Wittman 1993; Al-Radi and Seeden 1980; Hijara 1997; Mallowan and Rose 1935; Mottram 2010). There is also direct evidence of feasting from two areas at Domuztepe (Campbell et al. 1999; Kansa and Campbell 2004).

Storage

The other common strategy employed by agricultural societies to reduce the impact of periodic crop failure or food shortages is the storage of surpluses, in this case either in the form of raw or processed grain, dried pulses, fruits, nuts, and other plant and animal foods, or "on the hoof," in the form of animals (Colson 1979; Halstead and O'Shea

1989). Also common is the conversion of food surpluses into durable objects or social obligations, which, as discussed above, may be called upon and exchanged for food in times of need. While this might appear to be a more passive means of counteracting scarcity than those discussed above, storage in fact takes on its most critical role in the context of feasting, which is predicated largely on the accumulation, storage, and use of surpluses (Hayden 2001). In turn, feasts not only provide a public venue for the exchange of durable valuables and the cementing of alliances, but also encourage the production of surpluses in order to fulfill reciprocal obligations (Dietler 2001; Halstead and O'Shea 1989).

Dedicated storage buildings first become common throughout northern Syria and Mesopotamia during the Pre-Halaf and Proto-Halaf periods and are subsequently attested at numerous Halaf sites (Akkermans and Wittman 1993; Mallowan and Rose 1935; Merpert and Munchaev 1973, 1987; Molist et al. 2007). At the same time, many assemblages attest to the appearance of and dramatic increase in necked storage jars (Braidwood and Braidwood 1960; Le Mière and Nieuwenhuyse 1996; Mottram 2010). The contents of the Sabi Abyad buildings confirm that they functioned as stores for foodstuffs and for the types of durable valuables that might be exchanged for foodstuffs (Verhoeven 1999). More importantly, from the occurrence within them of hundreds of locally made and used clay sealings it is apparent that they also acted as record offices for administering contractual debts (Duistermaat 1996). While the Sabi Abyad excavators have interpreted these remains as evidence that the Proto-Halaf community was semi-nomadic, spending much of the year away from the settlement (Akkermans and Duistermaat 1997; Verhoeven 1999), an alternative explanation is that the storage of food and valuables, together with the means for identifying and regulating access to these items, were buffering devices against the risks arising from extended periods of environmental unpredictability (Colson 1979). Indeed, Sabi Abyad is not unusual in that many Pre- and Proto-Halaf villages contain large storage buildings (e.g., Merpert and Munchaev 1973; Molist 1998; Molist et al. 2007; Tsuneki et al. 1998). Moreover, the Sabi Abyad buildings find a compelling parallel in the much later TT6 Burnt House at Arpachiyah, which, in addition to the well-known polychrome plates and exotic craft items, contained numerous seals and clay sealings (Campbell 1992; Mallowan and Rose 1935). If all such buildings were a measure of the transhumant population, then we should certainly expect to find many more Pre- and Proto-Halaf sites than are known at present, especially temporary encampments. Storage therefore seems to be of critical importance in interpreting the foundation and structure of Halaf societies.

In this regard, attention should be drawn to the iconography on some Early Halaf vessels, which depicts what appear to be multilevel buildings with gabled roofs and upper storys constructed out of timber, reeds, and matting, and which are often shown with several birds perched along the central ridgepole (Figure 2.3). Going by an amulet found during the original Arpachiyah excavations and a clay model from the Sabi Abyad Proto-Halaf levels, the buildings were rectangular in plan and the walls sometimes supported by cross-bracing (Figure 3:9; Spoor and Collet 1996:Figure 8.11). One depiction on a sherd from Sabi Abyad implies that the ends of the projecting beams were sometimes adorned with carvings, animal skulls, or some other three-dimensional ornamentation,

FIGURE 2.3. Depictions on Early Halaf pottery of raised granaries/storage structures. Nos. 1–2, 4, 10: Sabi Abyad (after Akkermans 1989; Le Mière and Nieuwenhuyse 1996); 3: Fıstıklı Höyük (after Bernbeck et al. 2003); 5–6: Tell Halula (Mottram 2010); 7: Sakçe Gözü (after Garstang et al. 1937); 8–9: Arpachiyah (after Hijara 1978; Mallowan and Rose 1935); 11: Domuztepe (after Kansa et al. 2009).

perhaps the "ritual" objects recovered from the Proto-Halaf Burnt Village, although these are more likely to comprise the bodies of bird figures (Figure 3:2; Spoor and Collett 1996:Figures 8.6, 8.7). Another, painted on the center of a bowl from Arpachiyah, implies that access to the upper levels was via a ramp or ladder (Figure 3:8). Interestingly, such structures were not restricted to Halaf sites, as attested by a painted terracotta model from Al-'Ubaid in southern Mesopotamia (Woolley 1927:Plate XLVIII).

In the case of the structure shown on the Arpachiyah bowl, both Hijara (1978) and Ippolitoni-Strika (1990) have suggested that it was a sacred building, the latter at the same time cautioning against overlooking the possible existence of cult-buildings constructed of lightweight materials. Clearly, they were significant to community life, given the placement of the motif in the center of the bowl, with scenes of daily life revolving around it. I would suggest, however, that rather than constituting specifically "cult" buildings, they were in fact storage facilities. This does not mean that they were not imbued with symbolism, nor need it preclude a pivotal role in Halaf belief systems and practices given the recurrence of the same theme on pottery from sites across the Halaf range; however, it does mean that they were not specifically *religious* buildings, thus allowing for wider cross-cultural comparisons.

Morphologically, these buildings bear a close resemblance to the *liku* or yam storage houses built for high-ranking men in Trobriand island villages on the completion of successful harvest competitions (Schiefenhövel and Bell-Krannhals 1996:Figure 4:2). Interestingly, the large amounts of yams these men receive both from family and allies are used not only to feed their immediate and extended families, but also to host ceremonies and feasts important for the village as a whole. The buildings are also strikingly similar to some Indonesian raised rice granaries, and to granaries attested for the Japanese prehistoric Yayoi and Kofun periods (ca. 300 B.C.–A.D. 538) (Domenig 2003; Sato 1991; Schefold 2003:Figs. 4, 12; Wessing 2003:Figs. 1–2, Figure 4:1, 3). They are also similar to the *ambar* granaries found along the Black Sea in Turkey and, although made of different materials, to the *hórreos* and *espigueiros,* or raised stone and timber granaries, found in northern Spain and Portugal (Akar et al. 2003:Figure 2.7.4; Bradley 2005:Figure 1.1). In all these societies, granaries or staple food stores are charged with a rich symbolism, and often serve as the focus for important ceremonies and rituals, including funerary feasts. It is therefore likely that the structures depicted on the Early Halaf pottery represent storehouses with raised granaries, especially given their evident centrality to Halaf ritual and symbolism. Indeed, Akkermans (1993:61) has already suggested that Early Halaf Building 1, at Sabi Abyad supported more than one story owing to the thick walls and buttressed façade. Moreover, the remains of Proto-Halaf Building V, where the "ritual objects" were found, clearly indicate multiple levels of construction, including a superstructure made of timber beams and matting (Verhoeven and Kranendonk 1996).

In this respect it is also valuable to consider the most prominent decorative motifs of the Proto- and Early Halaf phases—the horizontal and standard crosshatch patterns. Wengrow (2001) has drawn attention to the fact that many characteristic Samarran and Halaf motifs correspond to the basic constructional elements used to make baskets.

FIGURE 2.4. Modern ethnographic examples of raised granaries and staple food stores analogous to those depicted on Early Halaf Pottery; 1: Toba Batak, Sumatra, rice granary (after Domenig 1980); 2: Trobriand *liku* yam house; 3: Baduy, Java, rice granary (after Schefold 2003); 4: Iron Age petroglyph of probable raised granary, Valcamonica, Italy (after Bradley 2005).

However, I would suggest that, rather than simply taking the designs associated with one medium and adapting them to another (Nieuwenhuyse 2007:215), Proto-Halaf and Halaf potters were referencing structural themes that were central to the construction of their worldview and to the production and reproduction of society. Not only were many of the commodities stored within the Sabi Abyad buildings originally contained within baskets (Duistermaat 1996), but so, too, were parts of the fundamental fabric of life, in the sense that the granaries and other architectural elements were constructed using the same techniques and materials. Going by the most complete images of the storehouses preserved on a jar from Domuztepe (Figure 3:11), even those parts of the buildings not made from reeds or matting were probably painted to give the appearance that they were, as were the storage jars standing at each corner. This relationship between the buildings, the ceramics, and the commodities stored within them suggest that the decoration symbolized or expressed meanings connected with the cycle of food production and consumption (Pikirayi 2007).

Absorbing Successful Climate Change Responses into the Social Fabric

Halaf "Big Men"

Going by the above evidence it seems likely that prolonged aridity during the period ca. 6300–6000 B.C. led Upper Mesopotamian Neolithic societies to make significant alterations to their subsistence, storage, and commensal practices and to the nature and extent of their social interactions. Also likely, given the appearance during the same period of *tholos* architecture and large-scale communal storage structures, are important changes to social organization, particularly given the potential for feasts to enable the enhancement of status and manipulation of social inequalities, and for group rivalry in relation to the negotiation of ritual and symbolic power (Chamberlin 2006; Helwing 2003). From this it might be assumed that Halaf societies incorporated some system of ranking, a notion that runs contrary to many current interpretations of Halaf society, which see Halaf groups as politically autonomous and highly egalitarian (e.g., Akkermans 1993; Frangipane 2007). However, the documented existence among contemporary unranked societies of systematic status differences such as between age sets or genders renders the term *egalitarian* both misleading and simplistic. It also adds ambiguity to the concept of ranking, which need not imply the existence of social stratification (Wason 1994). The key difference relates to whether or not some sectors of society had dominance and control over productive resources and were set on accumulation rather than distribution, the latter being critical to the survival of Late Neolithic societies. In societies that blend influential leadership with an egalitarian background, such as the Big Man societies of Melanesia, ranking tends not to be pervasive and is also transient and cannot be passed on to offspring (Wason 1994).

Big Man societies are essentially redistributive societies in which influence and recognition are more important than being wealthy. Leadership is based on effective influence within a relatively egalitarian society and is achieved through gift exchange and the ability to draw consensus (Wason 1994). In effect, a Big Man is an equal member of society who has a defined role that carries prestige but not wealth, and recognition but not status. Nevertheless, in Big Man societies there is also a pervasive ethos of competition within and between groups. When Big Men compete as peers, the stakes include the privilege, wealth, or even physical well-being of their respective social groups, not just the leaders' own status. Such a situation could well have spurred the progressive processes of emulation that Nieuwenhuyse (2007:220) maintains were the driving force behind stylistic innovations in the Late Neolithic painted wares.

Evidence that the accumulation of prestige played an important role in Proto- and later Halaf society is provided by various aspects of social and material culture, particularly the occurrence in burials or funerary deposits of objects that seem inherently valuable or in some way unusual in terms of their size, shape, material, or fineness of crafting (Merpert and Munchaev 1993a; Tsuneki et al. 1997; cf. Wason 1994:104). Other sumptuary items or indicators of prestige are likely to include the polychrome plates from the Arpachiyah Burnt House, given their quality, context, and evident display purposes (Campbell 1992).

One possible correlate of rank is mortuary treatment. Although it has been suggested that Halaf burials provide little or no evidence of ranking (Akkermans 1993:315), a likely exception is Grave 2 at Arpachiyah, in which four skulls were interred in ceramic vessels, in one instance in a repaired heirloom vessel decorated with exceptional scenes (Hijara 1978). This could parallel a situation documented in some Big Man societies, where the skull of the Big Man is made the focus of a shrine because his head is seen as a concentration of power and as a reservoir of his personal skills (Lillios 1999:247).

Other possible symbols of prestige and/or authority are the mace heads found on some sites (Braidwood and Braidwood 1960:Figure 65.11; Fuensanta and Charvàt 2005:Figure 4.4; Özbal et al. 2004:Figure 15.6; Korfmann 1982:Figure 19.3; Oppenheim and Schmidt 1943:Plate XXXVI.25). Their use as insignia of rank is suggested by a later Ubaid figurine from Eridu, which depicts a nude male holding a mace or scepter (Stein 1994:Figure 7). The presence of such items in burials or funerary contexts is consistent with a system of achieved status, since in Big Man societies issues of inheritance are geared more toward celebrating individual achievement rather than maintaining lineage ties and succession claims (Becker et al. 2007:Figure 27; Merpert and Munchaev 1993a:Figure 10.13.3). Personal objects, that is, objects associated with a person's unique traits or achievements, are often destroyed or broken when a person dies, whereas objects of production are passed on to heirs (Lillios 1999:248–249).

Relevant in this regard is the occurrence at several sites of what can only be described as heirloom vessels, some of which show signs of extensive use or repairs (Mottram 2010). All appear to date to the Early Halaf and most are either bowls or pouring vessels (see also Dooijes and Nieuwenhuyse 2009; Nieuwenhuyse 2007). This suggests that they perhaps represented ties to formative ritual traditions, especially those associated with societal production and reproduction and/or designed to support a sense of individual or community identity (Van Dyke and Alcock 2003). According to Lilios (1999), heirlooms are not common among hunter-forager societies, whereas among chiefdoms the transmission and inheritance of objects associated with ritual and production, or which are emblematic of economic success, social affiliations, and political rank, are of acute importance in ensuring succession. Thus, in Late Neolithic societies, in which succession was less of an imperative, heirlooms were possibly instrumental in the production of social memory, which constituted essential symbolic capital when power, economic resources, or key social alliances were at stake (Hendon 2000; Joyce 2000; Lillios 1999).

Halaf Village Structure and Degrees of Mobility

Current characterizations of the Halaf social landscape describe a situation in which each region contained a few large permanently occupied settlements, the majority of sites being small and occupied only over the short term (Akkermans 1993). Very often, this distinction is interpreted in terms of differences in subsistence strategies, although it is also suggested that some sites were established to exploit specific resources (Campbell 1992). Akkermans and Schwartz (2003:119) suggest that the breaks in sequences and frequent shifts of occupation observed at smaller sites such as Umm Qseir and Damishliyya argue for habitation on a seasonal basis or for one or two generations at most, while the people occupying these sites are often characterized as transhumant pastoralists or as hunters (Akkermans and Wittmann 1993). Recently, it has been implied that these developments were set in motion during the 8.2 event, as communities transformed from autonomous households practicing sedentary agriculture and herding, to more diversified populations combining both mobile pastoralists and sedentary agriculturalists (Akkermans et al. 2010; van der Plicht et al. 2011).

Clearly, Halaf societies, as inheritors of the legacy of the 8.2 event, were open to a wide variety of subsistence options, as is borne out by the diversity of plant and animal remains found on different sites; nevertheless, the current propensity to interpret Halaf societies as incorporating a high level of subsistence mobility probably has more to do with a trend in archaeology, arising out of postcolonial discourse, which seeks to redress an intellectual and theoretical imbalance—one that has placed a premium on the investigation of settlements and on the notion that sedentariness is the desired situation of all peoples—than it has to do, necessarily, with archaeological realities. While it is certainly an important investigative concern, the result is that now "nomads" tend to be seen everywhere in the archaeological record whether or not there is sound evidence for such (e.g., Akkermans and Duistermaat 1997; Bernbeck 2008; Bernbeck et al. 1999; Clayton 2004; Verhoeven 1999). Moreover, because residential mobility is often assumed to be economically or ecologically determined, the possibility that fluctuations in population size and village structure might be a function of social rather than economic factors is often overlooked (although see Akkermans and Schwartz 2003:152; Bernbeck 2008:63). Indeed, the roles played by social and ritual factors in determining mobility can be just as important as, and in some cases more important than, those played by ecology (Kent and Vierich 1989).

In the Halaf context, several specifics suggest that social factors were more important than ecological or economic factors in determining the degree of movement. Foremost of these is that most sites usually cited as evidence for residential mobility don't, in fact, reflect the patterning usually associated with regularly mobile groups, even those that may reside in a location for extended periods (Kent and Vierich 1989). Indeed, most small, supposedly short-term settlements, such as Fıstıklı Höyük, Khirbet esh-Shenef, and Umm Qseir, indicate sequences lasting sometimes up to 150 years rather than the cycles of annual seasonal abandonment and reoccupation associated with semi-nomadic pastoralism or seasonal transhumance. Any major internal shifts in the location of settlement seem to occur every 30 to 50 years; that is, after two to three generations. Moreover, there is no compelling evidence to link these shifts with seasonal or longer-term mobility associated with a particular form of subsistence.

No doubt, some sites were occupied on a short-term or seasonal basis. Arjoune, in the Orontes valley, is characterized by a series of pit dwellings (Parr 2003), as is Damishliyya, in the Balikh (Akkermans 1989). The likelihood that Halaf groups used tents as temporary dwellings has been suggested by Bernbeck (2008:55) and could well be the case given the motifs painted on the inside center of a plate from Tell Halaf (Figure 5:4). These show a striking resemblance to the black tents used by present-day Kurdish tribes in southeastern Turkey, which are anchored by wooden stakes attached to a short rope (Cribb 1991:Figure 10.8). Other "butterfly" or "comb" motifs seen on bowls and plates from Tell Halaf, Yarim Tepe, Arpachiyah, and elsewhere are possibly also tents, although they more closely resemble the circular huts and shelters constructed by many southern and West African groups (Figure 5:1–3, 5–6, 8–10). The fact that most of these motifs are found in the center of dishes or bowls signals their importance to the structuring of Halaf life and ideology. Indeed, much Halaf iconography seems to have at its core the *structures* of life. In this sense the ceramics became *containers* of life, playing host to the substances and spiritual forces that ensured the prosperity and reproduction of society as a whole (e.g., Forni 2007).

Nevertheless, there is no clear evidence that the majority of Halaf sites were similar to Arjourne and Damishliyya. For example, none of the Balikh sites that Akkermans (1993:181) suggests were temporary encampments has been firmly identified as such. Hole (1997:44) refers to one possible Halaf campsite in the foothills of the Jebel ʿAbd el-Aziz; however, overall the number of sites that could be claimed to be short-term pastoralist or herder campsites is very limited.

On the basis of various ethnographic and ethnoarchaeological studies it may be inferred that the founders of most Halaf villages intended staying for some time. Among various southern African groups, hut styles and investment in construction time are directly correlated with anticipated mobility (Kent 1991; Kent and Vierich 1989). Easy-to-construct grass huts occur at all habitation sites; however, more substantial mudbrick huts are found only at sites with an anticipated long-term occupation (Kent and Vierich 1989). As residential mobility decreases, other variables come into play, including an increase in the abundance and diversity of artifacts, the founding of specialized refuse dumps at a distance from hearths and houses, and an increase in the number and size of formal storage facilities. In short, the internal differentiation of settlements is far greater among less mobile groups (Kelly 1992; Kent 1991).

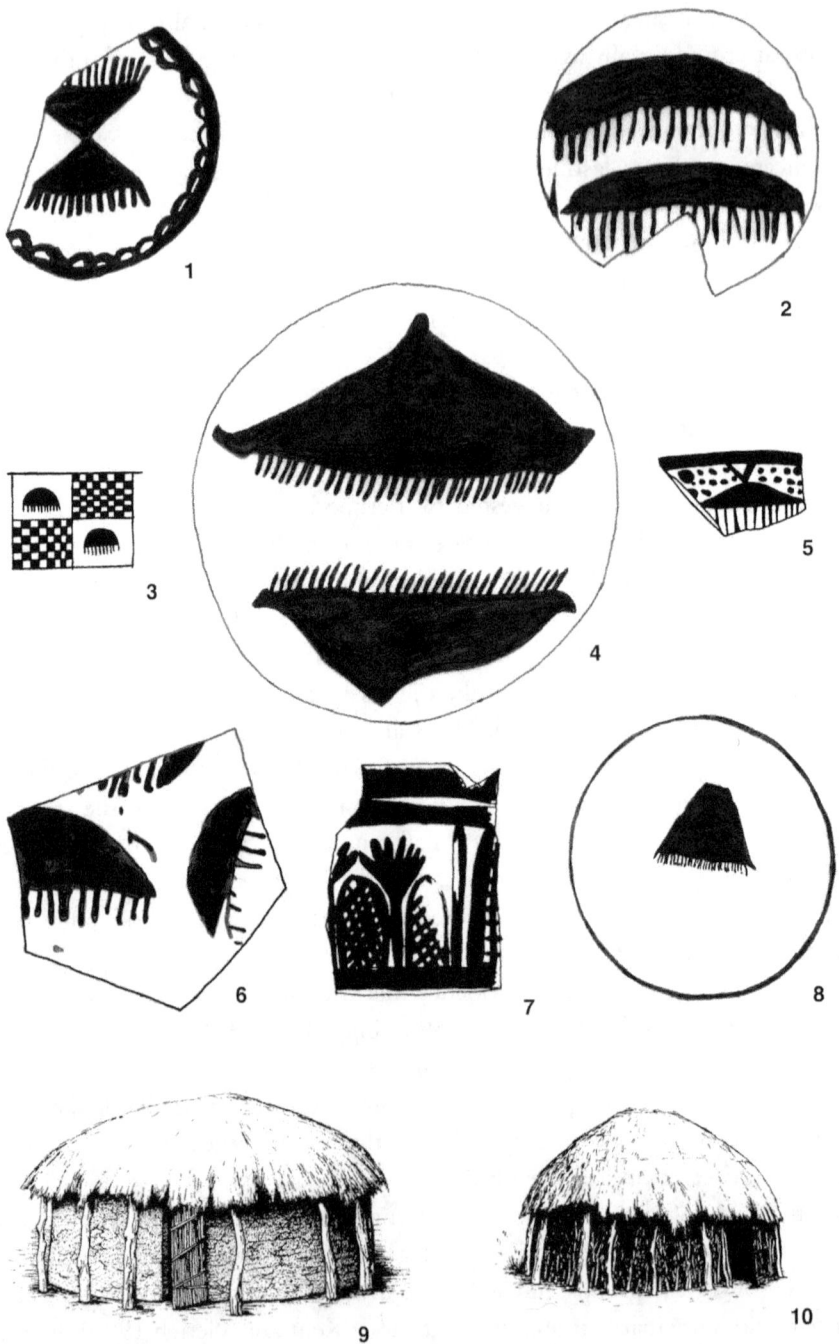

FIGURE 2.5. Depictions on Halaf pottery of probable habitation structures (Nos. 1–8) shown against analogous structures from southern Africa (Nos. 9–10) comprising both mudbrick and woven grass-and-branch huts; 1–2, 4, 8: Tell Halaf (after Oppenheim and Schmidt 1943); 5: Sakçe Gözü (after Garstang et al. 1937); 6: Yarim Tepe II (after Merpert et al. 1978); 3, 7: Arpachiyah (after Mallowan and Rose 1935); 9–10: Botswana (after Kent and Herbich 1989).

At several Halaf sites it is possible to identify initial founding stages characterized by minor infrastructure works and by pit dwellings that were presumably covered by tents or by some lightweight superstructure or reeds or matting supported by timber posts (e.g., Al-Radi and Seeden 1980; Bernbeck 2008; Du Plat Taylor et al. 1950; Tsuneki 1998a). Mallowan and Rose (1935:31) were probably not wrong, then, in identifying the motifs depicted on several Early Halaf sherds as reed or mat-covered huts (Figure 5:7).

Comparisons are sometimes drawn between Halaf villages and the circular hut compounds of various African groups among whom it is usual for huts to be occupied by only one, or at the most two, persons (Akkermans and Schwartz 2003; Breniquet 1987; Flannery 1972). As Breniquet (1987:237) notes, this arrangement is common among patrilineal polygynous societies wherein each compound typically accommodates an elder man, his wives, adult sons plus their wives and children, and unmarried daughters. Only about two-thirds of buildings are residential; the rest are kitchens, stables, or storage buildings.

Among the Luo of western Kenya, for instance, the regional settlement pattern is characterized by polygynous, patrilocal, three-generation extended families living in separate homesteads scattered over the landscape (Dietler and Herbich 1998). What is significant for the present discussion is that each homestead undergoes a generational cycle of foundation, growth, and abandonment, with the result that the landscape is composed of interspersed homesteads representing all stages of the cycle. When the owner of a house dies, the house must be left to deteriorate and fall down or else be pulled down. Homesteads are eventually abandoned after the death of the founding generation and converted to farmland by the sons of the original male head. The sons are also obligated by custom to move out of their father's homestead and found their own homesteads when their own sons are ready for marriage (Dietler and Herbich 1998).

It is not claimed that this explains precisely the Halaf situation; nevertheless, ethnographic models of this kind do raise the possibility that the settlement shifts observed on larger sites such as Sabi Abyad, and the distinct cycles of founding, construction, and abandonment identified on smaller sites such as Fıstıklı Höyük, were socially constituted, rather than a reflection of the subsistence economy (Bernbeck 2008). All the same, the social imperatives behind such shifts need not preclude their having originated in the settlement mobility necessitated by chronic drought conditions. Certainly, an explanation along these lines seems to fit better with the chronology of cycling observed on many Late Neolithic sites than other current theories. The key issue, then, as regards changes in architecture and settlement layout, is not to do with changes to the subsistence base, but with how sharing and more intensive social interactions during a period of economic uncertainty led to changes in the basic units of production, from autonomous households, possibly with a monogamous base, to polygynous extended households (Flannery 1972). Thus, the fact that *tholoi* are present but not prevalent during the Pre- and Proto-Halaf is because the form of social organization was then only in its formative stages, whereas by the Middle Halaf it was typical (Breniquet 1987). Interestingly enough, among polygynous societies, having multiple wives is often considered essential to being able to mount lavish feasts (Dietler 2001).

Conclusions

The above discussion highlights the central role strategies for coping with the effects of long-term climate change are likely to have played in the emergence of the cultural attributes we associate with the Halaf tradition. In particular, it demonstrates the likely importance of feasting and storage to the structuring of Late Neolithic societies and how these measures for minimizing risk are likely to have become embedded in social practice. The possibility that these same societies were more mobile than their predecessors has also been considered, although at the same time, doubt has been cast on the likelihood that this included a shift to a purely pastoral nomadic existence.

Rather than a collapse of the existing order, in the face of increasing aridity, the emergence of the Halaf ceramic style and associated sociocultural developments can thus be explained as an intelligent process of social adaptation to a harsher climate. In particular, the development of a broader regional consciousness would appear to have opened the way for the adoption of adaptation measures that would not otherwise have been feasible.

Acknowledgments

I would like to express my sincere thanks to Olivier Nieuwenhuyse for inviting me to contribute to this volume and to Peter Biehl for his forbearance with regard to the submission of this paper. My thanks go also to David Menere for his assistance with the production of the graphics.

References Cited

Al-Radi, S., and H. Seeden 1980 The AUB Rescue Excavations at Shams ed-Din Tannira. *Berytus* 28:87–126.

Akar T., M. Avci, and F. Dusunceli 2003 Barley: Post-Harvest Operations. In *Post Harvest Compendium*, edited by D. Mejía and E. Parrucci, FAO Information Network on Post-Harvest Operations. http://www.fao.org/inpho/content/compend/text/ch31.htm.

Akkermans, P. M. M. G. 1989 The Prehistoric Pottery of Tell Sabi Abyad. In *Excavations at Tell Sabi Abyad—Prehistoric Investigations in the Balikh Valley, Northern Syria*, edited by P. M. M. G. Akkermans, pp. 77–213. B.A.R. International Series 468, Oxford.

Akkermans, P. M. M. G. 1993 *Villages in the Steppe: Later Neolithic Settlement and Subsistence in the Balikh Valley, Northern Syria*. International Monographs in Prehistory, Ann Arbor, Michigan.

Akkermans, P. M. M. G., R. Cappers, C. Cavallo, O. Nieuwenhuyse, B. Nilhamn, and I. N. Otte 2006 Investigating the Early Pottery Neolithic of Northern Syria: New Evidence from Tell Sabi Abyad. *American Journal of Archaeology* 110:123–156.

Akkermans, P. M. M. G., and K. Duistermaat 1997 Of Storage and Nomads. The Sealings from Late Neolithic Sabi Abyad, Syria. *Paléorient* 22(2):17–44.

Akkermans, P. M. M. G., and G. M. Schwartz 2003 *The Archaeology of Syria. From Complex Hunter-Gatherers to Early Urban Societies (ca. 16,000–300 B.C.)*. Cambridge University Press, Cambridge.

Akkermans, P. M. M. G., J. van der Plicht, O. P. Nieuwenhuyse, A. Russell, A. Kaneda, and H. Buitenhuis 2010 Weathering Climate Change in the Near East: Dating and Neolithic Adaptations 8200 Years Ago. *Antiquity* 84: Online Project Gallery.

Akkermans, P. M. M. G., and M. Verhoeven 1995 An Image of Complexity: The Burnt Village at Late Neolithic Sabi Abyad, Syria. *American Journal of Archaeology* 99:5–32.

Akkermans, P. M. M. G., and B. Wittmann 1993 Khirbet es-Shenef 1991. Eine späthalafzeitliche Siedlung im Balikhtal, Nordsyrien. *Mitteilungen der Deutschen Orient-Gesellschaft* 125:143–166.

Alley, R. B., P. A. Mayewski, T. Sowers, M. Stuiver, K. C. Taylor, and P. U. Clark 1997 Holocene Climate Instability: A Prominent, Widespread Event 8200 yr Ago. *Geology* 25(6):483–486.

Barber, D. C., A. Dyke, C. Hillaire-Marcel, A. E. Jennings, J. T. Andrews, M. W. Kerwin, G. Bilodeau, R. McNeely, J. Southon, M. D. Morehead, and J.-M. Gagnon 1999 Forcing of the Cold Event of 8,200 Years Ago by Catastrophic Drainage of Laurentide Lakes. *Nature* 400:344–348.

Becker, J., T. Helms, M. Posselt, E. Vila, P. Aytac, J. Eckardt, and J. Malo 2007 Ausgrabungen in Tell Tawīla, Nordost-Syrien. *Mitteilungen der Deutschen Orient-Gesellschaft zu Berlin* 139:213–268.

Bernbeck, R. 2008 An Archaeology of Multisited Communities. In *The Archaeology of Mobility: Old World and New World Nomadism*, edited by H. Barnard and W. Wendrich, pp. 43–77. Cotsen Institute of Archaeology at University of California, Los Angeles.

Bernbeck, R., S. Pollock, and C. Coursey 1999 The Halaf Settlement at Kazane Höyük. Preliminary Report on the 1996 and 1997 Seasons. *Anatolica* 25:109–147.

Bernbeck, R., S. Pollock, S. Allen, A. G. Gessner, S. K. Costello, R. Costello, M. Foree, M. Y. Gleba, M. Goodwin, S. Lepinski, C. Nakamura, and S. Niebuhr 2003 The Biography of an Early Halaf Village: Fıstıklı Höyük 1999–2000. *Istanbuler Mitteilungen* 53:9–77.

Bollig, M. 2005 *Risk Management in a Hazardous Environment: A Comparative Study of Two Pastoral Societies*. Springer, New York.

Bradley, R. 2005 *Ritual and Domestic Life in Prehistoric Europe*. Routledge, Abingdon.

Braidwood, R. J., and L. S. Braidwood 1960 *Excavations in the Plain of Antioch, I: The Earlier Assemblages: Phases A–J. O.I.P. 61*. The Oriental Institute of the University of Chicago, Chicago.

Braun, D. P., and S. Plog 1982 Evolution of "Tribal" Social Networks: Theory and Prehistoric North American Evidence. *American Antiquity* 47:504–525.

Breniquet, C. 1987 Nouvelle hypothèse sur la disparition de la culture de Halaf. In *Préhistoire De La Mésopotamie. La Mésopotamie Préhistorique et l'Exploration Récent du Djebel Hamrin*, edited by J.-L. Huot, pp. 232–241. Éditions du C.N.R.S., Paris.

Brooks, N. 2006 Cultural Responses to Aridity in the Middle Holocene and Increased Social Complexity. *Quaternary International* 151:29–49.

Campbell, S. 1992 *Culture, Chronology, and Change in the Late Neolithic of North Mesopotamia*. Unpublished PhD dissertation, University of Edinburgh, Edinburgh.

Campbell, S. 2007 Rethinking Halaf Chronologies. *Paléorient* 33(1):103–136.

Campbell, S., E. Carter, E. Healey, S. Anderson, A. Kennedy, and S. Whitcher 1999 Emerging Complexity on the Kahramanmaraş Plain, Turkey: The Domuztepe Project, 1995–1997. *American Journal of Archaeology* 103:395–418.

Cavallo, C. 1996 The Animal Remains—A Preliminary Account. In *Tell Sabi Abyad: The Late Neolithic Settlement. Report on the Excavations of the University of Amsterdam (1988) and the National Museum of Antiquities Leiden (1991–1993) in Syria*, edited by P. M. M. G. Akkermans, pp. 475–520. Nederlands Historisch-Archaeologisch Instituut, Istanbul.

Chamberlin, M. A. 2006 Symbolic Conflict and the Spatiality of Traditions in Small-Scale Societies. *Cambridge Archaeological Journal* 16(1):39–51.

Clarke, M. J. 2001 Akha Feasting: An Ethnoarchaeological Perspective. In *Feasts: Archaeological and Ethnographic Perspectives on Food, Politics, and Power*, edited by M. Dietler and B. Hayden, pp. 144–167. Smithsonian Institution, Washington D.C.

Clayton, L. A. 2004 *The Technology of Food Preparation: The Social Dynamics of Changing Food Preparation Styles*. Unpublished Master's thesis, Binghamton University, State University of New York.

Collet, P., and R. H. Spoor 1996 The Ground-Stone Industry. In *Tell Sabi Abyad: The Late Neolithic Settlement. Report on the Excavations of the University of Amsterdam (1988) and the National Museum of Antiquities Leiden (1991–1993) in Syria*, edited by P. M. M. G. Akkermans, pp. 415–438. Nederlands Historisch-Archaeologisch Instituut, Istanbul.

Colson, E. 1979 In Good Years and in Bad: Food Strategies of Self-Reliant Societies. *Journal of Anthropological Research* 35(1):18–29.

Cribb, R. 1991 *Nomads in Archaeology*. Cambridge University Press, Cambridge.

Cruells, W., and O. Nieuwenhuyse 2004 The Proto-Halaf Period in Syria. New Sites, New Data. *Paléorient* 30(1):47–68.

Davidson, T. E. 1977 *Regional Variation within the Halaf Ceramic Tradition*. Unpublished PhD dissertation, University of Edinburgh.

Dietler, M. 2001 Theorizing the Feast: Rituals of Consumption, Commensal Politics, and Power in African Contexts. In *Feasts: Archaeological and Ethnographic Perspectives on Food, Politics, and Power*, edited by M. Dietler and B. Hayden, pp. 65–114. Smithsonian Institution, Washington D.C.

Dietler, M., and I. Herbich 1998 Habitus, Techniques, Style: An Integrated Approach to the Social Understanding of Material Culture and Boundaries. In *The Archaeology of Social Boundaries*, edited by M. Stark, pp. 233–263. Smithsonian Institution, Washington D.C.

Domenig, G. 2003 Consequences of Functional Change: Granaries, Granary-Dwellings, and Houses of the Toba Batak. In *Indonesian Houses. Volume 1: Tradition and Transformation in Vernacular Architecture*, edited by R. Schefold, P. J. M. Nas, and G. Domenig, pp. 61–97. KITLV, Leiden.

Dooijes, R., and O. P. Nieuwenhuyse 2009 Ancient Repairs in Archaeological Research: A Near Eastern Perspective. In *Holding It All Together: Ancient and Modern Approaches to Joining, Repairs, and Consolidation*, edited by J. Ambers, C. Higgitt, L. Harrison, and D. Saunders, pp. 8–12. Archetype Books, London.

Du Plat Taylor, J., M. V. Seton-Williams, and J. Waechter 1950 The Excavations at Sakce Gözü. *Iraq* 12:53–138.

Duistermaat, K. 1996 The Seals and Sealings. In *Tell Sabi Abyad: The Late Neolithic Settlement. Report on the Excavations of the University of Amsterdam (1988) and the National Museum of Antiquities Leiden (1991–1993) in Syria*, edited by P. M. M. G. Akkermans, pp. 339–401. Nederlands Historisch-Archaeologisch Instituut, Istanbul.

Evershed, R. P., S. Payne, A. G. Sherratt, M. S. Copley, J. Coolidge, D. Urem-Kotsu, K. Kotsakis, M. Özdoğan, A. E. Özdoğan, O. P. Nieuwenhuyse, P. M. M. G. Akkermans, D. Bailey, R.-R. Andeescu, S. Campbell, S. Farid, I. Hodder, N. Yalman, M. Özbaşaran, E. Biçakci, Y. Garfinkel, T. Levy, and M. M. Burton 2008 Earliest Date for Milk Use in the Near East and Southeastern Europe Linked to Cattle Herding. *Nature* 455:528–531.

Flannery, K. 1972 The Origins of the Village as a Settlement Type in Mesoamerica and the Near East: A Comparative Study. In *Man, Settlement, and Urbanism*, edited by P. J. Ucko, R. Tringham, and G. W. Dimbleby, pp. 23–53. Duckworth, London.

Forni, S. 2007 Containers of Life. Pottery and Social Relations in the Grassfields (Cameroon). *African Arts* 40(1):42–53.

Frangipane, M. 2007 Different Types of Egalitarian Societies and the Development of Inequality in Early Mesopotamia. *World Archaeology* 39(2):151–176.

Fuensanta, J. G., and P. Charvát 2005 Halafians and Ubaidians: The Case of Tilbes Höyük in Birecik (Southeastern Turkey). In *Ethnicity in Ancient Mesopotamia*, edited by W. H. Van Soldt, pp. 123–133. Nederlands Instituut voor het Nabje Oosten, Leiden.

Garstang, J. 1953 *Prehistoric Mersin: Yümük Tepe in Southern Turkey*. Clarendon Press, Oxford.

Grigson, C. 2003 Animal Husbandry in the Late Neolithic and Chalcolithic at Arjoune: The Secondary Products Revolution Revisited. In *Excavations at Arjoune, Syria*, edited by P. Parr, pp. 187–240. B.A.R. International Series 1134, Archaeopress, Oxford.

Halstead, P., and J. O'Shea 1989 Introduction: Cultural Responses to Risk and Uncertainty. In *Bad Year Economics: Cultural Responses to Risk and Uncertainty*, edited by P. Halstead and J. O'Shea, pp. 1–7. Cambridge University Press, Cambridge.

Hayden, B. 1996 Feasting in Prehistoric and Traditional Societies. In *Food and the Status Quest: An Interdisciplinary Perspective*, edited by P. Wilson, P. Wiessner and W. Schiefenhövel, pp. 127–147. Berghahn Books, New York.

Hayden, B. 2001 Fabulous Feasts: A Prolegomenon to the Importance of Feasting. In *Feasts: Archaeological and Ethnographic Perspectives on Food, Politics, and Power*, edited by M. Dietler and B. Hayden, pp. 23–64. Smithsonian Institution, Washington.

Helmer, D., L. Gourichon, and E. Vila 2007 The Development of the Exploitation of Products from Capra and Ovis (Meat, Milk, and Fleece) from the PPNB to the Early Bronze in the Northern Near East (8700 to 2000 B.C. cal.). *Anthropozoologica* 42(2):41–69.

Helwing, B. 2003 Feasts as a Social Dynamic in Prehistoric Western Asia—Three Case Studies from Syria and Anatolia. *Paléorient* 29(2):63–85.

Hendon, J. A. 2000 Having and Holding: Storage, Memory, Knowledge, and Social Relations. *American Anthropologist* 102(1):42–53.

Hijara, I. H. 1978 Three New Graves at Arpachiyah. *World Archaeology* 10(2):125–128.

Hijara, I. H. 1997 *The Halaf Period in Northern Mesopotamia*. NABU Publications, London.

Hole, F. 1997 Paleoenvironment and Human Society in the Jezireh of Northern Mesopotamia 20,000–6,000 BP. *Paléorient* 23(2):39–49.

Ippolitoni-Strika, F. 1990 A Bowl from Arpachiyah and the Tradition of Portable Shrines. *Mesopotamia* XXV:147–174.

Joyce, R. A. 2000 Heirlooms and Houses. In *Beyond Kinship: Social and Material Reproduction in House Societies*, edited by R. A. Joyce and S. D. Gillespie, pp. 189–212. University of Pennsylvania Press, Philadelphia.

Kansa, S. W., and S. Campbell 2004 Feasting with the Dead? A Ritual Bone Deposit at Domuztepe, Southeastern Turkey (c. 5550 cal B.C.). In *Behaviour behind Bones: The Zooarchaeology of Ritual, Religion, Status, and Identity*, edited by S. Jones O'Day, W. Van Neer, and A. Ervynck, pp. 2–13. Oxbow Books, Oxford.

Kansa, S., A. Kennedy, S. Campbell, and E. Carter 2009 Resource Exploitation at Late Neolithic Domuztepe. *Current Anthropology* 50:897–914.

Keeley, L. H. 1996 *War before Civilization*. Oxford University Press, Oxford.

Kelly, R. L. 1992 Mobility/Sedentism: Concepts, Archaeological Measures, and Effects. *Annual Review of Anthropology* 21:43–66.

Kent, S. 1991 The Relationship Between Mobility Strategies and Site Structure. In *The Interpretation of Archaeological Spatial Patterning*, edited by E. M. Knoll and T. D. Price, pp. 33–59. Plenum, New York.

Kent, S., and H. Vierich 1989 The Myth of Ecological Determinism—Anticipated Mobility and Site Spatial Organization. In *Farmers as Hunters: The Implications of Sedentism*, edited by Susan Kent, pp. 97–130. Cambridge University Press, Cambridge.

Korfmann, M. O. 1982 *Tilkitepe*. Ernst Wasmuth, Tübingen.

Le Mière, M., and O. Nieuwenhuyse 1996 The Prehistoric Pottery. In *Tell Sabi Abyad, The Late Neolithic Settlement. Report on the Excavations of the University of Amsterdam (1988) and the National Museum of Antiquities Leiden (1991–1993) in Syria*, edited by P. M. M. G. Akkermans, pp. 119–284. Nederlands Historisch-Archaeologisch Instituut, Istanbul.

Le Mière, M., and M. Picon 1994 Early Neolithic Pots and Cooking. In *Handwerk und Technologie im Altern Orient: Ein Beitrag zur Geschichte der Technik im Altertum*, edited by R.-B. Wartke, pp. 67–70. Philip von Zabern, Mainz.

Lillios K. T. 1999 Objects of Memory: The Ethnography and Archaeology of Heirlooms. *Journal of Archaeological Method and Theory* 6(3):235–262.

Mallowan, M. E. L., and J. C. Rose 1935 Excavations at Tell Arpachiyah, 1933. *Iraq* 2:1–178.

Mayewski, P. A., E. E. Rohling, J. C. Stager, W. Karlén, K. A. Maasch, L. D. Meeker, E. A. Meyerson, F. Gasse, S. van Kreveld, K. Holmgren, J. Lee-Thorp, G. Rosqvist, F. Rack, M. Staubwasser, R. R. Schneider, and E. J. Steig 2004 Holocene Climate Variability. *Quaternary Research* 62:243–255.

Mellaart, J. 1975 *The Neolithic of the Near East*. Thames and Hudson, London.

Merpert, N. Y., and R. M. Munchaev 1973 Early Agricultural Settlements in the Sinjar Plain, Northern Iraq. *Iraq* 35:93–113.

Merpert, N. Y., and R. M. Munchaev 1987 The Earliest Levels at Yarim Tepe I and Yarim Tepe II in Northern Iraq. *Iraq* 49:1–36.

Merpert, N. Y., and R. M. Munchaev 1993a Burial practices of the Halaf culture. In *Early Stages in the Evolution of Mesopotamian Civilization*, edited by N. Yoffee and J. J. Clark, pp. 207–223. University of Arizona Press, Tucson.

Merpert, N. Y., and R. M. Munchaev 1993b Yarim Tepe III. The Halaf Levels. In *Early Stages in the Evolution of Mesopotamian Civilization*, edited by N. Yoffee and J. J. Clark, pp. 163–205. University of Arizona Press, Tucson.

Migowski, C., M. Stei, S. Prasad, J. F. W. Negendank, and A. Agnon 2006 Holocene Climate Variability and Cultural Evolution in the Near East from the Dead Sea Sedimentary Record. *Quaternary Research* 66:421–431.

Mills, B. J. 2007 Performing the Feast: Visual Display and Suprahousehold Commensalism in the Puebloan Southwest. *American Antiquity* 72:210–239.

Minnis, P. E. 1996 Notes on Economic Uncertainty and Human Behaviour in the Prehistoric North American Southwest. In *Evolving Complexity and Environmental Risk in the Prehistoric Southwest*, edited by J. A. Tainter and B. Bagley Tainter, pp. 57–78. Addison-Wesley, Reading, Massachusetts.

Molist, M. 1998 Espace Collectif et Espace Domestique dans le Néolithique des IXème et VIIIème Millénaires B.P. au Nord de la Syrie: Apports du Site de Tell Halula (Vallée de l'Euphrate). In *Éspace Naturel, Espace Habité en Syrie du Nord (10e–2e millénaires av. J-C.)*, edited by M. Fortin and O. Aurenche, pp. 115–130. Maison de l'Orient Méditerranéen, Lyon.

Molist, M., J. Anfruns, F. Borrell, X. Clop, W. Cruells, A. Gómez, E. Guerrero, C. Tornero, and M. Saña 2007 Tell Halula (Vallée de l'Éuphrate, Syrie): nouvelles données sur les occupations Néolithiques. Notice Préliminaire sur les Travaux 2002–2004. In *Les Résultats du Programme de Formation à la Sauvegarde du Patrimonie Cultural de Syrie*, edited by J. A. Massih, pp. 21–52. Cultural Heritage, DGAM, Damascus.

Mottram, M. 2010 *Continuity Versus Cultural Markers: Results of the Controlled Surface Collection of Tell Halula, North Syria.* Unpublished PhD dissertation, The Australian National University, Canberra.

Munchaev R. M., and N. Y. Merpert 1973 Excavations at Yarim Tepe 1972. *Sumer* 29:3–16.

Nieuwenhuyse, O. 2000 Halaf Settlement in the Khabur Headwaters. In *Prospection Archéologique du Haut-Khabur Occidental (Syrie du N.E.). Vol. I*, edited by B. Lyonnet, pp. 151–260. Institut Français d'Archéologie du Proche-Orient, Beirut.

Nieuwenhuyse, O. 2007 Plain and Painted Pottery: The Rise of Neolithic Ceramic Styles on the Syrian and Northern Mesopotamian Plains. *Papers on Archaeology of the Leiden Museum of Antiquities 3*. Brepols, Turnhout.

Nieuwenhuyse, O. 2008 Feasting in the Steppe—Late Neolithic Ceramic Change and the Rise of the Halaf. In *Proceedings of the 5th International Congress on the Archaeology of the Ancient Near East*, edited by J. Cordoba, M. Molist, C. Pérez, I. Rubio, and S. Martínez, pp. 691–702. UAM Ediciones, Madrid.

Nieuwenhuyse, O. 2009 The "Painted Pottery Revolution": Emulation, Ceramic Innovation, and the Early Halaf in Northern Syria. In *Méthodes d'Approche des Premières Productions Céramiques: Étude de Cas dans les Balkans et au Levant*, edited by L. Astruc, A. Gaulon, and L. Salanova, pp. 81–91. Marie Leidorf, Rahden/Westphalia.

Oppenheim, M. F. von, and H. Schmidt 1943 *Die Prähistorischen Funde. Tell Halaf I*. Walter de Gruyter. Berlin.

Özbal R., F. Gerritsen, B. Diebold, E. Healey, N. Aydin, M. Loyet, F. Nardulli, D.Reese, H. Ekstrom, S. Sholts, N. Mekel-Bobrov, and B. Lahn 2004 Tell Kurdu Excavations 2001. *Anatolica* 30:37–107.

Parr, P. 2003 Concluding Remarks. In *Excavations at Arjoune, Syria*, edited by P. Parr, pp. 277–281. B.A. R. International Series 1134. Archaeopress, Oxford.

Perkins, A. L. 1949 *The Comparative Archaeology of Early Mesopotamia*. The Oriental Institute of the University of Chicago, Chicago.

Pikirayi, I. 2007 Ceramics and Group Identities: Towards a Social Archaeology in Southern African Iron Age Ceramic Studies. *Journal of Social Archaeology* 7(3):286–301.

Rautman, A. E. 1996 Risk, Reciprocity, and the Operation of Social Networks. In *Evolving Complexity and Environmental Risk in the Prehistoric Southwest*, edited by J. A. Tainter and B. Bagley Tainter, pp. 197–222. Addison-Wesley, Reading, Massachusetts.

Rohling, E. J. and H. Pälike 2005 Centennial-Scale Climate Cooling with a Sudden Cold Event Around 8,200 Years Ago. *Nature* 434:975–979.

Russell, A. 2010 *Retracing the Steppes: Zooarchaeological Analysis of Changing Subsistence Patterns in the Late Neolithic at Tell Sabi Abyad, Northern Syria, c. 6900 to 5900 B.C.* Unpublished PhD dissertation, University of Leiden, Leiden.

Saña Seguí, M. 1999 Arqueología de la Domesticación Animal: La Gestión de los Recursos Animales en Tell Halula (Valle del Éufrates-Siria) del 8.800 al 7.000 BP. *Treballs d'Arqueologia del Pròxim Orient, 1.* Universitat Autònoma de Barcelona, Barcelona.

Sato, K. 1991 Menghuni Lumbung. Beberapa Pertimbangan Mengenai Asal-Usul Konstruksi Rumah Panggung di Kepulauan Pasifik, *Antropologi* 49:31–47.

Schefold, R. 2003 The Southeast Asian-Type House: Common Features and Local Transformations of an Ancient Architectural Tradition. In *Indonesian Houses. Volume 1: Tradition and Transformation in Vernacular Architecture*, edited by R. Schefold, P. J. M. Nas, and G. Domenig, pp. 19–60. KITLV, Leiden.

Schiefenhövel, W., and I. Bell-Krannhals 1996 Of Harvests and Hierarchies: Securing Staple Food and Social Position in the Trobriand Islands. In *Food and the Status Quest: An Interdisciplinary Perspective*, edited by P. Wilson, P. Wiessner, and W. Schiefenhövel, pp. 235–251. Berghahn Books, Providence, Rhode Island.

Schiffer, M. B., and J. M. Skibo 1997 The Explanation of Artifact Variability. *American Antiquity* 62:27–50.

Shimoyama, A., and A. Ichikawa 2000 Appendix 2: Fatty Acid Analysis of Pottery Samples from Tell el-Kerkh. In Fourth Preliminary Report of the Excavations at Tell el-Kerkh (2000), Northwestern Syria, edited by A. Tsuneki et al. *Bulletin of the Ancient Orient Museum* 21:33–36.

Spoor, R. H., and P. Collet 1996 The Other Small Finds. In *Tell Sabi Abyad: The Late Neolithic Settlement. Report on the Excavations of the University of Amsterdam (1988) and the National Museum of Antiquities Leiden (1991–1993) in Syria*, edited by P. M. M. G. Akkermans, pp. 439–473. Nederlands Historisch-Archaeologisch Instituut, Istanbul.

Staubwasser, M., and H. Weiss 2006 Holocene Climate and Cultural Evolution in Late Prehistoric–Early Historic West Asia. *Quaternary Research* 66:372–387.

Stein, G. J. 1994 Economy, Ritual, and Power in 'Ubaid Mesopotamia. In *Chiefdoms and Early States in the Near East: The Organizational Dynamics of Complexity*, edited by G. J. Stein and M. S. Rothman pp. 35–46. Monographs in World Archaeology No. 18. Prehistory Press, Madison.

Tsuneki, A. 1998a Ending Remarks. In *Excavations at Tell Umm Qseir in Middle Khabur Valley, North Syria*, edited by A. Tsuneki and Y. Miyake, pp. 202–203. Department of Archaeology, Institute of History and Anthropology, University of Tsukuba, Tsukuba.

Tsuneki, A. 1998b Tholoi: Their Socio-economic Aspects. In *Excavations at Tell Umm Qseir in Middle Khabur Valley, North Syria*, edited by A. Tsuneki and Y. Miyake, pp. 164–176. Department of Archaeology, Institute of History and Anthropology, University of Tsukuba, Tsukuba.

Tsuneki, A., J. Hydar, Y. Miyake, S. Akkahane, T. Nakamura, M. Arimura, and S. Sekine 1997 First Preliminary Report of the Excavations at Tell el-Kerkh (1997), Northwestern Syria. *Bulletin of the Ancient Orient Museum* 18:1–40.

Tsuneki, A., J. Hydar, Y. Miyake, O. Maeda, T. Odaka, K.-I. Tanno, and A. Hasegawa 2000 Fourth Preliminary Report of the Excavations at Tell el-Kerkh (2000), Northwestern Syria. *Bulletin of the Ancient Orient Museum* 21:1–36.

Twiss, K. C. 2008 Transformations in an Early Agricultural Society: Feasting in the Southern Levantine Pre-Pottery Neolithic. *Journal of Anthropological Archaeology* 27:418–442.

Van Dyke, R. M., and S. E. Alcock 2003 Archaeologies of Memory: An Introduction. In *Archaeologies of Memory*, edited by R. M. Van Dyke and S. E. Alcock, pp. 1–13. Blackwell, Oxford.

Van der Plicht, J., P. M. M. G. Akkermans, O. Nieuwenhuyse, A. Kaneda, and A. Russell 2011 Tell Sabi Abyad, Syria: Radiocarbon Chronology, Cultural Change, and the 8.2 ka Event. *Radiocarbon* 53(2):229–243.

Van Zeist, W., and W. Waterbolk-Van Rooijen 1996 The Cultivated and Wild Plants. In *Tell Sabi Abyad: The Late Neolithic Settlement. Report on the Excavations of the University of Amsterdam (1988) and the National Museum of Antiquities Leiden (1991–1993) in Syria*, edited by P. M. M. G. Akkermans, pp. 475–520. Nederlands Historisch-Archaeologisch Instituut, Istanbul.

Verhoeven, M. 1999 *An Archaeological Ethnography of a Neolithic Community. Space, Place, and Social Relations in the Burnt Village at Tell Sabi Abyad, Syria*. Nederlands Historisch-Archaeologisch Instituut, Istanbul.

Verhoeven, M., and P. Kranendonk (with a contribution by N. Aten) 1996 The Excavations: Stratigraphy and Architecture. In *Tell Sabi Abyad: The Late Neolithic Settlement. Report on the Excavations of the University of Amsterdam (1988) and the National Museum of Antiquities Leiden (1991–1993) in Syria*, edited by P. M. M. G. Akkermans, pp. 25–118. Nederlands Historisch-Archaeologisch Instituut, Istanbul.

Wason, P. K. 1994 *The Archaeology of Rank*. Cambridge University Press, Cambridge.

Watson, P. J. 1983 The Halafian Culture: A Review and Synthesis. In *The Hilly Flanks and Beyond: Essays on the Prehistory of Southwestern Asia*, edited by T. C. Young Jr., P. E. L. Smith, and P. Mortenson, pp. 231–250. University of Chicago Press, Chicago.

Weiss, H., and R. S. Bradley 2001 What Drives Societal Collapse? *Science* 291:609–610.

Wengrow, D. 2001 The Evolution of Simplicity: Aesthetic Labour and Social Change in the Neolithic Near East. *World Archaeology* 33(2):168–188.

Weninger, B., E. Alram-Stern, E. Bauer, L. Clare, U. Danzeglocke, O. Jöris, C. Kubatzki, G. Rollefson, H. Todorova, and T. van Andel 2006 Climate Forcing Due to the 8200 cal yr BP Event Observed at Early Neolithic Sites in the Eastern Mediterranean. *Quaternary Research* 66:401–420.

Wessing, R. 2003 The Shape of Home: Spatial Ordering in Sundanese Kampung. In *Indonesian Houses. Volume 1: Tradition and Transformation in Vernacular Architecture*, edited by R. Schefold, P. J. M. Nas, and G. Domenig, pp. 427–460. KITLV, Leiden.

Wick, L., G. Lemcke, and M. Sturm 2003 Evidence of Lateglacial and Holocene Climatic Change and Human Impact in Eastern Anatolia: High-Resolution Pollen, Charcoal Isotopic and Geochemical Records from the Laminated Sediments of Lake Van, Turkey. *The Holocene* 13:665–675.

Wiersma, A. P., and H. Renssen 2006 Model-Data Comparison for the 8.2 ka BP Event: Confirmation of a Forcing Mechanism by Catastrophic Drainage of Laurentide Lakes. *Quaternary Science Reviews* 25:63–88.

Wilkinson, T. J. 2003 *Archaeological Landscapes of the Ancient Near East*. University of Arizona Press, Tucson.

Woolley, C. L. 1927 The Cemetery of Al-'Ubaid. In *Ur Excavations Vol. 1: Al-'Ubaid*, edited by H. R. Hall and C. L. Woolley, pp. 149–213. Oxford University Press, Oxford.

Zeder, M. A. 1996 The Role of Pigs in Near Eastern Subsistence: A View from the Southern Levant. In *Retrieving the Past: Essays on Archaeological Research and Methodology in Honour of Gus W. Van Beek*, edited by J. D. Seger, pp. 297–312. Eisenbrauns, Winona Lake.

CHAPTER THREE

The 8.2 Event in Upper Mesopotamia

Climate and Cultural Change

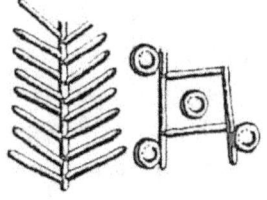

Olivier P. Nieuwenhuyse, Peter Akkermans,
Johannes van der Plicht, Anna Russell,
Akemi Kaneda

Abstract *In this paper we explore the socioeconomic repercussions of the so-called 8.2 event upon prehistoric communities in Upper Mesopotamia. This short-lived anomaly caused drought and somewhat cooler conditions for a period of about two centuries between 6200–6000 cal. BC. Various authors have argued that this precipitated dramatic economic disruption and social collapse among Late Neolithic societies. We argue, in contrast, that the debate has so far been based upon a poor archaeological record. We discuss settlement dynamics and demographic chance, shifts in material culture and animal exploitation patterns, using the well-documented case study of Tell Sabi Abyad (northern Syria). Inhabited between 7000–5500 cal. BC, this site shows dramatic changes over this long time span. However, the role of climate change as a prima causal factor remains to be discussed further.*

Then came the superflood on a faraway continent, and the sudden, howling, desiccating winds of 8200 B.P., after which everything came crashing down.

—E. Linden, *The Winds of Change*

INTRODUCTION

In 2006 a team led by Peter Akkermans and Hans van der Plicht began to explore the socioeconomic repercussions on prehistoric communities in the Near East of the so-called 8.2 event (Akkermans et al. 2010, 2011; van der Plicht et al. 2011).[1] The 8.2 event, as is well known, has been attributed to an abrupt disruption of the ocean thermohaline circulation (THC), most probably caused by the drainage of the huge postglacial Laurentide Lakes in front of the retreating Laurentide Ice Sheet (Barber et al. 1999; Bauer et al. 2004; Klitgaard-Kristensen et al. 1998; Teller et al. 2002; Wiersma and Renssen 2006). The event is thought to have brought about generally cold and dry conditions, causing significant drought in arid areas such as the Near East (Hoek and Bos 2007; Rohling and Pälike 2005; Weiss and Bradley 2001). The event was particularly severe in the Northern Hemisphere; oxygen records from Greenland ice-cores show an abrupt estimated cooling of an astonishing $6 \pm 2°C$ over Greenland (Alley et al. 1997). But its effects were felt worldwide, as indicated by a sudden and marked reduction in precipitation in Africa, the Near East, and Asia, and up to 2 degrees centigrade of cooling in the Near East at this time (Alley and Ágústsdóttir 2005; Bar Matthews et al. 1999; Enzel et al. 2003; Gasse 2000; Migowski et al. 2006; Renssen et al. 2002).

Indeed, such is the vigor of the 8.2 event—the largest anomaly to occur during the Holocene—that it has emerged in the perception of a much broader, nonscholarly audience as a dramatic metaphor for the doom that might befall our own society in the very near future. This abrupt climate event, or rather a somewhat popularized version of it, acted as a model for the great blockbuster *The Day after Tomorrow*, where the event repeats itself in just one week! Insights regarding this event emerging from the scientific community and the awareness that it might, perhaps, one day repeat itself even led to a serious White House policy report (Schwartz and Randall 2003), exploring various "what if?" scenarios.

Perhaps invigorated by such popular reception of a prehistoric climate event, several scholars have speculated on the responses to this event of the prehistoric communities that inhabited the Near East at the time. Many of our colleagues quickly reached a particularly grim consensus—the event initiated dramatic social disruption and wholesale cultural collapse. It induced region-wide socioeconomic downfalls, massive population migrations, increases in violence and warfare, and various other sorts of mayhem (Bar-Yosef 2001, 2006; Clare et al. 2008; Hassan 2009; Staubwasser and Weiss 2006; Weiss 2000; Weiss and Bradley 2001; Weninger et al. 2006). A long-flourishing Neolithic society came to its end as entire regions were abandoned en masse and destitute villagers became the world's first climate refugees, roving and "habitat tracking" (Staubwasser and Weiss 2006:378) toward the remaining river valleys.

What strikes us, as prehistorians working in the field, is the generally poor base of empirical data on which the various "collapse" scenarios for the 8.2 event are based. Several issues can be pointed out. Foremost, synchronizing climate change with specific episodes of prehistoric culture change remains a formidable challenge in the region (Düring, this volume; Maher et al. 2011). It cannot be emphasized enough that prehistoric cultures in the ancient Near East characteristically remain poorly dated. Their diachronic boundaries can all too often easily be shifted up and down by a couple of centuries to match the preferred climate change. Furthermore, as recent archaeological fieldwork makes abundantly clear, the later prehistory of the Near East was a time in which tremendous cultural innovations were continuously being made (Akkermans and Schwartz 2003; Akkermans et al. 2006; Carter and Phillip 2010; Nieuwenhuyse et al.

2013). This means that there is a priori a high probability of finding synchronicity between some climate event and some cultural innovation, as Maher, Banning, and Chazan point out (Maher et al. 2011:19; also Coombes and Barber 2005).

Well-documented sites with sound absolute dates for cultural sequences that show the possible repercussions or adaptations to the 8.2 event in situ are frustratingly rare. Most scholars therefore resort to contrasting larger spatial-temporal units, adopting a culture-historical comparative approach to explore broader changes before and after the event. This bears the risk of comparing apples and oranges, so to speak, and losing sight of small-scale patterns of change and innovation. Prehistoric culture groups in the Near East have typically been conceived of as spatially and temporally bounded entities, constructed in a polythetic fashion on the basis of limited sets of type fossils such as pottery or lithic tools (Bernbeck 2005, 2008; Nieuwenhuyse 2007). Although such constructs remain convenient shorthand for summarizing much diversity, they have often been treated as reified prehistoric actors, bouncing on and off the stage in response to changing environments. This culture model may not be adequate to explore possible responses to climate change. Instead, we should be open to investigate changes taking place in multiple domains and on multiple time scales simultaneously.

In the Near East, perhaps in contrast to the temperate regions of Europe (see the contribution of Gronenborn, this volume), the linkage between structurally increased aridity and societal downfall is often held to be self-evident (e.g., Issar and Zohar 2004), a tradition of thought going back to the pioneering work of Pumpelly (1905) and Childe (1929). After all, large parts of the region are ecologically marginal, and much of the region is characterized by a high level of vulnerability (see Clare et al. 2008). Indeed, in semi-arid regions such as the Near East, water availability is a key factor influencing human decision making, and an increase in aridity poses very serious threats (Brooks 2006; Catto and Catto 2004; deMenocal 2001). While increased rainfall allows civilizations to grow, even a slightly diminished rainfall can therefore only have had disastrous consequences, or so it is often thought. Many scholars working in the Near East see no obvious reason to question the causal role of climate as the "serial killer of civilizations" (Linden 2006:26). Yet what is overlooked in this equation is the fact that societies in the Near East have flourished and built great civilizations with ever-expanding vigor in spite of their marginal location; apparently, over the long term their propensity for adaptation exceeded their structural risk of collapse (Rosen 2007; Wilkinson 2003). Why should the 8.2 event necessarily result in collapse rather than instigating innovation, adaptation, and survival?

Some of the "collapse" scenarios argue from a rather narrow, mechanistic perspective. The main conceptual tool in many of these versions simply appears to be demonstrating synchronicity, accepting this as self-evident proof for causality. Yet, human societies are not some species of Mediterranean plankton, proliferating when it gets warmer, dying off when it gets cooler. While synchronicity is essential, it is not enough. One finds precious little discussion in the literature of possible alternative socioeconomic outcomes, and the inevitability of collapse is insufficiently questioned (Rosen 2007). Yet it is our task as archaeologists to point out how climate interacted with existing economies, social structures, and cultural traditions (Mackintosh et al. 2000; Minnis 1985; Reycraft and Bawden 2000; Wossink 2009). In the modern world, severe droughts combine with specific sociopolitical conditions to result in famine (Glantz 1994; van der Leeuw 2000). For past societies, then, we should explain how a particular period of aridity led to the specific cultural changes observed. Few scholars have taken up this much more difficult challenge. Dramatic scenarios of climate change in archaeology often have actors—culture, climate—but no plot.

In this contribution we discuss some of the social, economic, and ideological changes that manifested themselves at the time of the 8.2 event in Upper Mesopotamia. In archaeological terms, the era we are concerned with is the Late Neolithic (ca. 7000–5300 cal BC). The term *Upper Mesopotamia* refers to the northern part of the Fertile Crescent, the area covered today by southeastern Turkey, northern Syria, and northern Iraq. Climate models for this region predict a shift to dryer and cooler conditions during the 8.2 event (Wiersma and Renssen 2006). The modeling finds support in climate proxies that include speleothems and sea bottom sediments (Bar Matthews et al. 1999). How did people cope with these changes in their environment?

Our aims in this paper are twofold. First, we explore the case for collapse. What evidence can we muster for abandonment at the level of the village? Can we demonstrate demographic upheaval at the level of the wider region? We argue that there is no evidence for "collapse." Rather, the time of the 8.2 event was a time of significant social and economic transformation. Second, we briefly explore the nature of these transformations in order to evaluate the possible role of climate change as a causal agent. We focus on shifts in the exploitation of animals, and on changes in the production and consumption of containers.

Settlement Dynamics at Tell Sabi Abyad

In this paper, we take the Late Neolithic site of Tell Sabi Abyad as a case study. The almost two thousand years of continuous occupation at this location provide a perfect opportunity to gain a detailed insight into society before and after the event. The excavations have resulted in a sequence of uninterrupted settlement layers that shed light on not only the period of apparent climate change, but also on the long trajectory of cultural development that preceded the event, as well as the centuries that followed it. The rich information derived from these excavations allows a diachronic study of Late Neolithic society with a focus on both short and long-term changes in settlement, material culture, socioeconomic organization, and environment.

Tell Sabi Abyad is located in the upper Balikh valley of Northern Syria approximately 30 kilometers from the Syro-Turkish border (Figure 3.1). The Balikh Valley forms a narrow irrigated corridor through the arid Jazirah desert steppe (Copeland 1979). Tell Sabi Abyad is situated between isohyets of 300 mm (on the Turkish border) and 200 mm (on the Euphrates), making the area rather marginal for dry farming, with common crop failures (Van Zeist 1988; Wilkinson 1998). Even small changes in the amount of annual precipitation can have drastic results, as seen in the three-year drought that has recently

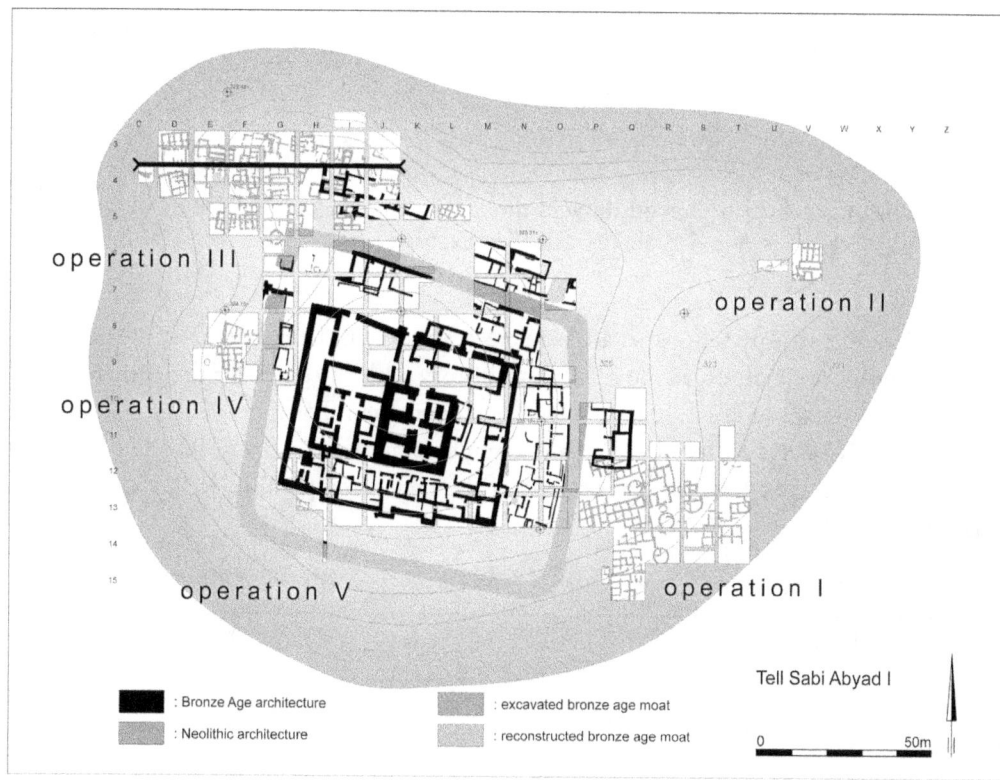

FIGURE 3.1. Map of Tell Sabi Abyad, showing the locations of areas of work (Operations). Solid line: the location of the section west-east shown in Figure 3.2.

hit Syria (Akkad 2009). This contemporary crisis may constitute a useful ethnographic analogue to those favoring the archaeological collapse scenario, as it led to dramatic crop failures and livestock loss, caused the abandonment of many a farming village and brought about much "habitat tracking" toward the major cities.

The mound of Tell Sabi Abyad has been shown to have a highly complex history of settlement. The main focus of this research is an area known as Operation III, on the northwestern slopes of the mound (Figure 3.1). The work in Operation III revealed four successive phases of deposition, which we have named Sequence A (ca. 7100–6200 BC), Sequence B (ca. 6200–5900 B.C.), Sequence C (ca. 5900–5800 B.C.), and Sequence D (ca. 5700–5500 B.C.) (Figure 3.2). The excavations in Operation III show that the earliest stratigraphic phase (sequence A) is comprised of at least 12 distinct levels starting during the Initial Pottery Neolithic (7000–6700 BC) and continuing through the Early Pottery Neolithic into the early stages of the Pre–Halaf Pottery stage (until ca. 6200 BC; see Nieuwenhuyse et al. 2010 for the terminology). Sequence B continues with a sequence of at least eight levels after ca. 6200 B.C. (Pre-Halaf and Transitional periods). This is followed by deposits dated to the Early Halaf (Sequence C) and the Middle Halaf (Sequence D) periods (Figure 3.3).

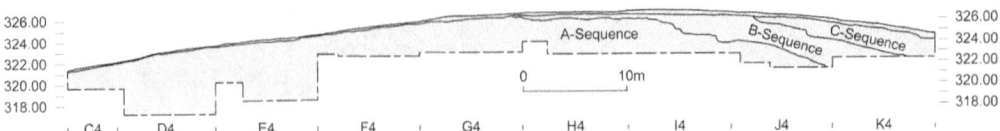

FIGURE 3.2. Schematic section through the excavations at Operation III at Tell Sabi Abyad. Sequences A and B are discussed in this paper.

Relevant for the present paper are sequences A and B, as these cover the time of the 8.2 event. In order to synchronize the settlement history with the dates for the climate event, a thorough program of absolute dating has been conducted. For the seventh to early sixth millennium BC layers of Tell Sabi Abyad, a grand total of more than 300 ^{14}C dates have been obtained thus far, making the site one of the best-dated prehistoric sites in the Near East (van der Plicht et al. 2011).[2]

Date cal. BC	Period	Tell Sabi Abyad I - operations					Tell Sabi Abyad II	Tell Sabi Abyad III
		I	II	III	IV	V		
5700	Middle Halaf			level C-1				
5800	Early Halaf	level 1						
		level 2						
5900		level 3	level 1	level C-2/8				
		level 4	level 2					
6000	Transitional	level 5 Burnt Village	level 3	level B-1		phase III		
		level 7	level 4	level B-2				
		level 8		level B-3				
				level B-4				
6100	Pre-Halaf	P-15 - 8		level B-5		phase II		
		P-15 - 9		level B-6				
				level B-7				
6200		P15 - 10		level B-8				
				level A-1				
6300				level A-2				
				level A-3	level 1	phase I		
6400	Early Pottery Neolithic	P-15 - 11		level A-4	level 2			
				level A-5				
6500				level A-6				
				level A-7				
6600				level A-8				
				level A-9				
6700				level A-10			level 1	trench H7
6800	Initial PN			level A-11			level 2	trench H8
				level A-12				
6900								
7000	Late PPNB						level 3	trench H9
7100							level 4	

FIGURE 3.3. The stratigraphy and absolute date of the prehistoric sequence excavated at Tell Sabi Abyad, showing the relationships between Operations I to V. Grey shade indicates the duration of the 8.2 ka event as observed in the Greenland ice cores (after van der Plicht et al. 2011).

Table 3.1
Tell Sabi Abyad. The Results of the Bayesian Analysis: Date Ranges for Operation III, Sequence A and B

Level	Nr. of ^{14}C dates	Date range (BC)
B3	2	6040–5995
B4	4	6050–6015
B5	6	6075–6040
B6	5	6095–6065
B7	6	6125–6080
B8	13	6180–6105
A1	23	6335–6225
A2	17	6385–6330
A3	6	6395–6375
A4	16	6455–6390
A5	9	6495–6455
A6	2	6505–6485
A7	14	6605–6495
A8	7	6630–6590
A9	4	6675–6620
A10	6	6750–6675
A11	4	6825–6740
A12	4	6865–6770

How does the stratigraphic sequence of the site correspond to the 8.2 event? The 8.2 event has been observed in many records elsewhere during the last decade, and many of these records are well dated (Wiersma 2008). We have already discussed the currently available evidence in detail elsewhere (van der Plicht et al. 2011:234–235). We summarize this discussion graphically in Figure 3.3, which plots the duration of the 8.2 event as observed in Greenland over the stratigraphy excavated at Tell Sabi Abyad. For the Greenland data we show only the 8.2 event as observed in ice cores, using the chronology of Thomas et al. (2007) for the combined Greenland ice cores, which is also consistent with the tree rings. This comparison shows that the 8.2 event synchronizes with the transition from Sequence A to Sequence B. The event began during the time represented by level A1 and fully manifested itself by level B8.

If we combine the vertical stratigraphy with the settlement plans spatially, this shows that Tell Sabi Abyad was far from deserted during or after the 8.2 event. People did not pack up and leave. Rather, what we see at this time is a shift in location preferences. During the entirety of Sequence A, people had built and rebuilt their dwellings on what is now the western half of the mound. Over the course of centuries this resulted in a steep buildup of sediments consisting of eroded loam buildings and copious amounts of occupational debris. After level A1 the occupation at the highest point of this centuries-old location came to an end. For the construction of new buildings people moved down the slopes of the older tell. Gradually, the village moved eastward, eventually partially covering the pre-event mound (Figure 3.2). This was a relocation of building activities synchronous with the 8.2 event, certainly. But this was not a very dramatic move, and it certainly did not involve wholesale site abandonment.

Archaeologists working in the southern Levant have recently documented a series of significant episodes of tell erosions, the so-called "rubble slides." Some of these may coincide with the 8.2 event (Gebel 2009; Rollefson 2009; Weninger 2009). They have discussed the possibility that these slides represent episodes of large-scale site abandonment induced by the event. The southern Levantine rubble slides phenomenon should not be confused with the formation of the B depositions at Tell Sabi Abyad. The deposits from Sequence B cover those of Sequence A, but largely this appears to have resulted from common processes of tell formation, resulting from building structures of loam, the ongoing deposition of village debris and the gradual erosion of abandoned buildings. While slope erosion certainly contributed to the formation of the B levels, so far there does not appear to have been anything unusual in this.

Apart from Operation III, post-event cultural deposits were excavated in Operation I, as well as in Operation II, on the southeastern and northeastern slopes of the mound, respectively (Akkermans 1993; Akkermans et al. 2006; Akkermans and Verhoeven 1995). In fact, it appears that the entire eastern half of Tell Sabi Abyad was inhabited in the centuries during and following the event (Figure 3.1). Although there are no stratigraphic connections linking Operation III and Operations I–II, the radiocarbon dates available for Operations I–II and the associated material culture make it clear that these areas existed side by side for several centuries. This confirms the image already gained from the work in Operation III: occupation at Tell Sabi Abyad shifted from west to east on basically the same site location, and it continued with no major disruption throughout the duration of the 8.2 event and afterward.

A Regional Perspective on Climate Change

The carefully dated sequence from Tell Sabi Abyad site makes it possible to look at settlement and demography from a wider, regional perspective. Here we shall briefly examine the evidence from a regional survey carried out in the Balikh Valley in the 1980s by the University of Amsterdam (Akkermans 1993). This survey yielded several hundreds of sites dating from the earliest human settlement in the Pleistocene into the modern period.[3] Subsequent work by the Oriental Institute yielded additional prehistoric sites and provided insights in the geomorphology of the landscape (Wilkinson 1998). We may use this evidence to plot pre-event and post-event site distributions in the valley. The Balikh lies in a semi-arid part of the Fertile Crescent, and the (modern) 220 mm rainfall isohyet runs through the central part of the valley. Certainly, a wholesale collapse induced by the 8.2 event should be visible in a notable discontinuity of human settlement. Can we demonstrate major demographic shifts such as predicted by the "cultural collapse" approach?

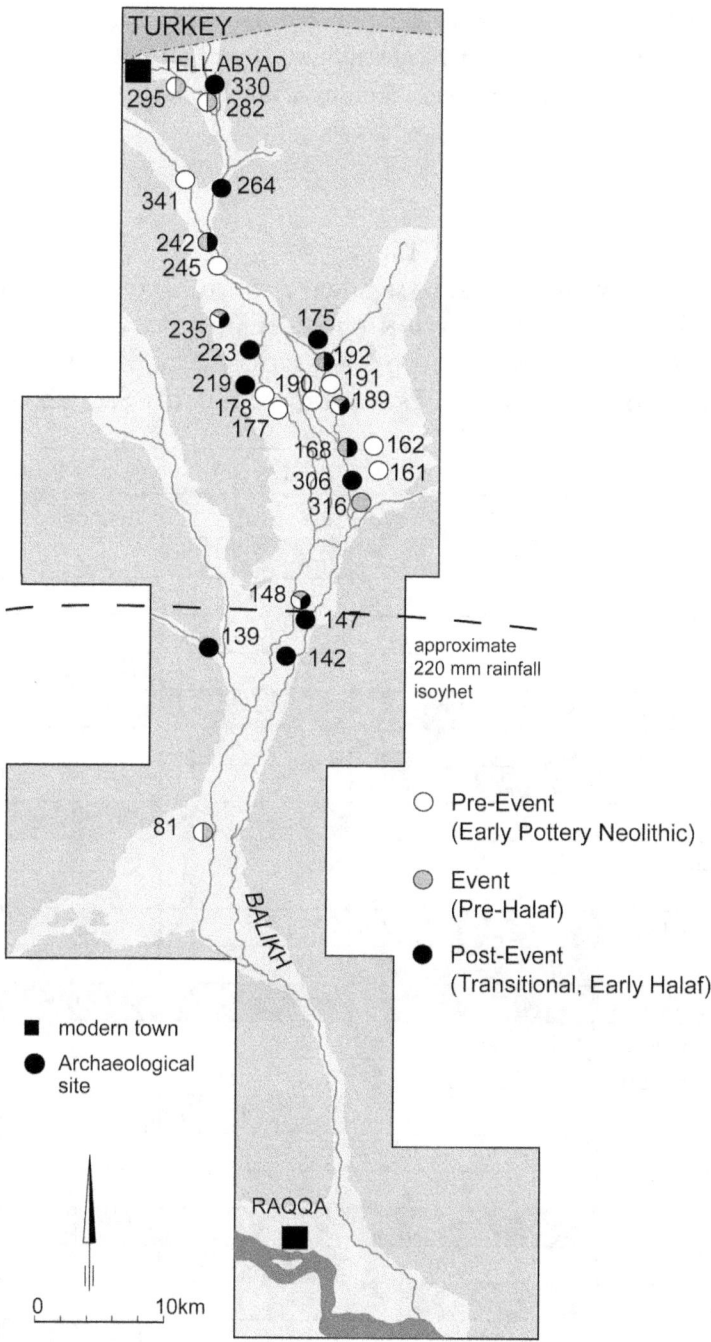

FIGURE 3.4. Distribution of Late Neolithic sites in the Balikh valley inhabited prior to, during and/or after the 8.2 event (University of Amsterdam survey).

In terms of culture-history the pre-event phase would correspond with what is termed the Early Pottery Neolithic (6800–6300 cal BC). The event itself would correspond broadly with what has been termed the Pre-Halaf (ca. 6300–6000). The two phases known as the Transitional Period (ca. 6000–5900 cal BC) and the Early Halaf period (ca.

5900–5700 cal BC) would cover the time immediately after the event (for definitions see Akkermans 1993; LeMiere and Nieuwenhuyse 1996; Nieuwenhuyse 2007). A total of 27 sites yielded evidence from one or more of these phases (Table 3.2). Figure 3.4 plots the distribution of these sites.

TABLE 3.2
UNIVERSITY OF AMSTERDAM SURVEY IN THE BALIKH VALLEY.
SITES LISTED WITH OCCUPATION IN THE EARLY POTTERY NEOLITHIC (PRE-EVENT), PRE-HALAF (8200) AND TRANSITIONAL TO EARLY HALAF PERIOD (POST-EVENT), SHOWING ESTIMATED MAXIMUM SITE EXTENT

Balikh Survey	Pre-event	Event	Post-event	
Site no. (BS)	Early Pottery Neolithic	Pre-Halaf	Transitional	Early Halaf
81	0,7	0,7		
139				0,8
142				0,6
147			0,5	5
148 – Tell Mounbateh	20	10	12,1	19,5
161	0,9			
162	1,6			
168		0,8	0,8	0,8
175 – Tell Hammam et-Turkman				1,5
177 – Tell Damishilyya I	0,5			
178	0,6			
189 – Tell Sabi Abyad I	2	3	3	2
190 – Tell Sabi Abyad II	0,2			
191 – Tell Sabi Abyad III	0,5			
192 – Tell Sabi Abyad IV		0,9		0,9
219				2,3
223				2
235 – Tell Wazgöl	0,8	0,8	1,4	1,4
242		0,3		0,3
245 – Tell Assouad	2,2			
264			—	—
282	0,6	0,6		
295	1,2	0,4		
306			0,8	0,8
316		0,5		
330				1
341	1			
Number of sites	15	11	8	16

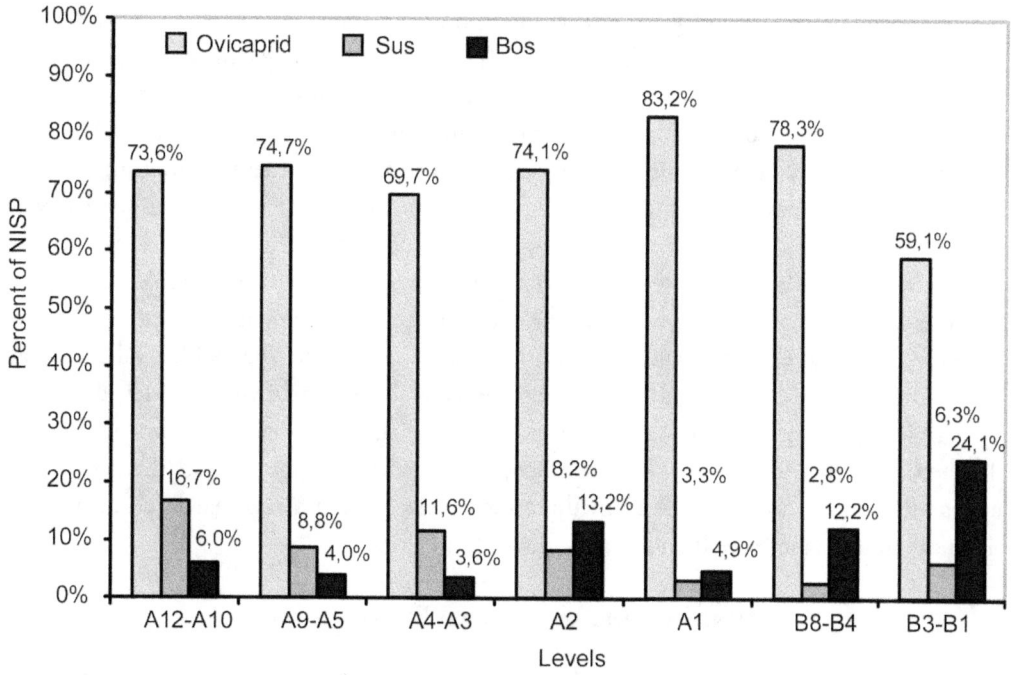

FIGURE 3.5. Tell Sabi Abyad, Operation III. Proportions of the main livestock species by level groupings (animal exploitation phases) (after Astruc and Russell 2013).

Of these prehistoric sites, seven Early Pottery Neolithic villages no longer show any signs of inhabitation during the Pre-Halaf period. Some of these, then, were perhaps abandoned at the onset of the arid conditions brought by the 8.2 event. On the other hand, four sites continued right into the event before they were abandoned, while three sites continued to be inhabited during and after the event. In addition to Tell Sabi Abyad (BS 189), these included Tell Mounbateh (BS 148) and Tell Wazgöl (BS 235). Interesting to note, four new sites were founded right at the start of the event, and a further three just after the event had ended. Significantly, throughout the event humans made continuous use of the marginal southern part of the valley, as indicated by the location of BS 81.

The reader will have noticed that the pre-event and post-event phases as plotted on the map were far from equal in duration. The monotonous character of the Early Pottery Neolithic ceramic assemblage and the slow pace of ceramic innovation during this long period simply do not allow further differentiation. In contrast, rapid pottery change and a strong technological and stylistic differentiation after 6200 cal BC enable a much more nuanced breaking-down of the post-event evidence. This leads to what is known as the "contemporaneity effect" (Pollock 1999), obscuring within-period settlement shifting in the longer periods. Such is the case, for instance, with BS 191 (Tell Sabi Abyad III), BS 192 (Tell Sabi Abyad II), and BS 245 (Tell Assouad), for which excavations show that these sites were abandoned during the Early Pottery Neolithic, long before the onset of the 8.2 event. Compensating for the contemporaneity effect would result in reducing the number of pre-event sites relative to the post-event period, evening out even more the differences in site numbers between the pre-event and post-event phases.

This does not come across as a significant process of abrupt site abandonment coinciding with a climate event. Rather, many, if not most, of these abandonments and new foundations may form part of the regular process of regular site relocation that was characteristic for the Late Neolithic in Upper Mesopotamia (Akkermans 1993, 2013; Bernbeck 2008; Nieuwenhuyse 2000; Nieuwenhuyse and Wilkinson 2008). Obviously, this image of continuity does not take into account possible changes in the role of these villages. But what it does make clear is that the 8.2 event in the Balikh valley so far cannot be associated with significantly increased site abandonment, the sudden desertion of marginal parts of the landscape or even a notable shift in village location preference. To the degree that settlement patterns reflect broader demographic trends, then, it would appear that the event had little immediate repercussions. On a regional level, at least, it did not induce wholesale demographic disruption.

Innovations in Animal Exploitation

The evidence with regard to the exploitation of wild and domesticated animals is essential in assessing the effects of climate change on the subsistence economy. Even moderate fluctuations in climate would be expected to show up, for instance, in changing relative frequencies of species, changes in body size over generations, or even in the elimination of entire species (Valtorta 2002; Tchernov 1982). In assessing changes in animal exploitation at Tell Sabi Abyad we have at our disposal the analysis of the faunal remains from Operation III, which cover the period ca. 6900–5900 cal BC, or the Early Pottery Neolithic to Early Halaf periods (Russell 2010). This adds to previous work on the material from Operation I, which covers the period ca. 6200–5900 cal BC, or the Pre-Halaf to Early Halaf periods (Cavallo 2000).

This work documents continuous changes in animal exploitation over the seventh and early sixth millennia. The long sequence may be grouped in seven successive "animal exploitation phases" (AEP), each phase characterized by a specific species composition and exploitation pattern (Russell 2010). Importantly, this periodization was made independent from, and does not completely overlap with, the culture-historical boundaries, which are largely based on the pottery. Thus, we may explore cultural changes across different domains—pottery containers versus animal uses—and at different time scales. We emphasize that the boundaries between successive animal exploitation phases were fluent and that shifts in the faunal patterns were always gradual and subtle. Human-animal relationships were continuously changing, both in terms of the wild animals exploited and the animal management strategies employed. The animal exploitation sequence does not suggest abrupt ruptures, dramatic shifts, or sudden mass extinctions of animals. This being said, clear shifts can be observed at around 6675 BC (Phase II), 6385 BC (Phase IV), and 6225 BC (Phase VI). Phase VI in particular is relevant here as it represents the shift from Sequence A to Sequence B, coinciding with the 8.2 event. For present purposes we highlight the management of ovicaprids and cattle. Over the long term, there were important shifts in the way people exploited these species.

Ovicaprids (domestic sheep and goat) dominated the assemblage in all levels. Sheep and goats clearly formed the basis of the animal-based subsistence system at Tell Sabi Abyad. Both postcranial fusion and mandibular tooth wear show roughly similar mortality profiles for each animal exploitation phase. In general, there were only subtle changes through time, with a few trends appearing. Levels A12 to A2 show a culling age of approximately two years of age with very few animals over three years of age present. In levels A1 to B4 (Pre-Halaf period), the main culling age shifts one year to around three years of age, with more animals living up to four years of age and older. In levels B3–B1 (Transitional Period) this trend continues, with even more animals living over four years of age but in these levels there is a relatively high neonatal mortality rate. These mortality profiles show that in the oldest levels the main drive for ovicaprid husbandry was meat production, the majority of animals being culled at the prime meat age of two years with only a small number of breeding stock maintained. Through time there appears to be a shift to a mixed economy of both meat and secondary product production. The hair or fleeces of the ovicaprids may also have become more important as the numbers of older animals also increased.

The data from Sabi Abyad, then, suggest a move from primarily meat to meat-to-meat plus milk and fiber production around 6225 BC in animal exploitation phase VI, but this formed part of a long-term trend. For the pre-event part of Sequence A (levels A12–A2), the mortality profiles of sheep (and goats) reflects the culling of animals when they are fully grown but still relatively young with tender meat. Milk production cannot be ruled out in these early levels however, as milk could have been taken on a small scale for human consumption, while the young livestock were reared until their first or second winter, when they were culled for meat (Halstead 1996). In the levels coinciding with the levels A1–B4, there is an increased emphasis on the production of milk with the lamb (or kid) still present so as to allow milk let-down and the culling of some females because of decreased milk yield or lamb production at two to four years of age. Indeed, residue analysis of pottery sherds from this period has produced the earliest evidence so far in the Near East of human milk consumption (Evershed et al. 2008). Fleece production can also be inferred from the presence of animals over four years of age. The emphasis on wool is also suggested by the synchronous introduction of spindle whorls in the Pre-Halaf period (Rooijakkers 2012). In the post-event levels (after level B3), the move to intensified secondary product production is even more clear, with milk and fleece production perhaps taking priority over meat production in sheep and goats.

In the marginal desert steppe of the Balikh valley, sheep and goats were the best species to exploit for secondary products. Both sheep and goats are good milk producers. While goats are generally thought to be better milk producers than sheep in terms of output (French 1970: 38), sheep milk has the highest fat content and as such is very nutritious (Ryder 1981). Milk also contains protein, minerals, vitamins, and carbohydrates (Outram and Mulville 2005). It is by its very essence a complete food. The judicious use of the sheep and goats for milk throughout the animals' lifespan would have had many benefits in terms of the regular nutrients available to the human population from this foodstuff, without decreasing the herd size as is necessary in accessing nutrition from meat. Milk produced by the lactating ovicaprids at Tell Sabi Abyad presumably made a significant addition to subsistence and helped the herders of Tell Sabi Abyad meet their nutritional needs.

Bos were found in all levels, cattle being the second most common species in the post-event levels, and the third most common species in the pre-event A levels. Sequence A documents a long-term process of gradually increasing control over, and eventually the complete domestication of, the aurochs (Russell 2010). In the pre-event levels, their mortality profiles do not reflect that of a wild, hunted population but rather a culturally controlled or proto-domestic cattle population. Herd security and meat production seem to have been the main foci of bos husbandry in these levels. This form of animal management continued through the final stages of Sequence A (levels A2 and A1), but by this time the animals can be considered fully domestic. In the levels coinciding with the climate events (levels B8–B4, or Pre-Halaf period) there was a switch to a more intensive meat production, a pattern that continued into the immediate post-event levels (Sequence C, or Early Halaf period).

Cattle at Sabi Abyad, then, appear to have been kept primarily for their meat. There is no evidence of their use for traction. Their mortality profiles do not suggest that milk production was a specific aim in the husbandry of these animals, although this certainly cannot be ruled out. In the Balikh valley during the time of the event, lush pasture would have been relatively sparse. Sheep and goats would have been the preferable animal for milk production as their nutritional needs would be met more easily than cattle (French 1970). As milk production relies on the production of young and the subsequent lactation period, sheep and goats would be a much more reliable source of dairy products, as well as being animals easier to milk (French 1970; Whittle 2003:93).

Some of these changes in animal exploitation may be argued to have been adaptive to more arid conditions. It is conceivable that deterioration in the local environment induced by the 8.2 event stimulated the increased reliance on ovicaprid herding and the intensification in using their secondary products. Other faunal shifts that might be related to more arid conditions include a short-lived reduction in pig husbandry and a decrease in the exploitation of wild animals in the levels coinciding with the climate event (levels B8–B1), a change that perhaps reflects shifts in the availability of certain species (Russell 2010). With regard to the domestication of cattle, however, the climate argument would be more difficult to sustain. Although cattle provided a lot of meat on the hoof, they were also much more costly and economically risky to keep because of their high food and water requirements and lower birth rates (Kansa et al. 2009). It would be difficult to argue that the final steps in their domestication were taken entirely as an adaptive response to more arid conditions caused by the 8.2 event.

As we have seen, these shifts in the management of ovicaprids and cattle did not come about suddenly, but instead were part of long-term changes in the exploitation of these vital animals. The main trend was an increasing reliance on domestic animals, particularly pastoral animals such as goat, sheep, and cattle. These long-term shifts cannot be interpreted exclusively as adaptation to sudden changes in the climate, although climate stresses may very well have accelerated already ongoing trends. Ways of living with animals are profoundly cultural (Ingold 2000); nonecological factors, too, must be taken into account. These may include changing hunting practices (Astruc and Russell 2013), and new cultural roles for domestic animals as tokens of wealth (Russell 1998). Cattle in particular are prime animals for competitive bouts of feasting (Russell 1998). Significantly, stylized images of cattle began to appear at about this time in the iconography, attesting to their enhanced cultural importance. People did not keep these animals exclusively as a means to escape starvation.

Containers and Climate Change

Containers—receptacles for holding and moving around liquid and solid goods—are a valuable proxy for investigating societal responses to climate change. In the Late Neolithic of the ancient Near East, containers of various materials, shapes, and sizes were used for a very broad range of activities, including storage, the preparation and consumption of food and drink, ritual, and social networking (Nieuwenhuyse 2007). If climatic stresses or other agents of change instigated large-scale, rapid socioeconomic readjustments, we expect these to be reflected in adaptive innovations in container production and consumption to cope with new circumstances.

In the most dramatic scenario—cultural collapse and large-scale demographic upheaval—we would expect a sudden, dramatic shift in pottery traditions, as the pre-event inhabitants of a site starved or moved out and new populations moved in. At Tell Sabi Abyad and other Late Neolithic excavated sites in Upper Mesopotamia the current evidence points to the opposite. Certainly, there were important changes in ceramic technology and style before, during, and after the climate event. However, these formed part of a long-term continuum of change and innovation (Nieuwenhuyse 2016). At no point in the sequence does the pottery suggest regional abandonment or massive replacement of a pre-event population of pottery producers by a post-event one.

This does not mean that we observe no changes coinciding with the 8.2 event. The final stage of the seventh millennium in particular saw massive changes in the production and usage of containers made of stone, plaster (so-called white ware) and pottery (Nieuwenhuyse 2016; Nilhamn and Koek 2013). Some of these changes, moreover, may be interpreted from a human-ecological perspective as being adaptive to increased aridity. Halstead and O'Shea (1989:34) identify four general cultural strategies as a response to increased aridity: mobility, diversification, physical storage and exchange (or "social storage"). These give expectations with regard to container production and consumption. Here we focus on the first two generalized strategies, innovating storage technologies and social storage. In terms of containers we may search for innovative technologies for making more durable, more efficient storage vessels, or an increasing size of preexisting types of storage containers. We may also explore the use of pottery as a medium for stylistic messaging, instrumental in maintaining affiliation with social networks above the level of the local community.

One very clear shift in pottery typology synchronizes well with the 8.2 event: an increased emphasis on pots for storage. The archetypal pottery storage container was the

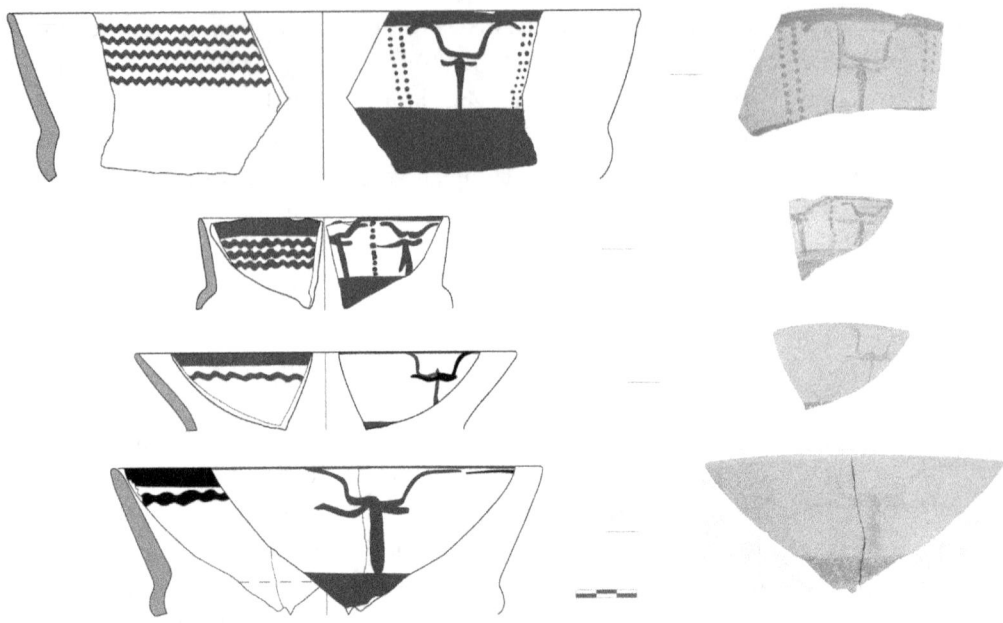

Figure 3.6. Tell Sabi Abyad, Operation III. Stylized images of the bull, so-called bucrania, painted on Halaf Fine Ware serving vessels known as "cream bowls."

jar: a closed shape with a distinct collar to facilitate the efficient closure of the vessel. In the pre-event levels at Tell Sabi Abyad (levels A12–A2, or Initial Pottery Neolithic and Early Pottery Neolithic), the range of vessel types included no large, voluminous pottery vessels with a neck at all; in the levels coinciding with the event and in all subsequent levels (level A1 and subsequently, or Pre-Halaf to Early Halaf), the pottery assemblage included a large proportion of pottery jars. Pottery jars make an abrupt first appearance by level A1, coinciding with Animal Exploitation Phase VI, or the start of the Pre-Halaf period. Moreover, size estimates of jars at Tell Sabi Abyad show that these containers rapidly increased in volume in the centuries during and after the climate event (Nieuwenhuyse 2016). The increased availability of strong, durable, and increasingly voluminous storage containers made of pottery would have been highly adaptive in times of food stress.

On second thought, however, the equation of storage pots with climate change turns out to be more complex. While from a blunt typological perspective, indeed, jars were not available prior to level A1, this does not mean that pre-event communities were not concerned with using pottery vessels for storage. Throughout the seventh millennium people plastered their pottery containers to make them less porous and more efficient for holding goods in waterproof, airtight conditions. At Tell Sabi Abyad the raw material used for plastering pottery containers appears to have been gypsum (Nieuwenhuyse 2016). Further, in the centuries preceding the development of pottery types with a collar (in levels A5–A2), Late Neolithic communities were using large, hole-mouth vessels with a so-called cordon, an applied band running horizontally over the exterior surface just below the rim. This facilitated efficient closure of the vessel, by covering it with a cloth or piece of leather, and fixing this cover with a rope below the cordon. What this suggests is that the innovations in pottery typology did not come about abruptly but formed part of a long-term trend toward a more efficient employment of pottery containers for storage.

An important issue concerns the availability of necessary potters' expertise for producing these large, voluminous pottery storage jars. No matter how "primitive" these plain, often irregularly shaped and coarsely finished vessels may look to us, in the final parts of the seventh millennium these represented cutting-edge technology. Making a large, hence heavy, pottery jar is not an option open to just any novice potter. Potters must have knowledge on the preparation of specially tempered clays (in the case of the Late Neolithic of Upper Mesopotamia: the right dose of coarsely chopped organic temper), they must understand the effects on the final product of alternative shaping methods, they had to know how to construct installations for firing the huge vessels, and they should have a minimum control over the conditions of firing. The excavations at Tell Sabi Abyad suggest that this package of specialist expertise accumulated gradually over the course of the seventh millennium BC (Nieuwenhuyse 2016). Prior to level A1, then, it may simply not have been possible to produce large pottery jars, whereas from level A1 on this option was open. A gradual building up of technological expertise, internal to Late Neolithic societies, may have contributed to the development of efficient pottery jars as much as other, external causal agents.

Let us now turn to what was undoubtedly the most conspicuous change in container production: the introduction of decorated pottery vessels. This introduction occurred after some five hundred years of relentlessly plain pottery, and was immediately followed by a steep rise of the proportion of decorated pottery. At Tell Sabi Abyad, the start of this phenomenon was by level A1. Figure 3.8 shows the relevant data from Tell Sabi Abyad,

FIGURE 3.7. Tell Sabi Abyad, Operation III. Large storage vessels from level A1.

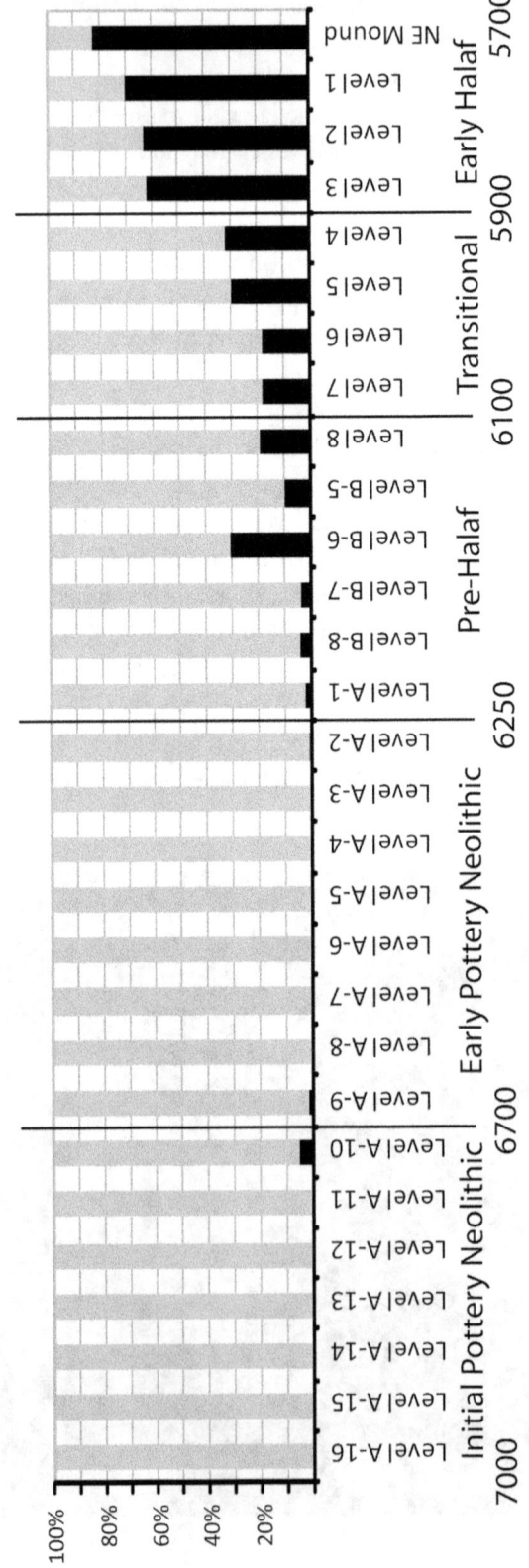

FIGURE 3.8. Tell Sabi Abyad, Operation III. The rising proportion of decorated ceramics from level A1 on.

but it is stressed that the phenomenon is observed at other excavated sites as well; this was a phenomenon participated in over a very wide region (Nieuwenhuyse 2007). Not only did people suddenly begin to decorate their pottery; the way they did so reflects an increasing awareness of developments beyond the local horizon. From the Mediterranean coast across southeastern Turkey and northern Syria into Iraq and beyond, people came to share broadly comparable traditions of intricately decorated high-quality Fine Ware vessels (Nieuwenhuyse 2007, 2013).

These supra-regional networks may be seen as an interaction sphere (Yoffee 1993) that united previously more localized Late Neolithic communities. Significantly, the emphasis in these innovations in pottery technology and decorative style lay on serving vessels: small, open vessels suitable for presenting and consuming food and drink. Size estimates suggest that elaborately decorated serving vessels were employed both in private and in collective contexts (Nieuwenhuyse 2007). Commensal activities became more ostentatious, more overtly emphasized at the end of the seventh millennium BC The pottery gained a new role in ritual occasions and in feasts, becoming an essential component of these activities (Nieuwenhuyse 2007). It can be argued that these painted styles reflect the emergence of new regulations concerning hospitality and proper ways of receiving and entertaining guests from outside the local village (Nieuwenhuyse 2013). From a human-ecological perspective, this would have been adaptive. These ritually regulated social networks would have offered a crucial "safety network," offering dispersed groups the option of reciprocal sharing in the form of a system of "delayed storage" (or "social storage") (O'Shea 1989). This would have assisted people surviving difficult times in an arid, marginal landscape (Bruins et al. 2003; Minnis 1985). The high degree of synchronicity with the 8.2 event is very suggestive and would argue for a contributing role of climate stresses in furthering these innovations in the social-cultural realm.

As with changes in animal exploitation and the availability of storage containers, however, a closer inspection shows that this important innovation was not as abrupt as the raw pottery counts (Figure 3.8) would suggest. If we look at the availability of decorated ceramic vessels in more detail, we find that decorated pottery types at Tell Sabi Abyad were introduced already by level A4, several centuries before the onset of the 8.2 event. In this Early Pottery Neolithic stage, corresponding to Animal Exploitation Phase IV, decorated vessels were excruciatingly rare, representing less than 1 percent of the ceramic assemblage as a whole. Moreover, in contrast to post-event developments, decorative style did not (yet) emphasize serving vessels, but cooking pots instead. Red-slipped or, very occasionally, painted cooking vessels are a rare but characteristic feature of the ceramic assemblage in levels A4 to A2 at Tell Sabi Abyad (Nieuwenhuyse 2016). Decorated pottery, then, was not "invented" during the 8.2 event; rather, people adapted preexisting practices to give new meaning to the medium of decorated pottery vessels.

Conclusions

In this paper we have explored some of the socioeconomic repercussions of the 8.2 event upon Late Neolithic communities inhabiting the semi-arid plains of Upper Mesopotamia.

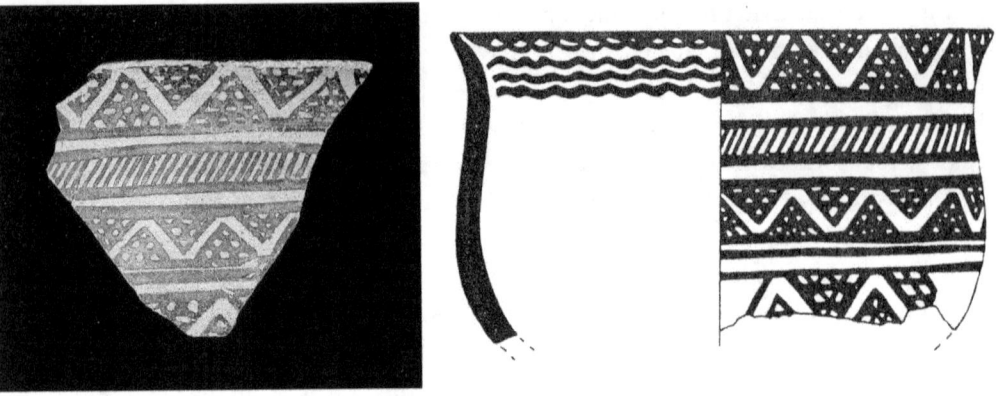

Figure 3.9. Tell Sabi Abyad, Operation I. Transitional period painted serving vessel "Samarra style" (after Akkermans 1993; Nieuwenhuyse 2007).

We have argued, as others have before us, that we can demonstrate a broad package of cultural change coinciding with this particular rapid climate change. We have also demonstrated, however, that the 8.2 event did not result in "collapse." There is no evidence that the village was abandoned, and even at the regional level we can so far not demonstrate any significant demographic disruption. There was no significant break in cultural traditions.

Rather, the archaeological record suggests rapid and profound cultural transformation. This transformation, then, to a large measure synchronizes well with the date and the duration of the 8.2 event. However, the precise timing of these innovations, or rather the moment they first began to manifest themselves, cautions against a neat equation of culture change with climate change. Some of these innovations seem to have been put in motion already in the centuries prior to the event; the moment of the rapid climate event coincides with an acceleration of ongoing trends. The long process of increasing control over wild populations of aurochs, for instance, culminated in the final domestication of this fierce animal at the end of the seventh millennium, but the process began much earlier. Other innovations, such as the introduction of bulky pottery storage jars, constituted a real novelty of sorts if we adopt a typological perspective (presence or absence of specific types), but upon closer perspective we find that they built upon previous developments in the centuries prior to the event.

The specific role of climate as a causal agent requires much further discussion. Several arguments speak in favor of it being a strong causal agent with profound, supraregional effects: the strong synchronicity of the various innovations, accelerations of ongoing trends and transformation of preexisting cultural practices observed; the breathtaking scope of these transformations, manifesting themselves in the economy, technologies, ritual practices, and village location preferences; the fact that, as far as we can presently gauge, very similar transformations occurred across the Upper Mesopotamian plains, not just at Tell Sabi Abyad. Environmental stresses resulting from a rapid climate change would constitute a plausible candidate.

However, we believe that climate was not the only factor. Additional or alternative causal factors, too, would have had the potential for effecting profound socioeconomic and cultural changes. For one, the development of the socioeconomic package known as the "secondary products revolution" should not be underestimated. From the view of human ecology, the intensified exploitation of secondary products and the development of semipastoral life ways in the final parts of the seventh millennium were adaptive (also Mottram this volume). They were a coping mechanism that made it possible for Late Neolithic communities to survive. Turning this argument around, however, intensifying the exploitation of secondary products would in itself have constituted a major causal factor instigating much further socioeconomic change (Sherratt 1981, 1983), transforming ways of using the landscape, offering new means for increasing social complexity, and stimulating the development of the supraregional networks that we see reflected in the ceramics. The discussion may quickly become stranded in an unhelpful "chicken and egg" debate: What came first? But we should not be too quick to isolate climate-induced environmental stresses as the sole agent for social change.

We have argued that archaeologists interested in climate change should seek to understand how the effects of climate stresses interacted with preexisting cultural practices, local contexts, and levels of technology. As to our case study, it would appear that at the time of the 8.2 abrupt climate event Late Neolithic communities inhabiting the Upper Mesopotamian steppes were to a large degree pre-adapted. We should not lose sight of the fact that they had already inhabited a marginal, arid environment quite successfully for several millennia. They developed a suite of practices that enabled them to survive, even flourish. This already included efficient storage technologies and relationships with domestic or proto-domesticated animals that allowed for an intensification of relying on their secondary products. When the adverse effects of the 8.2 event became manifest at the close of the seventh millennium, they were able to elaborate on these preexisting practices and adapt them to new circumstances. The reduced availability of wild resources documented at the time of the climate event would have limited the potential of alternative buffering techniques such as increasing mobility, leaving storage and social exchange as more viable options (O'Shea 1989:66).

Interestingly, once these changes were set in motion, there was no turning back. After the short-lived 8.2 event had ended, the new cultural practices persisted and became much further developed during what is known as the Halaf period. In a way, whatever the specific role of climate change may have been, the profound cultural transformations at the close of the seventh and early sixth millennium BC were a positive force. They still form the foundations of our modern society today.

Notes

1. Our contribution arises from the project *Abrupt Climate Change and Cultural Transformation*. This four-year project was funded by NWO (dossier 360-62-040) and hosted by the Faculty of Archaeology of Leiden University.
2. All 14C dates were measured at the Groningen Radiocarbon facility. The radiocarbon data presented in this paper are all based on charred botanical samples, not on bone remains.

3. The material collected from this survey is kept within the National Museum of Antiquities Leiden. The Late Neolithic ceramics are being prepared for their final publication.

References Cited

Akkad, D. 2009 Escaping the Drought. *Syria Today* 53 (September 2009):22–29.

Akkermans, P. M. M. G. 1993 *Villages in the Steppe*. University of Michigan Press, Ann Arbor.

Akkermans, P. M. M. G. 2013 Living Space, Temporality, and Community Segmentation: Interpreting Late Neolithic Settlement in Northern Syria. In *Interpreting the Late Neolithic of Upper Mesopotamia,* edited by O. P. Nieuwenhuyse, R. Bernbeck, P. M. M. G. Akkermans, and J. Rogasch, pp. 63–76. PALMA Series 9. Brepols, Turnhout.

Akkermans, P. M. M. G., and G. Schwartz 2003 *The Archaeology of Syria: From Complex Hunter-gatherers to Early Urban Societies (ca. 16000–300 B.C.)*. Cambridge University Press, Cambridge.

Akkermans, P. M. M. G., R. Cappers, C. Cavallo, O. P. Nieuwenhuyse, B. Nilhamn, and I. Otte 2006 Investigating the Early Pottery Neolithic of Northern Syria: New Evidence From Tell Sabi Abyad. *American Journal of Archaeology* 110:123–156.

Akkermans, P. M. M. G., J. van der Plicht, O. P. Nieuwenhuyse, A. Russell, A. Kaneda, and H. Buitenhuis 2010 Weathering Climate Change in the Near East: Dating and Neolithic Adaptations 8200 Years Ago. *Antiquity:* online project gallery. Electronic document, http://www.antiquity.ac.uk/projgall/plicht325/.

Akkermans, P. M. M. G., J. van der Plicht, O. P. Nieuwenhuyse, A. Russell, and A. Kaneda 2011 Cultural Transformation and the 8.2 ka event in Upper Mesopotamia. In *Ancient Society and Climate,* edited by S. Kerner, R. Dann, and P. Bangsgaard Jensen, pp. 95–110. Museum Tusculanum Press, Copenhagen.

Akkermans, P. M. M. G., and M. Verhoeven 1995 An Image of Complexity: The Burnt Village at Late Neolithic Sabi Abyad, Syria. *American Journal of Archaeology* 99(1):5–32.

Alley, R. B., P. A. Mayewski, T. Sowers, M. Stuiver, K. C. Taylor, and P. U. Clark 1997 Holocene Climate Instability: A Prominent, Widespread Event 8200 yr Ago. *Geology* 25(6):483–486.

Alley, R. B., and A. M. Ágústsdóttir 2005 The 8k Event: Cause and Consequences of a Major Holocene Abrupt Climate Change. *Quaternary Science Reviews* 24:1123–1149.

Astruc, L., and A. Russell 2013 Hunting and Lithic Specialization in the Balikh Valley: New Data from Tell Sabi Abyad (VIIth Millennium B.C.). In *Interpreting the Late Neolithic of Upper Mesopotamia,* edited by O. P. Nieuwenhuyse, R. Bernbeck, P. M. M. G. Akkermans, and Jana Rogasch, pp. 331–344. PALMA Series 9. Brepols, Turnhout.

Barber, D. C., A. Dyke, C. Hillaire-Marcel, A. E. Jennings, J. T. Andrews, M. W. Kerwin, G. Bilodeau, R. McNeely, J. Southon, M. D. Morehead, and J. M. Gagnon 1999 Forcing of the Cold Event of 8,200 Years Ago by Catastrophic Drainage of Laurentide Lakes. *Nature* 400:344–348.

Bar-Matthews, M. A. A., A. Kaufman, and G. J. Wasserburg 1999 The Eastern Mediterranean Paleoclimate as a Reflection of Regional Events: Soreq Cave, Israel. *Earth and Planetary Science Letters* 166:85–95.

Bar-Yosef, O. 2001 The World around Cyprus: From Epi-Paleolithic Foragers to the Collapse of the PPNB Civilization. In *The Earliest Prehistory of Cyprus,* edited by S. Swiny, pp. 129–164. American School of Oriental Research, Boston.

Bar-Yosef, O. 2006 L 'impact des changements climatiques du Dryas récent et de l'Holocène inférieur sur les sociétés de chasseurs-cueilleurs et d'agriculteurs du Proche-Orient. In *L'Homme Face au Climate*, edited by E. Bard, pp. 283–301. Odile Jacob, Paris.

Bauer, E., A. Ganopolski, and M. Montoya 2004 Simulation of the Cold Climate Event 8200 Years Ago by Meltwater Outburst from Lake Agassiz. *Paleoceanography* 19:1–13.

Bernbeck, R. 2005 Taming Time and Timing the Tamed. In *Proceedings of the Fifth International Congress on the Archaeology of the Ancient Near East*, edited by J. Córdoba, M. Molist, M. Carmen Pérez, I. Rubio, and S. Martinez, pp. 709–728. UAM, Madrid.

Bernbeck, R. 2008 An Archaeology of Multi-Sited Communities. In *The Archaeology of Mobility. Old World and New World Nomadism*, edited by H. Barnard and W. Wendrich, pp. 43–77. Cotsen Institute of Archaeology, Los Angeles.

Brooks, N. 2006 Cultural Responses to Aridity in the Middle Holocene and Increased Social Complexity. *Quaternary International* 151:29–49.

Bruins, H. J., J. J. Akong'a, M. M. E. M. Rutten, and G. M. Kessel 2003 Drought Planning and Rainwater Harvesting for Arid-zone Pastoralists: The Turkana and Maasai (Kenya) and the Negev Bedouin (Israel). NIRP Research for Policy Series 17. Nuffic, The Hague.

Carter, R. A., and G. Philip (editors) 2010 *Beyond the Ubaid: Transformation and Integration in the Late Prehistoric Societies of the Middle East*. Oriental Institute, Chicago.

Catto, N., and G. Catto 2004 Climate Change, Communities, and Civilizations: Driving Force, Supporting Play, or Background Noise? *Quaternary International* 123–125:7–10.

Cavallo, C. 2000 Animals in the Steppe—A Zooarchaeological Analysis of Later Neolithic Tell Sabi Abyad, Syria. BAR International Series 891. Archaeopress, Oxford.

Childe, V. G. 1929 *The Most Ancient East*. Alfred A. Knopf, New York.

Clare, L., E. J. Röhling, B. Weninger, and J. Hilpert 2008 Warfare in Late Neolithic/Early Chalcolithic Pisidia, Southwestern Turkey: Climate Induced Social Unrest in the Late 7th Millennium cal B.C. *Documenta Praehistorica* 35:65–92.

Coombes, P., and K. Barber 2005 Environmental Determinism in Holocene Research: Causality or Coincidence? *Area* 37(3):303–311.

Copeland, L. 1979 Observations on the Prehistory of the Balikh Valley, Syria, During the 7th to 4th Millennium B.C. *Paléorient* 5:251–275.

deMenocal, P. B. 2001 Cultural Responses to Climate Change During the Late Holocene. *Science* 292:667–673.

Enzel, Y., R. Bookman, D. Sharon, H. Gvirtzman, U. Dayan, B. Ziv, and M. Stein 2003 Late Holocene Climates of the Near East Deduced from Dead Sea Level Variations and Modern Regional Winter Rainfall. *Quaternary International* 60:263–273.

Evershed, R. P., S. Payne, A. G. Sherratt, M. S. Copley, J. Coolidge, D. Urem-Kotsu, K. Kotsakis, M. Özdoğan, A. Erim-Ozdogan, O. P. Nieuwenhuyse, P. M. M. G. Akkermans, D. Bailey, R. R. Andeescu, S. Campbell, S. Farid, I. Hodder, N. Yalman, M. Özbaşaran, E. Bıcakcı, Y. Garfinkel, T. Levy, and M. M. Burton 2008 Earliest Date for Milk Use in the Near East and Southeastern Europe Linked to Cattle Herding. *Nature* 7180:1–4.

French, M. H. 1970 *Observations on the Goat*. Food and Agricultural Organization of the United Nations, Rome.

Gasse, F. 2000 Hydrological Changes in African Tropics Since the Last Glacial Maximum. *Quaternary Science Reviews* 19:189–211.

Gebel, H. G. K. 2009 The Intricacy of Neolithic Rubble Layers. The Ba'ja, Basta, and 'Ain Rahub Evidence. *Neolithics* 2009(1):33–48.

Glantz, M. H. 1994 Drought, Desertification, and Food Production. In *Drought Follows the Plow*, edited by M. H. Glantz, pp. 7–32. Cambridge University Press, Cambridge.

Halstead, P. 1996 Pastoralism or Household Herding? Problems of Scale and Specialization in Early Greek Animal Husbandry. *World Archaeology* 28(1):20–42.

Halstead, P., and J. M. O'Shea 1989 Introduction: Cultural Responses to Risks and Uncertainty. In *Bad Year Economics: Cultural Responses to Risk and Uncertainty*, edited by P. Halstead and J. M. O'Shea, pp. 1–7. Cambridge University Press, Cambridge.

Hassan, F. A. 2009 Human Agency, Climate Change, and Culture: An Archaeological Perspective. In *Anthropology and Climate Change: From Encounters to Actions*, edited by S. A. Crate and M. Nuttal, pp. 39–69. Left Coast Press, Walnut Creek.

Hoek, W. Z., and J. A. A. Bos 2007 Early Holocene Climate Oscillations—Causes and Consequences. *Quaternary Science Reviews* 26:1901–1906.

Ingold, T. 2000 *The Perception of the Environment: Essays on Livelihood, Dwelling, and Skill*. Routledge, London.

Issar, A. S., and M. Zohar 2004 *Climate Change: Environment and Civilization in the Middle East*. Springer, New York.

Kansa, S. W., A. Kennedy, S. Campbell, and E. Carter 2009 Resource Exploitation at Late Neolithic Domuztepe. *Current Anthropology* 50(6):897–914.

Klitgaard-Kristensen, D., H. P. Sejrup, H. H. S. Johnsen, and M. Spurk 1998 A Regional 8200 cal. yr B.P. Cooling Event in Northwest Europe, Induced by Final Stages of the Laurentide Ice-Sheet Deglaciation? *Journal of Quaternary Science* 13(2):165–169.

LeMiere, M., and O. P. Nieuwenhuyse 1996 The Prehistoric Pottery. In *Tell Sabi Abyad the Later Neolithic Settlement. Report on the Excavations of the University of Amsterdam (1988) and the National Museum of Antiquities, Leiden (1991–1992)*, edited by P. M. M. G. Akkermans, pp. 119–284. NHAI, Istanbul.

Linden, E. 2006 *The Winds of Change. Climate, Weather and the Destruction of Civilizations*. Simon and Schuster, New York.

Maher, L. A., T. Banning, and M. Chazan 2011 Oasis or Mirage? Assessing the Role of Abrupt Climate Change in the Prehistory of the Southern Levant. *Cambridge Archaeological Journal* 21(1):1–29.

McIntosh, R. J., J. A. Tainter, and S. K. McIntosh 2000 Climate, History and Human Action. In *The Way the Wind Blows. Climate, History, and Human Action*, edited by R. J. McIntosh, J. A. Tainter, and S. K. McIntosh, pp. 1–44. Columbia University Press, New York.

Migowski, C., M. Stein, S. Prasad, J. F. W. Negendank, and A. Agnon 2006 Holocene Climate Variability and Cultural Evolution in the Near East from the Dead Sea Sedimentary Record. *Quaternary Research* 66:421–431.

Minnis, P. E. 1985 *Social Adaptation to Food Stress. A Prehistoric Southwestern Example*. University of Chicago Press, Chicago.

Nieuwenhuyse, O. P. 2000 Halaf Settlement in the Khabur Headwaters. In *Prospection archéologique dans le Haut-Khabur, vol. 1: méthodologie, paléolithique et néolithique*, edited by B. Lyonnet, pp. 151–260. IFAPO, Damascus.

Nieuwenhuyse, O. P. 2007 *Plain and Painted Pottery. The Rise of Neolithic Ceramics Styles on the Syrian and Northern Syrian Plains*. PALMA Series vol. 3. Brepols, Turnhout.

Nieuwenhuyse, O. P. 2013 The Social Uses of Decorated Ceramics in Late Neolithic Upper Mesopotamia. In *Interpreting the Late Neolithic of Upper Mesopotamia*, edited by

O. P. Nieuwenhuyse, R. Bernbeck, P. M. M. G. Akkermans, and J. Rogasch, pp. 135–146. PALMA Series 9. Brepols, Turnhout.

Nieuwenhuyse, O. P. 2016 *Relentlessly Plain. The Seventh Millennium Ceramic Assemblage from Tell Sabi Abyad (Northern Syria)*. Oxbow, Oxford.

Nieuwenhuyse, O. P., and T. J. Wilkinson 2008 Late Neolithic Settlement in the Area of Tell Beydar (NE Syria). *Subartu* 21:305–27.

Nieuwenhuyse, O. P., P. M. M. G. Akkermans, and J. van der Plicht 2010 Not So Coarse, nor Always Plain—The Earliest Pottery of Syria. *Antiquity* 84:71–85.

Nieuwenhuyse, O. P., R. Bernbeck, P. M. M. G. Akkermans, and J. Rogasch (editors) 2013 *Interpreting the Late Neolithic of Upper Mesopotamia*. PALMA Series 9. Brepols, Turnhout.

Nilhamn, B., and E. Koek 2013 Early Pottery Neolithic White Ware from Tell Sabi Abyad. In *Interpreting the Late Neolithic of Upper Mesopotamia*, edited by O. P. Nieuwenhuyse, R. Bernbeck, P. M. M. G. Akkermans, and J. Rogasch, pp. 289–296. PALMA Series 9. Brepols, Turnhout.

O'Shea, J. M. 1989 The Role of Wild Resources in Small-Scale Agricultural Systems: Tales from the Lakes and the Plain. In *Bad Year Economics. Cultural Responses to Risk and Uncertainty*, edited by P. Halstead and J. M. O'Shea, pp. 57–67. Cambridge University Press, Cambridge.

Outram, A. K., and J. Mulville 2005 The Zooarchaeology of Fats, Oils, Milk, and Dairying: An Introduction and Overview. In *The Zooarchaeology of Fats, Oils, and Dairying*, edited by J. Mulville and A. K. Outram, pp. 1–7. Oxbow Books, Oxford.

Pollock, S. 1999 *Ancient Mesopotamia*. Cambridge University Press, Cambridge.

Pumpelly, R. W. 1905 *Explorations in Turkestan: With an Account of the Basin of Eastern Persia and Sistan*. Carnegie Institute, Washington.

Renssen, H., H. Goosse, and T. Fichefet 2002 Modeling the Effect of Freshwater Pulse on the Early Holocene Climate: The Influence of High-Frequency Climate Variability. *Paleoceanography* 17(2):1–15.

Reycraft, R. M., and G. Bawden 2000 Introduction. In *Environmental Disaster and the Archaeology of Human Response*, edited by G. Bawden and R. M. Reycraft, pp. 1–10. University of New Mexico, Albuquerque.

Rohling, E. J., and H. Pälike 2005 Centennial-Scale Climate Cooling with a Sudden Cold Event Around 8,200 Years Ago. *Nature* 434:975–979.

Rollefson, G. O. 2009 Slippery Slope. The Late Neolithic Rubble Layer in the Southern Levant. *Neo-Lithics* 2009(1):12–18.

Rooijakkers, T. 2012 Spinning Animal Fibres at Late Neolithic Tell Sabi Abyad, Syria? *Paléorient*. 38(1–2):93–109.

Rosen, A. M. 2007 *Civilizing Climate. Social Responses to Climate Change in the Ancient Near East*. Rowman and Littlefield, New York.

Russell, A. 2010 *Retracing the Steppes. A Zooarchaeological Analysis of Changing Subsistence Patterns in the Late Neolithic at Tell Sabi Abyad, Northern Syria, c. 6900 to 5900 B.C.* Sidestone, Leiden.

Russell, N. 1998 Cattle as Wealth in Neolithic Europe: Where's the Beef? In *The Archaeology of Value. Essays on Prestige and the Processes of Valuation*, edited by D. W. Bailey, pp. 42–54. BAR International Series 730. Archaeopress, Oxford.

Ryder, M. J. 1981 Livestock Products: Skins and Fleeces. In *Farming Practice in British Prehistory*, edited by R. Mercer, pp. 182–209. Edinburgh University Press, Edinburgh.

Schwartz, P., and D. Randall 2003 An Abrupt Climate Change Scenario and Its Implications for United States National Security. Electronic document, http://www.ems.org/climate/pentagon_climate_change.html#report, accessed September 12, 2013.

Sherratt, A. 1981 Plough and Pastoralism. Aspects of the Secondary Products Revolution. In *Pattern of the Past. Studies in Honour of David Clarke*, edited by I. Hodder, G. Isaac, and N. Hammond, pp. 261–306. Cambridge University Press, Cambridge.

Sherratt, A. 1983 The Secondary Exploitation of Animals in the Old World. *World Archaeology* 15(1):90–104.

Staubwasser, M., and H. Weiss 2006 Holocene Climate and Cultural Evolution in Late Prehistoric—Early Historic West Asia. *Quaternary Research* 66:372–387.

Tchernov, E. 1982 Faunal Response to Environmental Changes in the Eastern Mediterranean during the Last 20,000 Years. In *Palaeoclimates, Palaeoenvironments, and Human Communities in the Eastern Mediterranean Region in Later Prehistory*, edited by J. L. Bintliff and W. van Zeist, pp. 105–129. BAR International Series 133. Archaeopress, Oxford.

Teller, J. T., D. W. Leverington, and J. D. Mann 2002 Freshwater Outbursts to Oceans from Glacial Lake Agassiz and Their Role in Climate Change During the Last Deglaciation. *Quaternary Science Reviews* 22:879–887.

Thomas, E. R., E. W. Wolff, R. Mulvaney, J. P. Steffensen, S. J. Johnsen, C. Arrowsmith, J. W. C. White, B. Vaighn, and T. Popp 2007 The 8.2 ka Event from Greenland Ice Cores. *Quaternary Science Reviews* 26:70–81.

Valtorta, S. E. 2002 Animal Production in a Changing Climate: Impacts and Mitigation. *Brody Lectures at the 16th International Congress on Biometereology*.

Van der Leeuw, S. 2000 Land Degradation as a Socionatural Process. In *The Way the Wind Blows: Climate History and Human Action*, edited by R. J. McIntosh, J. A. Tainter, and S. K. McIntosh, pp. 357–383. Columbia University Press, New York.

Van der Plicht, J., P. M. M. G. Akkermans, O. P. Nieuwenhuyse, A. Kaneda, and A. Russell 2011 Tell Sabi Abyad, Syria: Radiocarbon Chronology, Cultural Change, and the 8.2 ka Event. *Radiocarbon* 53(2):229–243.

Van Zeist, W. 1988 Some Notes on the Plant Husbandry of Tell Hammam et-Turkman. In *Hamman et-Turkman I*, edited by M. N. Van Loon, pp. 705–715. NHAI, Istanbul.

Weiss, H. 2000 Beyond the Younger Dryas: Collapse as Adaptation to Abrupt Climate Change in Ancient West Asia and the Eastern Mediterranean. In *Environmental Disaster and the Archaeology of Human Response*, edited by G. Bawden and R. M. Reycraft, pp. 75–98. University of New Mexico, Albuquerque.

Weiss, H., and R. S. Bradley 2001 What Drives Societal Collapse? *Science* 291:609–610.

Weninger, B. 2009 Introduction. *Neo-Lithics* 2009(1):5–11.

Weninger, B., E. Alram-Stern, E. Bauer, L. Clare, U. Danzeglocke, O. Jöris, C. Kubatzki, G. Rollefson, H. Todorova, and T. van Andel 2006 Climate Forcing Due to the 8200 cal yr B.P. Event Observed at Early Neolithic Sites in the Eastern Mediterranean. *Quaternary Research* 66:401–420.

Whittle, A. W. R. 2003 *The Archaeology of People: Dimensions of Neolithic Life*. Routledge, London.

Wiersma, A. P. 2008 *Character and Causes of the 8.2 ka Climate Event*. PhD thesis Free University of Amsterdam, Amsterdam.

Wiersma, A. P., and H. Renssen 2006 Model-Data Comparison for the 8.2 ka B.P. Event: Confirmation of a Forcing Mechanism by Catastrophic Drainage of Laurentide Lakes. *Quaternary Science Reviews* 25: 63–88.

Wilkinson, T. J. 1998 Water and Human Settlement in the Balikh Valley, Syria: Investigations from 1992–1995. *Journal of Field Archaeology* 25(1):63–87.

Wilkinson, T. J. 2003 *Archaeological Landscapes of the Near East*. University of Arizona Press, Tucson.

Wossink, A. 2009 *Challenging Climate Change. Competition and Cooperation among Pastoralists and Agriculturalists in Northern Mesopotamia (c. 3000–1600 B.C.)*. Sidestone, Leiden.

Yoffee, N. 1993 Mesopotamian Interaction Spheres. In *Early Stages in the Evolution of Mesopotamian Civilization: Soviet Excavations in Northern Iraq*, edited by N. Yoffee and J. J. Clark, pp. 257–270. University of Arizona Press, Tucson.

CHAPTER FOUR

The Aftermath of the 8.2 Event

Cultural and Environmental Effects in the Anatolian Late Neolithic and Early Chalcolithic

*Patrick T. Willett, Ingmar Franz,
Ceren Kabukcu, David Orton, Jana Rogasch,
Elizabeth Stroud, Eva Rosenstock, Peter F. Biehl*

Abstract *In the Konya Plain in Central Anatolia the transition from the Late Neolithic (LN) to the Early Chalcolithic (EC) occurs circa the beginning of a period of regional drying conditions. This transitional period also roughly corresponds to the so-called 8.2 cal B.P. Climatic Event (or simply "8.2 event"), which has been suggested as a possible cause for suspected drought on a wide scale, increased seasonality and fluctuating weather conditions, particularly in the Near East and North Africa between 6400 and 5800 cal B.C. On the Konya Plain, the prehistoric settlement of Çatalhöyük spans the LN/EC transition before final abandonment ca. 5600 cal B.C., providing an ideal case study for evaluating the impact of climate and environment change on cultural systems during the seventh millennium. This paper will present regional proxy data relating to the environment alongside results from excavations at Çatalhöyük West and discuss the possible effects of the 8.2 event on the settlement.*

INTRODUCTION

This article aims to assess possible effects of the global climate change that occurred during the 8.2 event around 6200 cal B.C. (6400–5800 cal B.C.: Wiersma and Renssen 2006) in the Central Anatolian context and whether and how humans responded to them. The two key hypotheses that are being tested are that the change in climate and environment was at least one cause for the discontinuation of settlements at the end of the seventh millennium (Clare et al. 2009) on one hand and for people to move westward into Western Anatolia and Southeast Europe, and eventually across Europe (see Düring in this volume; Clare and Weninger 2014; Weninger et al. 2005) on the other. Çatalhöyük, the largest and most thoroughly sampled Neolithic site in the Konya Plain, with its sequence spanning the seventh and the beginning of the sixth millennium offers a microcosm that may help us unlock some of the aspects of these questions (Biehl and Rosenstock 2009; Biehl 2012; Biehl et al. 2012). At Çatalhöyük, settlement shifted from the East to the West Mound at around 6100 cal B.C., forming two adjacent consecutively occupied settlements. They offer an exceptional chance to analyze possible human responses to the 8.2 event on a micro-scale and may give us the possibility of answering the question of why and how the shift from the East Mound to the West Mound took place. Once we understand the regional process, we can try to determine the broader effects the shift had on the Near East and Europe.

The site (Figure 4.1) covers 32 acres near the modern town of Çumra ca. 50 km to the southeast of the city of Konya. There are two large mounds—the 21 m tall Neolithic East Mound, and the 7 m tall Chalcolithic West Mound—separated now by a dry riverbed, formerly a branch of the Çarşamba River. Following initial excavations by James Mellaart in the 1960s on both mounds (Mellaart 1965, 1967), extensive fieldwork at Çatalhöyük East was renewed under Ian Hodder in 1993 (Hodder 1996, 2007, 2013, 2014), and later several projects extended the investigations onto the West

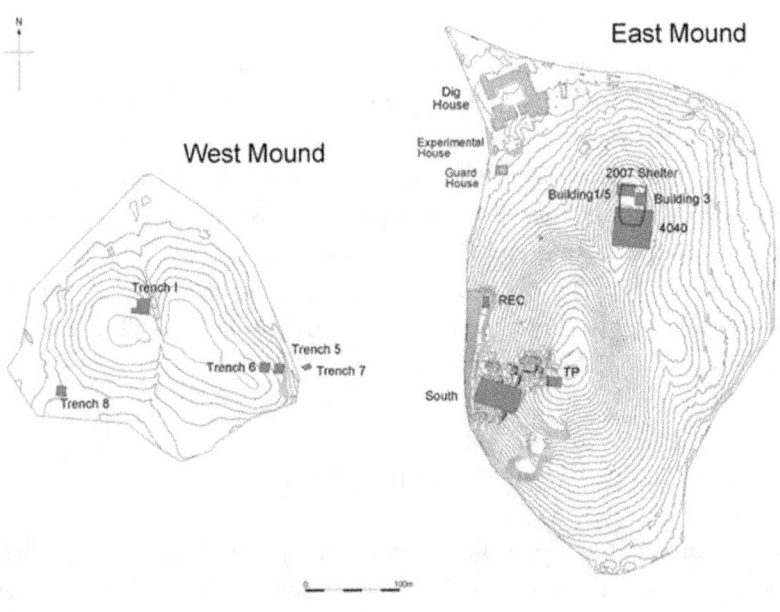

FIGURE 4.1. Topographic map of Çatalhöyük—Konya, Turkey,

Mound (Biehl et al. 2012; Erdoğu 2009; Gibson and Last 2003). On the East Mound, earliest occupation occurs ca. 7100 cal B.C. in the Aceramic Neolithic (AN) and continues through the Early Neolithic (EN) and Late Neolithic (LN) until ca. 5900 cal B.C. (Bayliss et al. 2015; Marciniak et al. 2015). Mellaart's level VI, or South.N/O in the new phasing system (Farid 2008:20), marks the peak and end of what can be termed the "classic" EN settlement. While David French (1967:175, chart 2) had attributed the whole East Mound sequence to the EN, Mellaart (1965:155, 1975:15 table 1) had regarded only levels I and 0 as LN. After the start of the new excavations, the recognition of incised decoration on pots in level VI as well as of occasional red-painted sherds from level III onward (Schoop 2005:107) has pushed the EN/LN border deeper into the stratigraphy of the site. The most recent suggestions emphasize the notion of a major change in architecture, artifact spectrum, and archaeobiological evidence at Mellaart level V or P respectively at ca. 6450 cal B.C. (Hodder 2013, 2014), proposing a start of the LN in this level (Marciniak et al. 2015). Another major disruption is seen at around 6200 cal B.C., marking a second phase within the LN (Marciniak et al. 2015). A deep sounding (Trench 7) into the earliest preserved deposits at the fringe of the West Mound has retrieved a series of ^{14}C dates that push the beginning of the occupation in that area of Çatalhöyük to ca. 6100 cal B.C., thus allowing for a substantial overlap of two centuries with the final levels of occupation on the East Mound (Orton et al., in prep). While our knowledge of the exact character of this transitional complex between the LN and the EC periods remain fragmentary due to the sparse preservation on the eroded top levels of the East Mound and the buried bottom levels of the West Mound, newer radiocarbon dates date levels with a developed EC material culture on the West Mound to a core time slot of ca. 6100 to 5600 cal B.C. (Orton et al., in prep) after which settlement occupation not only ceases in Çatalhöyük, but also entirely in the Konya Plain (Baird 2002, 2006).

In the following, we will first outline the traits of the "classic" EN and the evidence for the LN of Çatalhöyük East, summarize our limited insights into the transitional LN/EC on both mounds and describe the "classic" EC of the West Mound in order to assess the relative importance of the aforementioned turning points in the sequence at ca. 6450, 6200, and 6100 cal B.C. Additionally, the evidence for environmental change in Central Anatolia during this sequence is summarized and possible chronological links with the cultural sequence are noted. Finally, explanations for the observed changes are discussed.

Çatalhöyük in the Neolithic

The EN at Çatalhöyük is characterized by small, rectangular buildings, which were tightly packed with abutting walls, forming a hive-like complex in an agglomerating settlement structure. The lack of doorways at ground level and the rarity of avenues and alleyways between buildings have been interpreted in terms of transit via the rooftops (Hodder 2013). Recent findings have spurred a discussion on second stories at Çatalhöyük East (Hodder 2009). Houses were usually only large enough to accommodate small numbers of inhabitants, storage space was fairly restricted, and the dead were commonly buried below the floor (Hodder 2013). It has been estimated that at its peak, the settlement was home to anywhere from 3,500 to 8,000 residents (Cessford 2005a), though more conservative estimates of ca. 1,000 to 1,500 have been put forward (Marciniak and Kuijt, forthcoming).

Agriculture at Çatalhöyük centered on wheats, barleys, and pulses. Large numbers of glume wheats (emmer, einkorn, and "new type" wheat), as well as free-threshing wheat (*Triticum aestivum*) and naked barley (*Hordeum vulgare var. nudum*) have been found (Bogaard et al. 2013). Pulse taxa include peas *(Pisum sativum)*, lentils (*Lens culinaris*), bitter vetch (*Vicia ervilia*), and chickpea (*Cicer arietinum*) (Bogaard et al. 2013). Wild foods were also used, including the storage of seeds from a wild mustard (*Descurainia sophia*) and wild almond (*Amygdalus* sp.) (Bogaard et al. 2013). With regard to the animal economy, in the earlier levels domestic caprines predominate in terms of numbers, making up roughly 70 percent of the animal bone assemblage, though wild cattle—at ca. 15–20 percent of remains—must have contributed the majority of the meat consumed (Russell et al. 2013a). The remainder of the faunal material is comprised of various taxa of wild game including equids and pigs. Around Hodder level P, or roughly 6450 cal B.C., there is a marked shift in the composition of the faunal assemblage. Reliance on caprine herding apparently increased, as evidenced by a substantial rise in the number of sheep and goat remains, both as a ratio and in absolute terms (as inferred from density of bone finds within middens, Russell et al. 2013a:216–217). There is a drastic decline in the consumption of wild game of all taxa, with the majority of cattle being domestic, and pigs all but disappearing after ca. 6500–6400 cal B.C. It is also during this period as wild cattle, and hunted game in general, declined in consumption as a food source that they reached their peak in symbolic depiction, both through the medium of murals displaying wild taxa and in terms of special deposits and architectural installations (Russell et al. 2013a:250).

The pottery assemblage assigned to these later occupation phases of the East Mound comprises mostly simple bowls (serving and storage vessels) and deep egg-shaped jars (cooking vessels). Residue analyses have shown that in contrast to the earlier phases on site, bowls were also used for cooking at times. Additionally, the recovered quantity of pottery is higher than in the earlier phases. In the later phases, decoration on pottery is sometimes also applied in the form of incised lines or incised figurative representations, red slip covering the whole vessel, and red painted lines on bowls. Also more frequently so-called S-profiled vessels and necked vessels beside more C-shaped or straight profile vessels can be observed in the pottery assemblage. Vessels with so-called basket-handles also are commonly found in later occupation phases, especially in the levels probably even later than Mellaart's levels III to 0 (Farid 2008:26) from the area excavated by Arkadiusz Marciniak and Lech Czerniak (TP, "Team Poznań" in Figure 1) (Yalman et al. 2013). Pottery was most likely mainly built-up with the so-called sequential slap construction (Rosemary Joyce, pers. communication 2014) and with the help of molds (presumably baskets), which is attested by regularly observable concave bases on bowls (Duygu Tarkan-Özbudak, personal communication 2014). The firing temperature of the pottery lay around 800 °C (Doherty and Tarkan-Özbudak 2013).

The pottery (Figure 4.2) could be assigned to five main fabric groups: sandy fabrics, marly-white fabrics, silty fabrics, gritty dark fabrics, and red paste fabrics. They again were assigned to two so-called lines. The definitions of these lines combine the provenance of the inclusions with the darkness of the pottery. Part of the "light line" is the pottery with sandy fabrics, marly-white fabrics, and silty fabrics, which are typical of vessels for serving and storage purposes. They are dominated by local finer calcareous Çarşamba river sediment inclusions. Part of the "dark line" is the pottery with gritty dark fabrics and red paste fabrics, which are typical of the egg-shaped cooking jars. They are dominated by nonlocal gritty non-calcareous volcanic May river inclusions (Doherty and Tarkan-Özbudak 2013; Tarkan-Özbudak 2013).

So-called potstands or firedogs also begin to appear in the later occupation phases of Çatalhöyük East Mound. They are horn-shaped devices of sun-dried clay which most likely were used to support the egg-shaped cooking jars in hot embers or fire from the sides to prevent them from falling over (Franz 2012), and could thus have replaced earlier cooking techniques based on heated clay balls (Hodder 2006). Also milk fats are attested in small quantities of the pottery and thus show that ovicaprid or cattle milk had become a part of the economy, though still on a small scale (Evershed et al. 2008).

Preliminary calculations show that in the excavation areas of South, TP, TPC, NORTH, and IST (Figure 4.1) at least 3,022,149 liters were excavated in 20 years, which again contained at least 468,041 grams of pottery. This means that in these areas the mean density of pottery per excavation volume is only 0.16 g/l (Tarkan—personal communication).

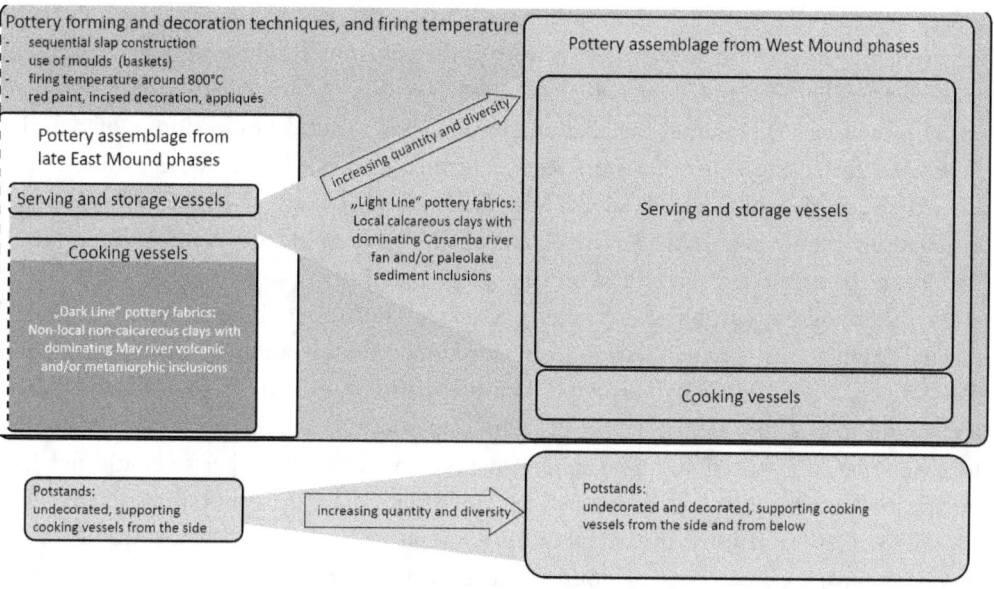

FIGURE 4.2. Pottery forms, decoration techniques, and firing temperatures at Çatalhöyük.

Çatalhöyük in the Early Chalcolithic

Just across the former Çarşamba River, settlement of the Çatalhöyük West Mound began by around 6100 cal B.C. After limited explorations in the 1960s, James Mellaart (1962, 1965) regarded the settlement as belonging to the EC period based on the pottery he encountered, and though it is now clear that much of the change that culminates on the site was initiated in the upper phases of the East Mound around 6450 cal B.C., the West Mound represents the only phase of Çatalhöyük with a fully developed EC material culture. At a size of 8 hectares and height of 7 meters, Çatalhöyük West Mound is one of the largest sites of the period in Anatolia (Baird 2006), and its overlapping occupation with the East Mound makes it an ideal case study for examining the transition from the Late Neolithic. Renewed excavations began in 1998 and have been carried out by a number of teams working in different areas. Preliminary ^{14}C dates determined from building infill in Trenches 5 and 7 mainly fall between 6000–5800 cal B.C. (Biehl et al. 2012; Orton et al., in prep).

One of the most notable departures from what is typical on the East Mound is in the buildings. Differences between West Mound and East Mound buildings become apparent when examining details of the house biographies, while formally the buildings share many characteristics: on both mounds they were built from earth, rectangular, with flat roofs. However, one notices that West Mound buildings have thicker walls, and—the most obvious feature—large internal buttresses. Small buttresses are found occasionally in upper levels on the East Mound (e.g., Czerniak and Marciniak 2008; Regan et al. 2008:65–66,74; Tung 2012:23; Yeomans and Sadarangani 2009:34) as features for stabilizing walls and structuring internal space, but only on the West Mound are they found as a regular feature inside buildings. With their now substantial size, buttresses structure the interior in a way that is very different from East Mound buildings, and make for smaller and more compartmentalized interiors. Another very visible difference is the overall lack of internal installations inside West Mound houses: none of the platforms and fire installations that were regular features in East Mound houses were found, with the exception of a hearth in B.25 in Trench 1 (Gibson and Last 2003). The only installations found in Trench 5 were a few smaller features on the floor of Building 98 that could be associated with food storage or preparation.

This evidence, combined—thicker walls, large buttresses, little internal space, lack of vital installations—makes a strong case for West Mound buildings having been double-storied. So far, features interpreted as the remains of a second story were preserved and found in B.25 in Trench 1 and B.78 in Trench 8, where the upper story is thought to have walls painted red and a larger internal area because this upper story did not have buttresses (Erdoğu 2008). If it can be concluded that all or most buildings on the West Mound had a second story, then the buildings found in Trench 5, as shown in Figure 4.3, represent only the lower half of houses whose living floor was not preserved. Given the reduced and fragmented internal space, presumably the lack of daylight, and the nature of the installations found in the B.98 basement, most of these lower stories were most likely used as areas for storage and potentially food processing that did not require heat.

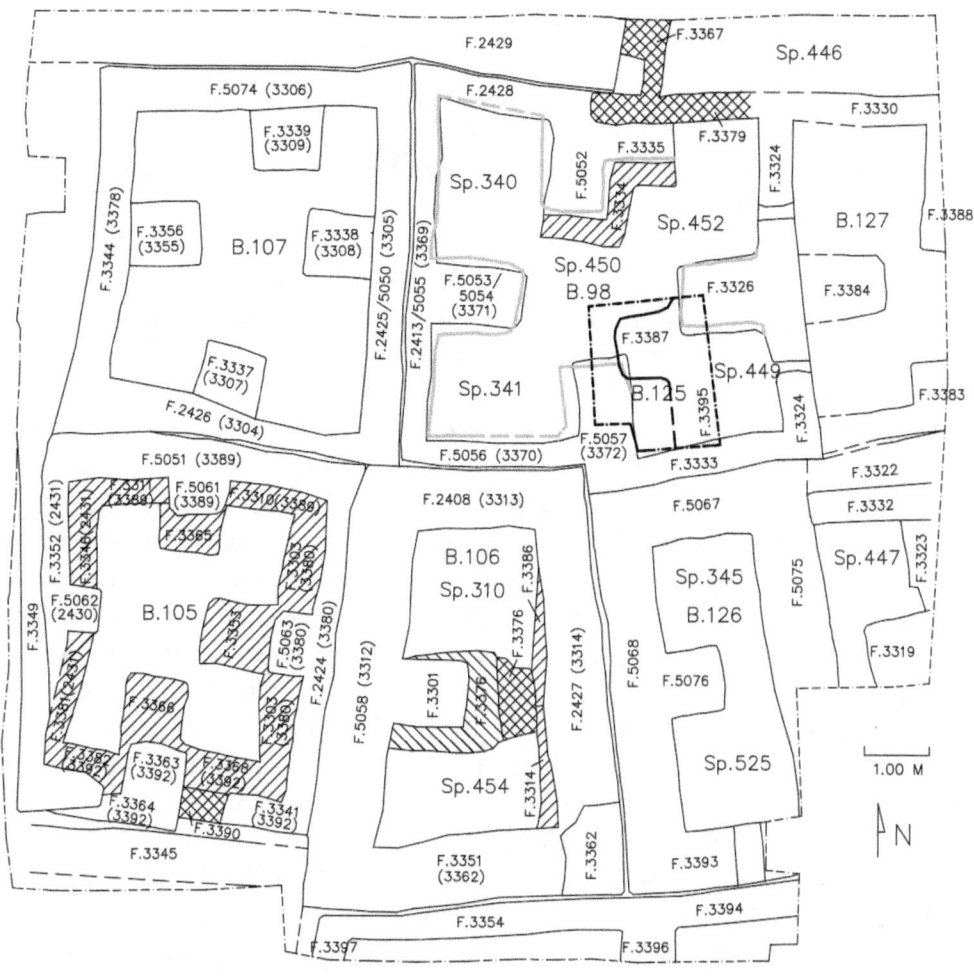

FIGURE 4.3. Excavation plan of Çatalhöyük West Mound Trench 5.

Adding a second story to the house on the one hand seems like a logical continuation of the development noted on the later East Mound, when buildings began to become larger and more complexly structured (Hodder 2013:22). On the other hand, two-storied buildings make for a new and very clear separation of living space and storage space.

More changes become visible when considering details of the building biographies. Starting with construction, it is interesting to note that builders on the West Mound seem to have reduced the number of larger pieces of wood needed for construction. The huge buttresses guarantee that the ceiling between the stories could have been built with pieces of wood of only moderate size. However, if all second stories lacked buttresses like that of B.78, the roof still would have needed larger beams. The composition of the construction materials used for the earthen walls seems to have changed, with West Mound walls having a more greyish color and lower sand content than the reddish and sandy bricks used in, for example, the TP area (Marciniak et al. 2013:79) and North area (e.g. Tung 2012:29), both representing the uppermost levels on the East Mound. Also, ways of forming and laying bricks had changed, with West Mound buildings showing a dazzling variety of earth-building techniques (Biehl et al. 2011:46) not present in that density and quality in the upper East Mound neighborhoods.

At the other end of buildings' lives, the elaborate abandonment rituals that transformed East Mound buildings into parts of the sphere of memory and ancestors (Cessford and Near 2005; Matthews 2005) were on the West Mound replaced by a more practical approach that probably included leaving houses to erode, removing plaster from walls and floors for reuse, and using abandoned houses as places for refuse disposal (Biehl et al. 2012; Rogasch et al. in press). Another piece of evidence for a changing social role of the house is the lack of sub-floor burials on the West Mound.

Only fractional data is available on settlement layout from the West Mound, since the overall area excavated there was limited. From what is known, it seems that settlement returned to the strict agglomeration that had characterized the middle layers of the East Mound, but was abandoned toward the end of the Neolithic sequence (Hodder 2013:22). Another interesting factor in West Mound settlement organization is the lack of the unroofed spaces regularly interspersed between buildings that had been a vital part of the classic East Mound settlement, and even became extended to real courtyards with cooking facilities in the late East Mound sequence (Hodder 2013:22, 27). Again, the limited area excavated on the West Mound leaves much room for such unroofed spaces to have existed somewhere in the settlement, but none was found within the excavated trenches thus far and also not at the closest roughly contemporary (Thissen 2002:16 Figure 1) site of Can Hasan I (French 1998).

In terms of pottery, a drastic increase in both quantity and elaboration appears on the West Mound (Figure 4.2). So far, calculations of the total excavation volume and the pottery quantities from Trenches 5 to 7 show that, in eight years, a total volume of at least 158,586 liters was excavated, which yielded at least 1,254,618 grams of pottery. This means that the mean density here for pottery per excavation volume is 8 g/l. Comparing this value with the calculations from the East Mound shows that on the West Mound, 50 times more pottery can be expected in the settlement remains than on the East Mound.

The pottery assemblage from Trenches 5–7 is dominated by light-colored serving and storage vessels painted with red geometric decoration patterns resembling basketry decoration, although sometimes incised decorations and appliqués are also observable. Undecorated two-lugged cooking vessels, which often show an ellipsoid circumference and an S-profile, are only a small part of the assemblage. In general, the vessels show a wider range of shapes and sizes than the vessels from the East Mound. Besides the predominant basketry-like decoration style, other regularly observable characteristics are ellipsoid rim and base circumferences, concave bases, basket impressions on bases, and basket-handles. Together these characteristics indicate that this pottery is not only imitating basket containers, but also is formed with the help of basket molds. Potstands are regularly found besides pottery and also show a wider range of sizes and shapes, and often are decorated with incised geometric patterns similar to the decoration on pottery. Their variations in sizes and shapes indicate a broad range of use (Franz 2008, 2009, 2010, 2011, 2012, 2013).

In general, the pottery is built up with slaps of calcareous clay and is fired around 800 °C under oxidizing conditions (Franz and Ostaptchouk 2012). Preliminary analyses indicate that the assemblage shows only fabrics made of local Çarşamba river fan or paleo-lake sediments. These fabrics are very similar to the fabrics of the "light line" of the East Mound pottery, and can also be divided into silty, sandy, and marly fabrics (Camizuli 2008; Doherty 2013, Tarkan-Özbudak 2013). Remarkably, the "dark line" fabrics of the East Mound pottery seem not at all present in the West Mound pottery production.

In terms of the subsistence economy, there is a general continuity with the later phases of occupation on the East Mound. The West Mound archaeobotanical assemblage reflects many of the trends found within the East Mound assemblage. A continued reliance on the crops of the East Mound is seen, with einkorn, (*Triticum monococcum*), emmer (*Triticum dicoccum*), and "new type" glume wheats along with free-threshing wheat still found. (Bogarrd et al. 2013; Stroud 2013). Pulses are dominated by lentil and to a lesser extent pea (*Pisum sativum*), but all crop taxa found on the East Mound have been found on the West Mound (Bogarrd et al. 2013; Stroud 2013). Hulled barley (*Hordeum vulgare* var. *vulgare*) becomes more frequent over time, continuing the increase in this crop type, a trend that started in the later levels of the East Mound (Bogarrd et al. 2013; Stroud 2013). Nuts of almond, terebinth (*Pistacia*), and hackberry (*Cetlis*) are also found within the West Mound assemblage, corresponding with the charcoal record and continuing the trend of wild food procurement seen on the East Mound (Stroud 2013). The archaeobotanical assemblage shows continuity in the use of cereals, pulses, and wild plants from East Mound to the West Mound.

The faunal assemblage reveals a similar pattern of continuity with the later East Mound phases. Sheep and goats continue to increase as a percentage, along with a slight increase in equids, decline in cattle—probably primarily domestic by this point, although their sheer rarity hampers metrical analysis—and a near disappearance of (wild) pig. There is also an apparent disappearance of cattle symbolism in regard to architectural installations and special deposits on the West Mound, though a perforated clay appliqué "miniature bucranium" was recovered from B. 106 (Franz 2009:49 Figure 53), while a symbolic role for caprines may be hinted at by two carefully prepared frontlets—pairs of horn cores attached to each other by trimmed-down frontal bones—found in a Trench 5 building infill, one each of domestic sheep and goat (Orton 2011; Russell et al. 2013b:11). Analyses of the pottery for milk fats and proteins is still underway, but preliminary faunal data suggest a shift in herding patterns (Russell et al. 2013b:7–8) that will need to be interpreted in the context of evidence for wider changes in caprine husbandry in Chalcolithic central Anatolia (Arbuckle et al. 2009).

Proxy data for environmental reconstruction in Central Anatolia in the late seventh and early sixth millennia B.C.

There is ample global evidence that the early to mid Holocene climatic optimum with its higher precipitation was interrupted by rapid climatic change between ca. 6400 and 5800 cal B.C. called the "8.2 event" according to the paleoclimatologists' custom of using calibrated B.P. dates instead of the cal B.C. format common in archaeology (for recent overviews see Dean et al. 2015, Morrill et al. 2014). Assessment of the exact extent to which the 8.2 event affected the Northern Hemisphere in a macro- and microregional perspective, however, is highly dependent on the availability of local climate proxies and their interpretation as to whether the event brought fresh and humid conditions or high seasonal contrasts characterized by droughts to Anatolia. As geomorphological indicators, pollen profiles, or anthracological evidence can also display human interaction with the environment (see below), more direct proxies such as sea surface temperatures derived from planktonic foraminifera data in sea cores or climate-sensitive stable isotopes derived from speleothems in caves are needed as control data.

For Anatolia, however, only the sea core LC21 in the Eastern Mediterranean east of Crete has yielded data (Rohling et al. 2002), and they point toward a cold spell occurring between 6400 and 5800 cal B.C. (Clare and Weninger 2014). The same is true for Sofular Cave in Northern Anatolia, where evidence for drier conditions around 6200 cal B.C. comes from sampled speleothems (Göktürk et al. 2011) and may provide the first hints that Central Anatolia was also affected by the event in a similar way, though the cave is separated from the region by the Pontic Mountains.

The prehistoric geomorphology of the Konya Plain, which is a large basin formed in the Late Quaternary, has been subject to thorough investigation over the past few decades, including the Konya Basin Palaeoenvironmental Research Programme (KOPAL) (Roberts et al. 1999), and for a summary of more recent studies conducted by a multidisciplinary team, see Charles et al. (2014). Results of previous research have demonstrated that the basin was at times during the Late Quaternary entirely occupied by a single extensive freshwater or brackish lake. The existence of this lake is responsible for a great number of landscape features that persist today, and also much of the > 400 m of lake marl and other lacustrine and fine grain sediments that fill the basin. At the start of the Holocene, though, the Konya basin was largely drained but remained moist, at least in parts probably experiencing seasonal flooding, and supporting both persistent riparian and dry woodlands. The Çarşamba and May rivers, which enter the closed Konya Basin, deposited significant alluvium in the areas surrounding Çatalhöyük. These sediments create ideal conditions for cereal cultivation, and the locations of Neolithic settlements in Anatolia often correlate with such formations (Cutting 2005:26). Çatalhöyük is not an exception in this regard, despite variable drainage conditions, as has been demonstrated by strontium isotope analyses of plant remains (Bogaard et al. 2013). Cultivation of local alluvium by the inhabitants of Çatalhöyük seems to have withstood both flooding and very dry conditions (Charles et al. 2014:78).

Drastic fluctuation in the water regime has been shown for the period from 6000–5000 cal B.C., just following the date range posited for drying conditions caused elsewhere by the 8.2 event. While the presence of water-abraded EC pottery deposited in a river channel south of the site, however, points to the existence of a still-flowing river (Roberts et al. 1996:39), it has been suggested that the environment of the Konya Plain was characterized by a complete drying out of lakes, displaying no single known lake or marsh deposit for the entire 1,000 year span (Fontugne et al. 1999:585). This period also encompasses the gradual shifting of the Çatalhöyük settlement from East to West and the final abandonment of the site.

Among the available pollen profiles (see Düring 2011:15 Figure 1.3 for an overview), the Beyşehir profile does not stem from the lake itself, but from a marsh nearby, and is most likely only representative for a local catchment area (Roberts 1990:54). We are therefore left with Karamik Bataklığı in the northern Lakes District, Eski Acıgöl in Nevşehir Province, a crater lake in an ancient volcano ca. 200 km Northeast of Çatalhöyük between the Konya Plain and Cappadocia, and Akgöl Adabağ in the Ereğli plain as the only usable pollen profiles from Central Anatolia if we want to look for possible signals of the 8.2 event in the pollen record. At Karamik Bataklığı, steppe vegetation changed to continuous forest and forest-steppe sometime after ca. 6800 cal B.C. (Bottema and van Zeist 1981). The sequence from Eski Acıgöl also shows increasing arboreal pollen with cedar (*Cedrus*) as an indicator of at least 600 mm of rainfall until ca. 6000 cal B.C. After 6000 cal B.C., however, increasing cereal pollen as a sign of intensified crop cultivation coincides with signs for forest degradation. A concomitant decline in rotifer percentages (*Rotatoria*) and an increase in green algae (*Pediastrum*) attest an accompanying increase of nutrients in the lake (Woldring 2002; Woldring and Bottema 2001). Eutrophication is known as a common consequence of intensive land use in both plant cultivation and animal husbandry (Hillbrand et al. 2014). The micro-charcoal displays an increase in the period after the 8.2 event, indicating a higher frequency of wildfires that could point to a dry spell, although human burning activities for forest clearance cannot be ruled out (Turner et al. 2008). Akgöl Adabağ shows increasing arboreal pollen until 7448–6606 cal B.C. (data calibrated using Oxcal 4.1. with the Intcal09 curve). After that, however, we see a short episode of decrease in arboreal pollen (Woldring 2002:63; Figure 2 "sample 32") whose exact date and interpretation remain unclear and thus should also be checked for a possible signal of either the 8.2 event or human-induced degradation.

Reconstruction of woodland environments during the occupation at Çatalhöyük using anthracological evidence from the site has added greater dimension to the picture. It demonstrates the long-term resilience of local wetlands and riparian woodlands with the continued use of species such as willow/poplar, elm, reeds, and sedges (Asouti and Kabukcu 2014). Wood charcoal macro-remain analyses from assemblages derived from the East Mound levels South R, S and T as well as TP levels display low frequencies of oak (*Quercus*) charcoals, whereas remains of ash (*Fraxinus*), juniper (*Juniperus*), elm and hackberry (*Ulmaceae*) and willow/poplar (*Salicaceae*) wood charcoal are ubiquitous. Those from building infills at Çatalhöyük West reveal that a majority of the fuel wood used in the settlement comprised juniper and elm/hackberry wood. In addition, a wide range of other taxa were also present including deciduous oak, willow/poplar, ash, terebinth (*Pistacia*), maple (*Acer*), tamarisk (*Tamarix*), almond (*Amygdalus*), wild apples and/or hawthorn (*Maloideae*), wild plums and cherries (*Prunus*), and wormwood (*Artemisia*) (Kabukcu 2015). This range of taxa was also identified in charcoal assemblages derived from primary depositional contexts including hearths, thus confirming their regular use as fuel-wood (Kabukcu 2015).

The continuity in the use of juniper and elm/hackberries as the dominant fuel wood taxa from the LN to the EC, alongside the continued presence of maple, almond, terebinth, hawthorn, and wild plum/cherry charcoal, suggests that the local semi-arid open woodlands and riparian vegetation were not impacted in any significant way by climatic aridity or human activities throughout the seventh-sixth millennia cal B.C. This observation finds additional support in the integrated charcoal and pollen evidence from central Anatolian sites dating to the early-mid Holocene (Kabukcu 2015; Asouti and Kabukcu 2014). Furthermore, the increasing alluvial deposition in the Konya plain through this period likely resulted in the more widespread distribution of better drained, high quality alluvial soils around the site which were able to sustain more productive riverine woodlands in the immediate vicinity of the site dominated by taxa such as elm, ash, and willow/poplar. This is also suggested by local pollen evidence, which also points to the prevalence of elm during the EC period (Eastwood et al. 2007).

THE LATE SEVENTH AND EARLY SIXTH MILLENNIUM AT ÇATALHÖYÜK IN THE CONTEXT OF THE "SECOND NEOLITHIC AND PAINTED POTTERY REVOLUTION"

Key to the greater discussion, the climate of the later half of the seventh millennium was likely a major contributor to the dispersal of Neolithic societies. The 8.2 event could be an additional motivation for the abandonment of old sites and the founding of new ones in this period, but it has to be kept in mind that all sites with ^{14}C dates from Western Anatolia have yielded evidence that the expansion of the Neolithic to the west was already underway at 6200 cal B.C.: most sites start shortly after 6500 cal B.C. (Thissen 2010), which makes it more likely that the changes visible in Çatalhöyük level V or P around 6450 cal B.C. go with the westward expansion (Rosenstock 2014:232 Figure 10). Therefore, the theory of a causal connection of the 8.2 event with the westward expansion requires additional research in Central Anatolia, where it is not yet well understood. This abrupt event has been described as the largest climate anomaly of the Holocene yet to date and has been put forward by Lee Clare (Clare et al. 2009) as the cause of widespread social upheaval, reorganization, and major cultural shifts in the prehistoric Near East where it was marked by annual temperature reductions of up to two degrees centigrade and a severe decline in precipitation (see Nieuwenhuyse et al. in this volume). The environment proxies outlined above seem to paint a complex picture of the late seventh millennium though, comprised of stability in the local woodland ecology at Çatalhöyük mixed with long-term drying trends in the Konya Plain as well as signals for considerable human impact in the first half of the sixth millennium. Archaeobotanical evidence also supports the notion of stability, but as research on this period has centered mostly on material from crop fields, which are commonly buffered from climate change in the short term, this position may eventually be appended.

There are many new developments that take shape in the seventh millennium, subsumed under the heading "Second Neolithic Revolution," as a term introduced by Düring (2011:122–125) for the intensification of resource use. In addition to new domesticates such as barley and free-threshing wheat, cattle were also added to the repertoire during this time (Arbuckle 2013; Kislev 1999; Russell et al. 2005), as well as dairy products (Evershed et al. 2008). Farming strategy improves, incorporating intensive methods and taking advantage of secondary products such as manure (Bogaard 2005). Agricultural innovations such as these expanded the range of landscapes suitable to Neolithic farming, allowing new areas outside of the steppe to be cultivated, or increased the appeal of productive economies in places where it had previously been undesirable (Schoop 2005), speeding along the process of neolithization westward. From there it spread relatively rapidly toward the Aegean and the Lower Balkans, arriving in Greece by 6400 cal B.C. with features similar to those known from Western Anatolia (Lichter 2005). This intensification could also have been the cause of the proliferation of sites in the Konya Plain in the Early Chalcolithic (Baird 2002): as painted pottery is already present in the Late Neolithic, some of the sites surveyed here that were attributed to the EC could already date to the LN of the late seventh millennium. In this case, the changes in subsistence in Çatalhöyük ca. 6450 cal B.C., the dispersal of sites into the Konya plain and the expansion of Neolithic settlements into the west and northwest of Turkey and the first Neolithic settlements in southeastern Europe would be contemporary phenomena, and it would be tempting to causally relate the economic changes to this dispersal (Schoop 2005).

Looking at the developments in the pottery assemblages from the East Mound and the West Mound in the time frame 6200–6000 cal B.C., we can see a persistence in the use of local clays for making pottery by using the so-called sequential slap construction technique and basket molds for forming the vessels. Also, the decoration techniques and the firing temperature stay the same. The major change that is observed is the massively increased manufacture of painted light-colored pottery made of local calcareous clays and the simultaneous decrease in manufacture of dark-colored undecorated cooking pots made of nonlocal non-calcareous clays. All changes, in fact, already begin in the late East Mound pottery assemblage before 6200 cal B.C. The West Mound pottery assemblage is the final basketry-imitating and blooming stage of Neolithic pottery manufacture at Çatalhöyük. This introduction of "baskets made of clay" could also have had an impact on the manufacture of proper basketry over time, most likely replacing them. This would perfectly explain the much larger quantities of pottery at Çatalhöyük between 6000–5500 cal B.C.

To bring the pottery from Çatalhöyük into a wider regional context, it is not a far reach to look toward the Pre-Halaf and early Halaf periods (ca. 6200–5700 cal B.C.) in Mesopotamia. This period is also characterized by the emergence of large quantities of painted pottery. Olivier Nieuwenhuyse (2007; see also Nieuwenhuyse in this volume) called the rapid increase of painted vessels in the pottery assemblage in Northern Syria in the Transitional phase between the Pre-Halaf and Early Halaf phase the "Painted Pottery Revolution." Alongside the increase in the quality of the pottery, the diversity also increased. First, the pottery in the Pre-Halaf Phase (ca. 6200 cal B.C.) is characterized by mostly plant-tempered pottery with simple profiles, which could be seen in addition to slips and impressed, incised, and appliqué decorations. Later in the Transitional Phase (ca. 6100 cal B.C.) the dark painted mineral-tempered fine wares with more complex profiles appear and spread, which gradually replace the older wares, and the other decoration techniques. Finally, in the Early Halaf Phase (ca. 5900 cal B.C.), painted fine ware with elaborate profiles dominates the pottery assemblage. This kind of painted Halafian ware is known from sites in northern Syria, southeastern Turkey, and northern Iraq. The typical Early Halafian bowl is a carinated painted bowl with a light-colored surface. From a technological point of view, this kind of pottery is not a new invention, but it seems that the firing techniques did advance (Nieuwenhuyse 2009). Reasons for the preference of painted decoration over other decoration techniques could have been "the prestige attached to the new technology of dark-on-light painted vessels; another may have been the superior versatility of painting for creating increasingly intricate designs" (Nieuwenhuyse 2009:88).

Comparing the conceptual, stylistic, and technological characteristics of Çatalhöyük pottery with the contemporary Halafian pottery shows many similarities and parallels. An alternative to the interpretation of these wares as "prestige products" could therefore be, although the Halafian pottery looks different from the pottery from Central Anatolia, that both are part of the same phenomenon observed in Çatalhöyük: pottery imitates basketry, and the increasing quantities of pottery together with the increasing diversity indicate a shift away from organic basketry vessels to mineral basketry-esque vessels roughly between 6200 and 5500 cal B.C. The different pottery styles follow the various regional centuries-old basketry traditions, and hint at the great variety of Neolithic basketry. The replacement of organic basketry with "baskets made of clay" is most likely explainable by a multicausal approach for each region or site. Three differently valued reasons could have been: the lack of suitable resources for basket making (which seems not very likely, as the grasses and reeds needed for making basketry may be found near all water sources), a more time-efficient production process for producing transportable containers (Brown 1989), and an increased dependency on pottery with time (Hodder 2012). These quite practical reasons indicate that the painted pottery seems to have not been a "prestige product" but just an easy-to-make replacement for baskets.

Discussion and Conclusions

Many of the changes witnessed in the economic and social organization of the Çatalhöyük settlement are clearly initiated in the middle of the East Mound sequence, starting ca. 6450 cal B.C., which continue to develop into the Early Chalcolithic material culture found on the West Mound. This also includes the heavy reliance on sheep and goat herding, and the waning relative importance of cattle, and wild cattle in particular, alongside other hunted game. One possible interpretation could be a drying environment in the proximity of the settlement beginning at this time, as both wild cattle and pig are better suited for damper conditions and at least partial woodland cover. On the other hand, the coincident decline in equids, which are typical semi-arid steppic and open grassland species, make this evidence somewhat ambiguous—although there is a slight uptick in their representation on the West Mound. Anthracological evidence from Çatalhöyük speaks to the long-term stability and resilience of local riverine woodlands and their constituent taxa and direct impacts of the 8.2 event on the environment at Çatalhöyük remain unclear to date. Seemingly directly coincident with the 8.2 event though, the appearance of painted pottery marks a substantial shift in society at Çatalhöyük. A wide-ranging new symbolic repertoire takes hold at the site ca. 6100 cal B.C., with the settling of the West Mound. Buildings return to the agglomerated form observed in the "classic" East Mound phases, but with a reorganized internal structure and use of space in buildings, incorporating internal buttresses and probably second stories as typical features. There was a changed preference for hulled barley over naked barley on the West Mound. Later drying conditions in the Konya Plain, beginning ca. 6000 cal B.C., evident in the sedimentation record, may have affected the local water regime, possibly perturbing the flow of the Çarşamba River and causing the shifting of the settlement to the west (Figure 4.4).

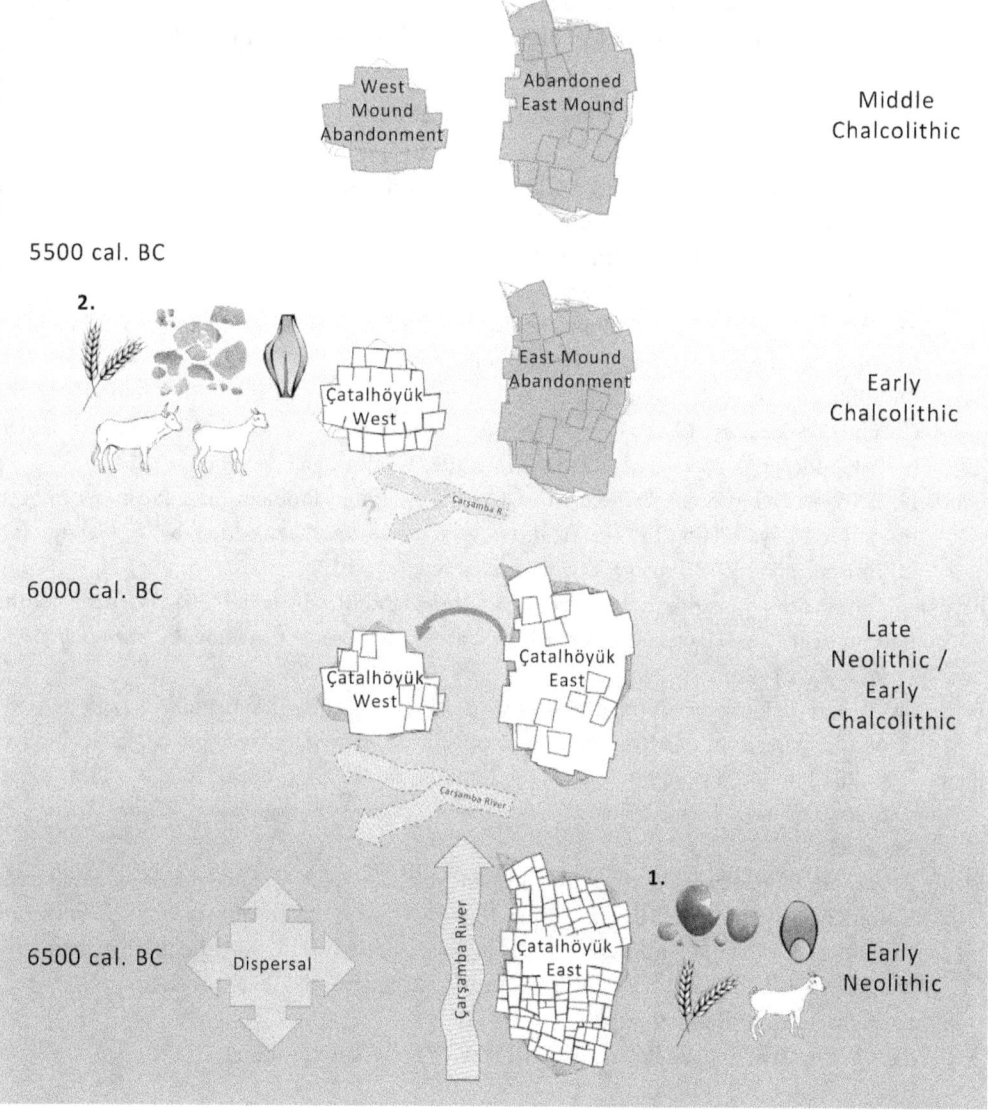

FIGURE 4.4. Diachronic change at Çatalhöyük from the late seventh to sixth millennium. 1Clockwise from top left: "classic" East Mount pottery, staple domesticates of naked barley (preferred), sheep/goat, and wheat. 2Clockwise from top center: typical West Mound painted pottery, staple domesticates of hulled barley (preferred), sheep/goat, cattle, and wheat.

The final abandonment of the West Mound at Çatalhöyük, as well as the whole Konya Plain, from ca. 5500 cal B.C., seems to be another event which is difficult to find clear evidence for in the environmental, climatic, and archaeological data. It may be the result of an overexploitation of the environment in the wake of the intensification of land use brought about by the "Second Neolithic Revolution" (Düring 2011:122–125). In contrast to earlier notions that saw a major hiatus between the East and West Mounds at Çatalhöyük (French 1967:175 chart 2), the evidence for continuity at Çatalhöyük from the Late Neolithic into the Early Chalcolithic demonstrates that instead the end of the Early Chalcolithic at the middle of the sixth millennium was the time that fundamentally changed the life of the people of Çatalhöyük as well as of the Konya Plain.

References Cited

Arbuckle, B. 2013 The Late Adoption of Cattle and Pig Husbandry in Neolithic Central Turkey. *Journal of Archaeological Science* 40(4):1805–1815.

Arbuckle, B., A. Öztan, and S. Gülçur 2009 The Evolution of Sheep and Goat Husbandry in Central Anatolia. *Anthropozoologica* 44(1):129–157.

Asouti, E., and C. Kabukcu 2014 Holocene Semi-arid Oak Woodlands in the Irano-Anatolian Region of Southwest Asia: Natural or Anthropogenic? *Quaternary Science Reviews* 90:158–182.

Baird, D. 2006 The History of Settlements and Social Landscapes in the Early Holocene in the Çatalhöyük Area. In *Çatalhöyük Perspectives: Themes from the 1995–1999 Seasons,* edited by I. Hodder, pp. 55–74. MacDonald Institute, Cambridge, England.

Baird, D. 2002 Early Holocene Settlement in Central Anatolia: Problems and Prospects as Seen from the Konya Plain. In *The Neolithic of Central Anatolia,* edited by F. Gerard and L. Thissen, pp. 139–159. Ege Yayinlari, Istanbul.

Bayliss, A., F. Brock, S. Farid, I. Hodder, J. Southon, and R. E. Taylor 2015 Getting to the Bottom of It All: A Bayesian Approach to Dating the Start of Çatalhöyük. *Journal of World Prehistory* 28:1–26.

Berger, J.-F., and J. Guilaine 2009 The 8200 cal BP Abrupt Environmental Change and the Neolithic Transition: A Mediterranean Perspective. *Quaternary International* 200:31–49.

Biehl, P. F. 2012 Rapid Change versus Long-Term Social Change during the Neolithic-Chalcolithic Transition in Central Anatolia. *Interdisciplinaria archaeologica Natural Sciences in Archaeology* 3:75–83.

Biehl, P. F., and E. Rosenstock 2009 Von Çatalhöyük Ost nach Çatalhöyük West: Kulturelle Umbrüche an der Schwelle vom 7. zum 6. Jt. v. Chr. in Zentralanatolien. In *Zurück zum Gegenstand, Festschrift für Andreas E. Furtwängler,* edited by Ralph Einicke, Stephan Lehman, Henryk Löhr, Gundula Mehnert, Andreas Mehnert, and Anja Slawisch, pp 471–482. Langenweißbach, Beier & Beran.

Biehl, P. F., J. Rogasch, and E. Rosenstock 2011 West Mound Excavations, Trench 5. *Çatalhöyük Archive Report* 2011:38–48.

Biehl, P. F., I. Franz, S. Ostaptchouk, D. Orton, J. Rogasch, and E. Rosenstock 2012 One Community and Two Tells: The Phenomenon of Relocating Tell Settlements at the Turn of the 7th and 6th Millennia in Central Anatolia. In *Socio-Environmental Dynamics over the Last 12,000 Years: The Creation of Landscapes,* edited by J. Müller, pp. 53–66. Offa, Kiel.

Biehl, P., J. Rogasch, and E. Rosenstock 2012 West Mound Trench 5 Excavations 2012. *Çatalhöyük Archive Report* 2012:76–102.

Bogaard, A. 2005 "Garden Agriculture" and the Nature of Early Farming in Europe and the Near East. *World Archaeology* 37:177–196.

Bogaard, A., M. Charles, A. Livarda, M. Ergun, D. Filipovic, and G. Jones 2013 The Archaeobotany of Mid-Later Occupation Levels at Neolithic Çatalhöyük. In *Humans and Landscapes of Çatalhöyük: Reports from the 2000–2008 Seasons. Çatalhöyük Research Project Series, Volume 8*, edited by I. Hodder. Monographs of the Cotsen Institute of Archaeology, University of California at Los Angeles, Los Angeles.

Boyer, P., N. Roberts, and D. Baird 2006 Holocene Environment and Settlement on the Carsamba Alluvial Fan, South-Central Turkey: Integrating Geoarchaeology and Archaeological Field Survey. *Geoarchaeology* 21:675–698.

Brown, J. A. 1989 The Beginnings of Pottery as an Economic Process. In *What's New? A Closer Look at the Process of Innovation*, edited by S. E. van der Leeuw pp. 203–224. Unwin Hyman, London.

Camizuli, E. 2008 Clay Provenance of Neolithic and Chalcolithic Ceramics from Çatalhöyük (Turkey). In *Çatalhöyük 2008 Archive Report* 289–339. http://www.catalhoyuk.com/downloads/Archive_Report_2008.pdf (February 2015).

Cessford, C. 2005a Estimating the Neolithic Population of Çatalhöyük. In *Inhabiting Çatalhöyük: Reports from the 1995–1999 Seasons*, edited by I. Hodder, pp. 323–328. McDonald Institute for Archaeological Research, Cambridge.

Cessford, C. 2005b Absolute Dating at Çatalhöyük. In *Changing Materialities at Çatalhöyük: Reports from the 1995–1999 Seasons*, edited by I. Hodder, pp 65–100. McDonald Institute for Archaeological Research, Cambridge.

Cessford, C., and J. Near 2005 Fire, Burning, and Pyrotechnology at Çatalhöyük. In *Çatalhöyük Perspectives: Reports from the 1995–1999 Seasons*, edited by I. Hodder, pp. 171–182. British Institute at Ankara, London.

Charles, M., C. Doherty, E. Asouti, A. Bogaard, E. Henton, C. L. Spencer, C. B. Ruff, P. Ryan, J. Sadvari, and K. Twiss 2014 Landscape and Taskscape at Çatalhöyük: An Integrated Perspective. In *Integrating Çatalhöyük: Themes from the 2000–2008 Seasons, Volume 10*, edited by I. Hodder, pp. 71–90. Monographs of the Cotsen Institute of Archaeology, University of California at Los Angeles, Los Angeles.

Clare, L., E. J. Rohling, B. Weninger, and J. Hilpert 2008 Warfare in Late Neolithic\Early Chalcolithic Pisidia, Southwestern Turkey. Climate Induced Social Unrest in the Late 7th Millennium cal BC. *Documenta Praehistorica* XXXV:65–92.

Clare, L., and B. Weninger 2014 Absolute Chronology and Rapid Climate Change in Central and West Anatolia. In *The Neolithic in Turkey Vol. 6: 10500–5200 BC: Environment Settlement, Flora, Fauna, Dating, Symbols of Belief, With views from North, South, East and West*, edited by M. Özdoğan, N. Basgelen, and P. Kuniholm, pp. 1–65. Archaeology and Art Publications, Istanbul.

Cutting, M. V. 2005 *The Neolithic and Early Chalcolithic Farmers of Central and Southwest Anatolia*. Archaeopress, Oxford.

Czerniak, L., and A. Marciniak 2008 The Excavations of the TP (Team Poznań) Area in the 2008 Season. *Çatalhöyük Archive Report* 2008:73–82.

Dean, J. R., M. D. Jones, M. J. Leng, S. R. Noble, S. E. Metcalfe, H. J. Sloane, D. Sahy, W. J. Eastwood, and N. Roberts 2015 Eastern Mediterranean hydroclimate over the late glacial and Holocene, reconstructed from the sediments of Nar lake, central Turkey, using stable isotopes and carbonate mineralogy. *Quaternary Science Reviews* 124:162–174.

Doherty, C. 2013 Sourcing Çatalhöyük's Clays. In *Substantive Technologies at Çatalhöyük. Reports from the 2000–2008 Çatalhöyük Research Project Series Volume 9*, edited by I. Hodder. BIAA Monograph 48:51–66.

Doherty, C., and D. Tarkan-Özbudak 2013 Pottery Production at Çatalhöyük: A Petrographic Perspective. In *Substantive Technologies at Çatalhöyük. Reports from the 2000–2008 Seasons. Çatalhöyük Research Project Series Volume 9* edited by I. Hodder. BIAA Monograph 48:179–188.

Düring, B. S. 2011 *The Prehistory of Asia Minor*. Cambridge University Press, New York.

Eastwood, W. J., N. Roberts, and P. Boyer 2007 Pollen Analysis at Çatalhöyük. In *Excavations at Çatalhöyük: The 1995–1999 Seasons*, edited by I. Hodder, pp. 573–580. McDonald Institute Monographs, Cambridge.

Erdoğu, B. 2008 Trench 8. *Çatalhöyük Archive Report* 2008:105–109.

Erdoğu, B. 2009 Ritual Symbolism in the Early Chalcolithic Period of Central Anatolia. *Journal for Interdisciplinary Research on Religion and Science* 5:129–151.

Farid, S. 2008 A Review of the Mellaart Ievei System and the Introduction of a New Phasing System at Çatalhöyük 2008. In *Çatolhöyük 2008 Archive Report*:15–28. http://www.catalhoyuk.com/archive_reports/Archive_Report_2008.pdf. Last updated 2008 (20.05.2011).

Fontugne, M., C. Kuzucuoglu, M. Karabiyikoglu, C. Hatté, and J. F. Pastre 1999 From Pleniglacial to Holocene: A ^{14}C Chronostratigraphy of Environmental Changes in the Konya Plain, Turkey. *Quaternary Science Reviews* 18:573–591.

Franz, I. 2008 West Mound Trenches 5 & 7—Trench 5-7 Pottery. In *Çatalhöyük 2008 Archive Report* 97–100. http://www.catalhoyuk.com/downloads/Archive_Report_2008.pdf (February 2015).

Franz, I. 2009 West Mound Trench 5—Trench 5 Pottery. In *Çatalhöyük 2009 Archive Report* 42–50. http://www.catalhoyuk.com/downloads/Archive_Report_2009.pdf (February 2015).

Franz, I. 2010 Ceramics Archive Report 2010—West Mound Trench 5-7 Pottery. In *Çatalhöyük 2010 Archive Report* 77–90. http://www.catalhoyuk.com/downloads/Archive_Report_2010.pdf (February 2015).

Franz, I. 2011 West Mound Pottery, Trench 5-7. In *Çatalhöyük 2011 Archive Report* pp. 79–90. http://www.catalhoyuk.com/downloads/Archive_Report_2011.pdf (February 2015).

Franz, I. 2012 Trench 5-7 Pottery Archive Report. In *Çatalhöyük 2012 Archive Report* pp. 262–271. http://www.catalhoyuk.com/downloads/Archive_Report_2012.pdf (February 2015).

Franz, I. 2013 West Mound pottery, 2013. In *Çatalhöyük 2013 Archive Report* pp. 198–203. http://www.catalhoyuk.com/downloads/Archive_Report_2013.pdf (February 2015).

Franz, I., and S. Ostapchouk 2012 Illuminating the Pottery Production Process at Çatalhöyük West Mound (Turkey) around 8000 cal BP. In *Naturwissenschaftliche Analysen vor- und frühgeschichtlicher Keramik II Methoden, Anwendungsbereiche, Auswertungsmöglichkeiten*. Dr. Rudolf Habelt GmbH, Bonn.

French, D. 1967 Excavations at Can Hasan: Sixth Preliminary Report, 1966. *Anatolian Studies* 17:105–178.

French, D. 1998 Canhasan I: Stratigraphy and Structures. BIAA, London.

Gibson, C., and J. Last 2003 An Early Chalcolithic Building on the West Mound at Çatalhöyük. *Anatolian Archaeology* 9:12–14.

Göktürk, O. M., D. Fleitmann, S. Badertscher, H. Cheng, R. L. Edwards, M. Leuenberger, A. Fankhauser, O. Tüysüz, J. Kramers 2011 Climate on the Southern Black Sea Coast during the Holocene: Implications from the Sofular Cave record. *Quaternary Science Reviews* 30:2433–2445.

Hillbrand, M., B. van Geel, A. Hasenfratz, P. Hadorn, and J. Nicolas Haas 2014 Non-Pollen Palynomorphs Show Human- and Livestock-Induced Eeutrophication of Lake Nussbaumersee (Thurgau, Switzerland) since Neolithic Times (3840 BC). *The Holocene* 24:559–568.

Hodder, I. 1996 On the Surface: Çatalhöyük 1993–95. *Çatalhöyük Reseach Project Series Volume 1*. McDonald Institute for Archaeological Research / British Institute of Archaeology at Ankara Monograph No. 22.

Hodder, I. 2006 *The Leopard's Tale*. Thames & Hudson, New York.

Hodder, I. 2007 Excavating Çatalhöyük: South, North, and KOPAL Area Reports from the 1995–99 Seasons. *Çatalhöyük Reseach Project Series Volume 3*. McDonald Institute for Archaeological Research / British Institute of Archaeology at Ankara Monograph.

Hodder, I. 2009 2009 Season Review. *Çatalhöyük 2009 Archive Report*: 1–4. http://www.catalhoyuk.com/downloads/Archive_Report_2009.pdf. Last updated 2009 (25.10.2012).

Hodder, I. 2012 *Entangled: An Archaeology of the Relationships of Humans and Things*. Wiley-Blackwell, Chichester-Malden,.

Hodder, I. 2013 Dwelling at Çatalhöyük. In *Humans and Landscapes of Çatalhöyük: Reports from the 2000–2008 Seasons*, edited by I. Hodder, pp. 1–29. Cotsen Institute of Archaeology Press, Los Angeles.

Hodder, I. 2014 Çatalhöyük: The Leopard Changes Its Spots. A Summary of Recent Work. *Anatolian Studies* 64:1–22.

Kabukcu, C. 2015 Prehistoric Vegetation Change and Woodland Management in Central Anatolia: Late Pleistocene-Mid Holocene Anthracological Remains from the Konya Plain. PhD Thesis. University of Liverpool, Dept. of Archaeology, Classics and Egyptology.

Kislev, M. E. 1999 Agriculture in the Near East in the Seventh Millennium BC. In *Prehistory of Agriculture: New Experimental and Ethnographic Approaches* edited by P. C. Anderson, pp. 51–55. Institute of Archaeology, Los Angeles.

Kuzucuoğlu, C. 2002 The Environmental Frame in Central Anatolia from the 9th to the 6th Millennia cal BC. In *The Neolithic of Central Anatolia. Internal Developments and External Relations during the 9th–6th Millenia Cal BC. Proceedings of the International CANeW Table Ronde Istanbul, 23–24 November 2001*, edited by F. Gérard and L. Thissen, pp. 33–58. Ege Yayinlari, Istanbul.

Lemcke G., and Sturm, M. 1997 $\delta^{18}O$ and Trace Element Measurements as Proxy for the Reconstruction of Climate Changes at Lake Van (Turkey): Preliminary Results. In *Third Millennium BC Climate Change and Old World Collapse*, edited by N. Dalfes, G. Kukla, H. Weiss, pp. 653–678. Springer, Berlin.

Lichter, C. 2005 Western Anatolia in the Late Neolithic and Early Chalcolithic: The Actual State of Research. In *How Did Farming Reach Europe? Anatolian-European Relations from the Second Half of the 7th through the First Half of the 6th Millennium cal BC*, edited by C. Lichter, pp. 59–74. Zero Prod., Istanbul.

Marciniak, A., P. Filipowicz, E. Johansson, and A. Mickel 2013 The Excavations of the TPC Area in the 2013 Season. *Çatalhöyük Archive Report* 2013:74–94.

Marciniak, A., M. Z. Barański, A. Bayliss, L. Czerniak, T. Goslar, J. Southon, and R. E. Taylor 2015 Fragmenting Times: Interpreting a Bayesian Chronology for the Late Neolithic Occupation of Çatalhöyük East, Turkey. *Antiquity* 89:154–176.

Matthews, W. 2005 Life-Cycle and Life-Course of Buildings. In *Çatalhöyük Perspectives: Reports from the 1995–1999 Seasons*, edited by I. Hodder, pp. 125–150. British Institute at Ankara, London.

Mellaart, J. 1962 Excavations at Çatal Hüyük: First Preliminary Report, 1961. *Anatolian Studies* XI:39–75.

Mellaart, J. 1975 *The Neolithic of the Near East*. Thames and Hudson, London.

Mellaart, J. 1965 Çatal Hüyük West. *Anatolian Studies* XV:135–156.

Mellaart, J. 1967 *Çatalhöyük, A Neolithic Town in Anatolia*. Thames and Hudson, London.

Morrill, C., D. M. Anderson, B. A. Bauer, R. Buckner, E. P. Gille, W. S. Gross, and A. Shah 2013 Proxy Benchmarks for Intercomparison of 8.2 ka Simulations. *Climate of The Past* 9:423–432. doi:10.5194/cp-9-423-2013.

Nieuwenhuyse, O. 2007 *Plain and Painted Pottery*. Brepols, Turnhout.

Nieuwenhuyse, 2009 The Painted Pottery Revolution: Emulation, Ceramic Innovation and the Early Halaf in Northern Syria. In *Méthodes d'approche des premières productions céramiques: étude de cas dans les Balkans et au Levant. Table-ronde de la Maison de l'Archéologie et de l'Ethnologie (Nanterre, France) 28 février 2006*, edited by Atruc et al. Internationale Archäologie 12:81–91.Verlag Marie Leidorf, Rahden.

Orton, D. 2011 West Mound Animal Bones 2011, Trench 5. In *Çatalhöyük 2011 Archive Report*, pp. 49–51. http://www.catalhoyuk.com/downloads/Archive_Report_2011.pdf (February 2015).

Orton, D., I. Franz, J. Rogasch, E. Rosenstock, P. F. Biehl, and A. Bogaard in prep., A tale of two tells: dating the Catalhoyuk West Mound. *European Journal of Archaeology*.

Özbaşaran, M. 2012 Aşıklı. In *The Neolithic in Turkey, New Excavations & New Research: Central Turkey*, edited by M. Özdoğan, N. Başgelen and P. Kuniholm, pp. 135–158. Arkeoloji ve Sanat Yayinlari, Istanbul.

Pross, J., U. Kotthoff, U. C. Müller, O. Peyron, I. Dormoy, G. Schmiedl, S. Kalaitzidis, and A. M. Smith 2009 Massive Perturbation in Terrestrial Ecosystems of the Eastern Mediterranean Region Associated with the 8.2 kyr B.P. Climatic Event. *Geology* 37:887–890.

Regan, R., F. Sadarangani, and J. Taylor 2008 Building 75, Middens and Buildings 80 & 70. *Çatalhöyük Archive Report*:62–73.

Roberts, N., P. Boyle, and R. Parish 1996 Preliminary Results of Geoarchaeological Investigations at Çatalhöyük. In *On the Surface: Çatalhöyük 1993–1995. Çatalhöyük Reseach Project Series Volume 1*, edited by I. Hodder. McDonald Institute for Archaeological Research / British Institute of Archaeology at Ankara Monograph No. 22.

Roberts, N., S. Black, P. Boyer, W. J. Eastwood, H. I. Griffiths, H. F. Lamb et al. 1999 Chronology and Stratigraphy of Late Quaternary Sediments in the Konya Basin, Turkey: Results from the KOPAL project. *Quaternary Science Reviews* 18:611–630.

Rogasch, J., J. Brady, I. Franz, D. Orton, S. Ostaptchouk, C. Piliougine, P. Ryan, E. Stroud, E. Rosenstock, and P. F. Biehl In press. All Rubbish? The Formation of Deposits in Buildings at Çatalhöyük West, Turkey (ca. 5900–5800 BC). In *The Archaeology of Polution: The Creation of Landscapes*, edited by J. Müller and P. F. Biehl. Habelt, Bonn.

Rohling, E. J., P. A. Mayewski, A. Hayes, R. H. Abu-Zied, and J. S. L. Casford 2002 Holocene Atmosphere-Ocean Interactions: Records from Greenland and the Aegean Sea. *Climate Dynamics* 18:587–593.

Rosenstock, E. 2014 Like a Su Böreği: Settlements and Site Formation Processes in Neolithic Turkey. In *The Neolithic in Turkey: 10500–5200 BC: Environment Settlement, Flora, Fauna, Dating, Symbols of Belief, With views from North, South, East and West. vol. 6*, edited by M. Özdoğan, N. Başgelen, and P. Kuniholm, pp. 223–263. Archaeology and Art Publications, Istanbul.

Russell, N., L. Martin, and H. Buitenhuis 2005 Cattle Domestication at Çatalhöyük Revisited. *Current Anthropology* 46:101–108.

Russell, N., K. C. Twiss, D. C. Orton, and G. A. Demirergi 2013a More on the Çatalhöyük Mammal Remains. In *Humans and Landscapes of Çatalhöyük: reports from the 2000–2008 Seasons*, edited by I. Hodder, pp. 213–258. Cotsen Institute, Los Angeles.

Russell, N., K. C. Twiss, D. C. Orton, and G. A. Demirergi 2013b Changing Animal Use at Neolithic Çatalhöyük, Turkey. In *Archaeozoology of the Near East*, edited by X. B. de Cupere, V. Linseele, and S. Hamilton-Dyer, pp. 45–68. Ancient Near Eastern Studies Supplement. Peeters, Leuven.

Schoop, U. D. 2005 *Das Anatolische Chalkolithikum*. Albert Greiner, Remshalden.

Stroud, E. 2013 The Archaeobotany Report, West Mound. *Çatalhöyük Archive Report*:160–169.

Thissen, L. 2002 Appendix I: The CANeW 14C Database, Anatolia 10,000–5000 cal BC. In *The Neolithic of Central Anatolia*, edited by F. Gérard and L. Thissen, pp. 299–337. Ege Yayınları, Istanbul.

Thissen, L. 2010 The Neolithic-Chalcolithic Sequence in the SW Anatolian Lakes Region. *Documenta Praehistorica* 37:269–282.

Tung, B. 2012 Excavations in the North Area. *Çatalhöyük Archive Report* 2012:9–35.

Turner, R., N. Roberts, and M. D. Jones 2008 Climatic Pacing of Mediterranean Fire Histories from Lake Sedimentary Microcharcoal. *Global and Planetary Change* 63:317–324.

Weninger B., E. Alram-Stern, E. Bauer, L. Clare, U. Danzeglocke, O. Jöris, C. Kubatzki, G. Rollefson, H. Todorova 2005 Die Neolithisierung von Südost-europa als Folge des abrupten Klimawandels um 8200 cal BP. In *Klimaveränderung und Kul-turwandel in neolithischen Gesellschaften Mitteleuropas,6700–22200 v. Chr.*, edited by D. Gronenborn Römisch-German. Zentralmuseum Tagungen 1. Verlag des Römisch-Germanischen Zentral-museums, Mainz:75–117.

Wiersma, A. P., H. H. Renssen, H. H. Goosse, and T. T. Fichefet 2006 Evaluation of Different Freshwater Forcing Scenarios for the 8.2 ka BP Event in a Coupled Climate Model. *Climate Dynamics* 27:831–849.

Woldring, H. 2002 Climate Change and the Onset of Sedentism in Cappadocia. In *The Neolithic of Central Anatolia: Internal developments and external relations during the 9th–6th Millennia cal BC*, edited by F. Gérard and L. Thissen, pp. 59–66. Ege, Istanbul.

Woldring H., and S. Bottema 2001 The Vegetation History of East-Central Anatolia in Relation to Archaeology: The Eski Acigöl Pollen Evidence Compared with the Near Eastern Environment = L'histoire de la végétation de l'Anatolie est-centrale en relation avec l'archéologie: le témoignage de pollen provenant de l'Eski Acigöl comparé à l'environnement proche-oriental. *Palaeohistoria* 43–44:1–34.

Yalman, N., D. Tarkan-Özbudak, and H. Gültekin 2013 The Neolithic Pottery of Çatalhöyük: Recent Studies. In *Substantive Technologies at Çatalhöyük. Reports from the 2000–2008 Seasons: Çatalhöyük Research Project Series Volume 9*, edited by I. Hodder. BIAA Monograph 48:143–178.

Yeomans, L., and F. Sadarangani 2009 Building 87, Building 85, external spaces Sp.365, Sp.370, Sp.132 and Sp.369. *Çatalhöyük Archive Report* 2009:32–38.

CHAPTER FIVE

Managing Risk through Diversification in Plant Exploitation during the Seventh Millennium B.C.

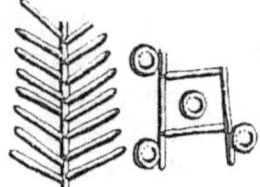

The Phytolith Record at Çatalhöyük

Philippa Ryan, Arlene Rosen

Abstract *Many large sites collapsed toward the end of the Pre-Pottery Neolithic, yet a small number—including Çatalhöyük in Central Anatolia—survived into the sixth millennium B.C. A period of possible abandonment at the end of Çatalhöyük East previously has been linked with the 8.2 ka (ca. 6250 cal B.C.) climate event (Weninger et al. 2006); however, recent excavations at Çatalhöyük demonstrate continuous occupation between the East (Neolithic) and West (Chalcolithic) Mounds. It is possible that some Neolithic economies were more resilient than others. This paper presents phytolith (plant-cell microfossils) evidence for plant use at Çatalhöyük, and from this discusses the role of plants in risk-buffering systems and what this can tell us about the long-term survival of the settlement.*

INTRODUCTION

Site abandonments during the Late PPNB to Late Neolithic have been discussed in relation to human impact on the environment (Rollefson and Köhler-Rollefson 1992), the failure of social strategies (Kuijt 2000), and the result of climate changes (Weninger et al. 2006). While many large sites collapsed toward the end of the Pre-Pottery Neolithic, a few, including Çatalhöyük, survived into the sixth millennium B.C. Managing agricultural risk would have been crucial and it is possible that some Neolithic economies were more resilient than others. Diversification methods to buffer against lean years could include using a variety of crops and livestock, spatial strategies in field locations, or wild foods (Halstead and O'Shea 1989; Halstead 1989; O'Shea 1989). Other buffering possibilities included food storage strategies as well as food sharing among households

(Demirergi et al. 2014; Halstead and O'Shea 1989; Marston 2011). Previously it has been suggested that there was a gap between the seventh millennium BC East Mound at Çatalhöyük and the sixth millennium BC West Mound, and that perhaps this was linked to the 8.2 ka event (Weninger et al. 2006)—however, it is now clear that occupation was continuous. In order to understand any possible impact of the 8.2 ka event (ca. 6250 cal B.C.) on Neolithic societies, it is necessary to determine how it might or might not have affected Neolithic subsistence economies.

There is evidence for the 8.2 ka event (ca. 6250 cal B.C.) from ice-cores, while various proxy records from around the globe indicate anomalies that range from a sudden short-term event to a span of from 400–600 years (Alley et al. 1997; Rohling and Pälike 2005). There is a "spike" in the carbon and oxygen isotope data from the Soreq Cave speleothems (northern Israel) that may potentially be correlated to the 8.2 ka event (Bar-Matthews et al. 1997). A possible arid phase 8,200 years ago is also suggested in the Dead Sea sedimentary record (Migowski et al. 2006). Wiersma and Renssen (2005) argue proxy data suggests cooling occurred mainly in the Northern Atlantic, indicating the event was most likely caused by meltwater, and that this corresponds temporally with climatic models of disruptions to summer monsoon weather systems.

Independent nonarchaeological vegetation records are important for understanding the potential impact of climate change on past resource availability. The relatively small body of evidence for the impact of climate change on vegetation histories due to the 8.2 ka event in the Near East contrasts with the far greater amount of such information for the Younger Dryas (e.g., Hajar et al. 2008; Moore and Hillman 1992; Rossingnol-Strick 1993; Wright and Thorpe 2003). The 8.2 ka event was about half the amplitude of the Younger Dryas (Alley et al. 1997), and it is possible these differences in magnitude are reflected in the poorer vegetation record for change. At present, there is no clear record for an 8.2 ka event in the main Central Anatolian pollen cores (Cappers et al. 2002; Roberts et al. 2001; Woldring 2002a&b; van Zeist and Bottema 1991:73). There is evidence for changes in the pollen record that coincides with the 8.2 ka event at Lake Van in Eastern Anatolia, however rather than aridity, greater humidity is indicated (Wick et al. 2003). Nevertheless, Wick et al. (2003) suggest the humidity may somehow relate to the 8.2 ka event. More recently, European climate studies (Peyron et al. 2011) have explored a potential explanation for such humidity, as discussed below.

Although the impact of the 8.2 ka event (ca. 6250 cal B.C.) on vegetation in the pollen record from the Near East is unclear, there is increasingly more evidence from European studies. Pross et al. (2009) argue the pollen evidence from Greece suggests the 8.2 ka event would have dramatically impacted settlers in the northeastern Mediterranean. Previously, Rossingnol-Strick identified increases in drought-tolerant taxa at Tenaghi Philippon (Greece) at 8000 or 7500 cal B.P. (ca. 6050 or 5550 cal B.C) for about 400 years, and at Xinias I at approximately 8000 cal B.P. (ca. 6050 cal B.C.) (Rossingnol-Strick 1993, 1995). A clear signal for the 8.2 ka event has now been recognized in pollen histories from Tenaghi Philippon (Greece) in association with a new seasonal pattern of dry winters and wet summers, and a weaker signal for the event at Lake Accesa (Italy) with wet winters and summers (Peyron et al. 2011). In the Balkans there is evidence for changes in plant biomass at several locations during the 8.2 ka event, with a clearer record in some areas than others and with some showing increases in humidity and others in aridity (Sadori et al. 2011). Variable changes in seasonality might explain contradictory records for aridity and humidity (Peyron et al. 2011).

ÇATALHÖYÜK AND ITS NEOLITHIC ENVIRONMENT

Situated on the Konya Plain of Central Turkey, the seventh millennium Neolithic East Mound at Çatalhöyük and the later sixth millennium Chalcolithic West Mound are based on either side of the now underground Çarsamba River. The earliest East Mound site levels are dated to 7400–7100 cal B.C. (Cessford 2005), and the dates from the latest Neolithic levels in the TP (Team Poznań) Area around 6250–6000 cal B.C.

The Konya Plain's current environment is semi-arid, however local as well as regional paleoclimatic research indicates greater precipitation during the Neolithic (Fontugne 1999; Kuzucuoğlu 2002; Roberts et al. 2001; Rossignol-Strick 1993, 1999). Some continuity is suggested by a continuation of a moister marsh/lake phase on the Konya Plain during the sixth millennium cal B.C. (Fontugne et al. 1999; Woldring 2002a&b). As already mentioned above, there is no clear signal for the 8.2 ka event in the main Central Anatolian pollen record. The Konya Basin Paleoenvironments project (KOPAL) geoarchaeological investigations suggest the area around Çatalhöyük (East) may have been subject to spring flooding (Boyer et al. 2006; Roberts et al. 2007; Roberts and Rosen 2009). Additional cores taken since 2007 suggest there were also areas of dryer land close to the site suited to agriculture, and a mosaic of wet and dry environments (Doherty et al. 2007, 2008; Charles et al. 2014). A mosaic of wet and dry land microenvironments during the Neolithic is also suggested by the macrobotanical record (seeds, fruits) and bird remains (Bogaard et al. 2013; Fairbairn et al. 2005; Russell and McGowan 2005).

Although agriculture was well established from the start of Çatalhöyük's occupation, analyses from in situ charred remains in storage contexts from burnt buildings have suggested that in addition to cereals, wild plants including nuts, fruits, crucifers, and possibly wild grasses were a more regular part of subsistence practices than previously thought (Bogaard et al. 2009; Fairbairn et al. 2007). Although there is no storage evidence for wetland tubers and nutlets, these types of plant foods may have been gathered and used directly, and the use of *Bolboschoenus* syn. *Scirpus* as a food source has been shown ethnographically (Wollstonecroft et al. 2008). During the seventh millennium B.C. at Çatalhöyük there seems to have been considerable continuity in many datasets, however, with some changes occurring in later levels across various material culture and environmental datasets (Asouti 2013; Düring 2006; Russell et al. 2013). Some variations occurred around Hodder's site phases South P and 4040 G (approximately 6500 cal B.C. based on the latest available dating and equivalent to Mellaart Level V). These included the appearance of new categories of outdoor open spaces with fire-spots as well as large external ovens (Bogaard et al. 2014; Regan et al. 2008). There were further distinct changes in architecture and material culture from the TP Area between 6250–6000 cal B.C. (Marciniak and Czerniak 2008) and into the West Mound. At the end of the seventh millennium B.C., a possible occupation gap between the East and West mounds was attributed to climate change (Weninger et al. 2006, but see Biehl and Rosenstock this vol.); however, there is now evidence of continuity of occupation between the East and West Mounds (Biehl et al. 2012; this volume). For much of the seventh millennium cal B.C. Çatalhöyük was situated alone in the broader landscape, but between 6000–5500 cal B.C. there was a general regional shift to a more evenly distributed settlement pattern (Baird 2002). Such changes in settlement patterns between the seventh and sixth millennia may have been due to a range of social and economic factors (Asouti 2009).

THE PHYTOLITH EVIDENCE

Phytolith analyses provide unique information about domestic and wild plant exploitation strategies. In monocotyledons (grasses, sedges, and palms), soluble silica is taken up by plants from groundwater and deposited within and between epidermal plant cells (Piperno 2006:5). Depending upon environmental conditions, such as aridity and clay-rich poorly drained soils, entire areas of epidermal tissue can become silicified forming "silica-skeletons" (Rosen 1992). Phytoliths are particularly abundant in grasses and sedges and to a lesser extent certain woody dicotyledons (trees and shrubs). Phytoliths are released into soils through burning and decay, allowing the study of plant use in charred and non-charred contexts. Phytoliths frequently indicate different plant parts, enabling the spatial study of crop processing, as well as the use of leaves and stems for weaving, construction, and fuel (Ryan 2011).

At Çatalhöyük, sediment samples were collected from floors and features within buildings, such as storage bins and hearths, as well as rubbish areas (middens) and external fire-spots. Samples were analyzed at both the scale of the excavation unit (bulk), and additionally smaller microscale samples were taken from within these units—for example, horizontally across occupation surfaces and vertically from microstratigraphy visible within some midden sections. Samples were taken from the various excavated areas across the East Mound. These include the 4040 Area excavations, which were designed to investigate social geography through the excavation of adjacent houses and middens (Farid 2004). The South Area has excavated sections through to the base of the mound, and samples from this area yielded information on temporal trends in plant use. The TP Area, which is close to, but not yet stratigraphically linked to the South Area, has the latest East Mound occupation levels (ca. 6250–6000 cal B.C.).

Methods used to process sediments are described in Rosen (2005). Samples were also analyzed from macroscopically visible silica skeletons that preserved traces of basketry (matting, baskets, and cordage), plants used as brick temper, and other layers of non-burnt plant materials. Two methods were used to investigate potential cereal-growing environments: (1) numbers of cells in conjoined wheat husk silica skeletons were counted following Rosen and Weiner (1994), and (2) following a recent method developed by Jenkins et al. (2011), proportions of grass short cells to dendritic long cells (from husks) were calculated from phytolith assemblages extracted from bin contexts. Since these grass single cells are not identifiable to genera, bin sediments were tested, because these were more likely to contain cereal grasses (or grass weeds growing within the same habitat), while other contexts may contain various grasses from multiple taphonomic pathways (such as dung).

Here we discuss results with a focus on indications for changing patterns of plant exploitation, and their relationship to archaeological and climate evidence. Additionally, we consider overall subsistence strategies to examine agricultural risk buffering. Further detailed results from samples analyzed are presented in Rosen (2005) and Ryan (2013 and 2011).

Domesticated and Wild Plant Foods

Cereal agriculture, in particular wheat (*Triticum* sp.), was evident in results from all sample contexts, including from middens, floors, ovens, and storage bins. Samples from storage bins can act as the most direct indicators of diet, and data are summarized in Figure 5.1. In addition to wheat, sediments from storage bins included phytoliths from barley, wild grasses, as well as possible phytoliths from nuts. In a comparison between phytoliths and charred plant remains present in bin units from Building 52 (4040 Area), certain dicotyledonous phytoliths occurred in highest proportions where charred almonds were found (Bogaard et al. 2009). These phytoliths also occurred in large proportions in

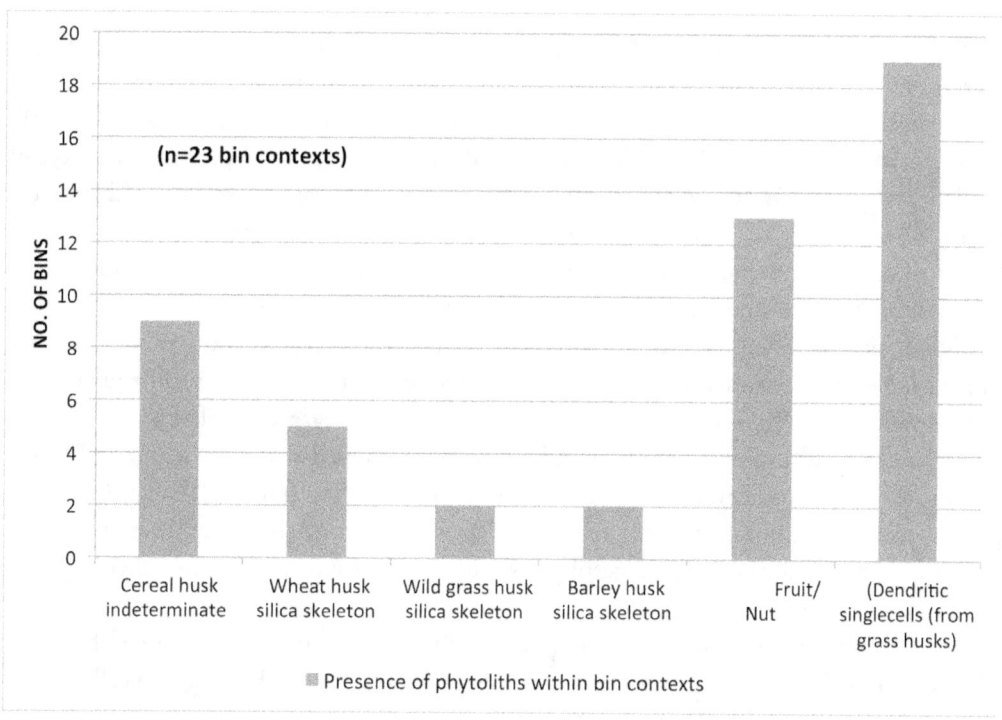

Figure 5.1. Histogram showing phytoliths from storage bins (East Mound).

samples from several bins in non-burnt buildings such as Building 65 in the South Area. It is possible that the incorporation of wild plant foods among stored provisions acted as a buffer against lean years and this approach may have played a role in the long-term success of economic strategies.

In Building 63 in the Istanbul Area, sediments from two bins containing charred barley notably did not contain barley phytoliths. This is not surprising since barley found in the macrobotanical record (until the latest East Mound levels) is all naked barley (*H. vulgare var. nudum*), the six-row form that occurs throughout the sequence. In addition there is a two-row naked barley from midway through the East Mound occupation (Bogaard et al. 2013). Such grains are stored naked without the chaff parts (lemma, palea) that produce diagnostic phytoliths. Some barley phytoliths found in other bins may come from light chaff detritus; although light chaff is removed from naked barley in the early stages of crop processing, some fragmentary debris may nevertheless have made its way onto the site. It is also likely that some of the barley phytoliths present in other bin contexts may be from wild weedy species. This is consistent with analyses from a wider range of contexts (floors, ash, rubbish) showing that barley phytoliths had a high correlation with wild weed grasses and a very low co-occurrence wheat phytoliths (Rosen 2005).

BARLEY: CHANGING PATTERNS OF RISK MANAGEMENT?

Barley husk silica skeletons were found more frequently in sediments from fire-installations (ovens, hearths) and fire-spots in later East Mound site levels (from South S onward and throughout TP). Figure 5.2 shows the average numbers of barley husk phytoliths per gram sediment from fire-spots, ovens and hearths, and Figure 5.3 an example of a barley husk silica skeleton. Increasing quantities and frequencies of barley husk phytoliths in these contexts most likely reflect the introduction of a new barley variety, two-row hulled barley (*H. distichum*). Phytoliths from hulled barley are more likely (than those from naked barley) to be found in domestic contexts such as storage bins and hearths because parts of the husk that produce diagnostic phytoliths (the lemma and palea) stay attached to the grain after harvest. In the macro-remains, naked six-row barley was found throughout the East Mound, and two-row naked barley was also present. Two-row hulled barley was found from the latest Neolithic levels as well as the subsequent sixth millennium West Mound (Bogaard and Charles 2010; Bogaard et al. 2013). There is also charred plant evidence for the uptake of two-row hulled barley in the Lake District to the West of Central Anatolia, during a similar time frame (6400–6100 cal B.C.), and in the sixth millennium at other sites in Central Anatolia (Asouti and Fairbairn 2002; Martinoli and Nesbitt 2003) suggesting a shift toward using hulled barley that occurred at a broad regional scale.

There are several possible explanations for the presence of two-row hulled barley in the latest East Mound site phases and subsequent West Mound. Ethnographically, wheat and barley can be planted together to reduce risk since barley is less water sensitive in comparison to wheat (Halstead and Jones 1989; Jones and Halstead 1995). Two-row hulled barley is better adapted to aridity than six-row barley (Riehl 2009), and it may

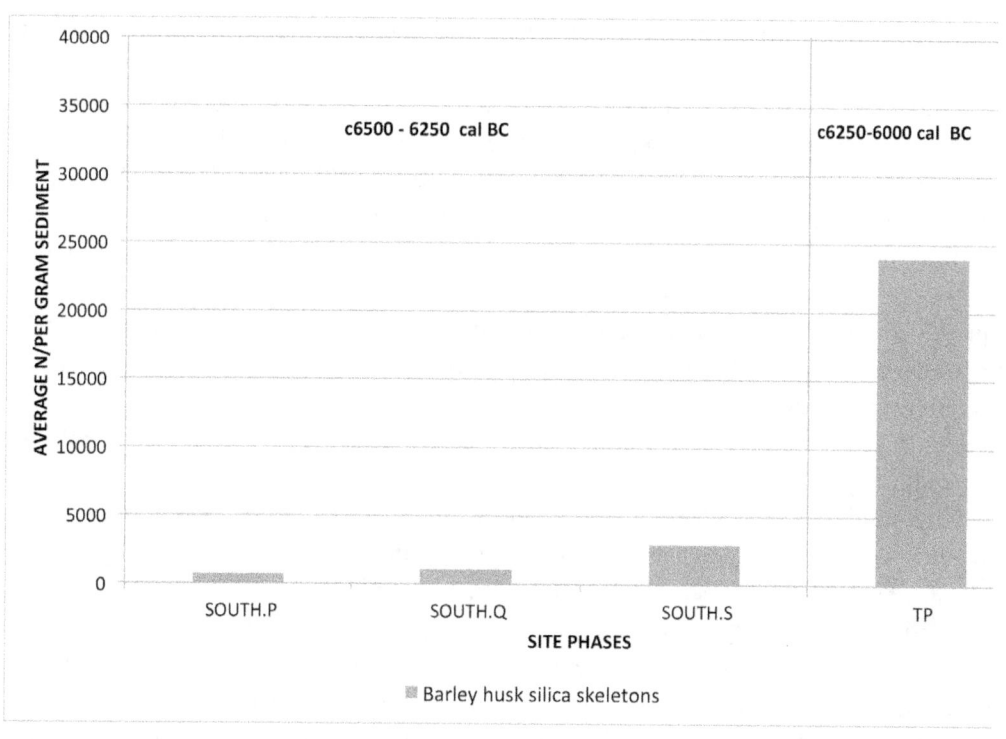

FIGURE 5.2. Histogram showing *Hordeum* sp. (barley) husk silica skeletons over time (East Mound).

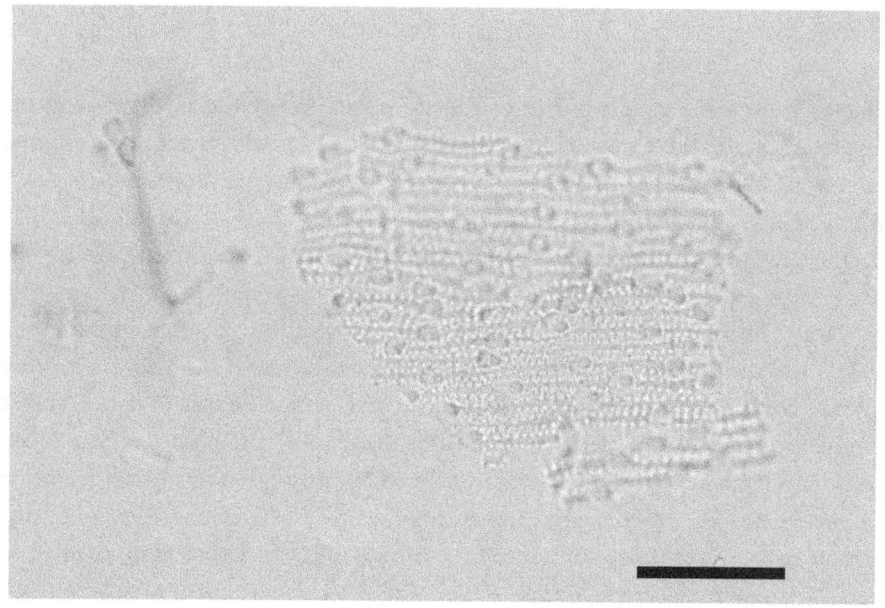

FIGURE 5.3. Image of *Hordeum* sp. (barley) silica skeleton (East Mound), scale 100 microns.

be significant that the introduction of this cereal coincides within a similar timeframe to the 8.2 ka event. Hulled barley may have been used for a new purpose, for instance as animal fodder. Chaff from hulled barley can add roughage to fodder, while in contrast naked barley is more commonly ethnographically eaten as food rather than fed to animals (Dorian Fuller, personal communication 2009). Hillman et al. (1997) point out that grazing animals avoid grasses with greater amounts of long brittle awns as they can be flesh penetrating, and from this perspective six-row barley is a less suitable fodder crop than two-row. Interestingly, possible cereal straw phytoliths were found more frequently in later East mound site phases as well as in the samples analyzed from the West mound—potentially reflecting an association with the new type of barley and its use; the presence of the whole plant would be consistent with the use of the crop as animal fodder. It is also possible that the new barley could act as either human or animal food depending upon the overall success of a given harvest (Halstead and Jones 1989; Jones and Halstead 1995). Hulled barley is problematic as a food cereal because it is difficult to remove the lemma and palea from the grains, and is generally avoided by modern farmers. However it is possible, and ethnographic research in northern Sudan (undertaken in 2014 by Philippa Ryan, unpublished) has shown that hulled barley was grown as a human food cereal until the mid-twentieth century. Although there are various different possible reasons for the uptake of this cereal, the use of a new barley species indicates a change within cropping practices that increased diversification hence enhancing resilience in the farming system, and which occurred around the time of the 8.2 ka event. Nevertheless, at Çatalhöyük such a shift may have been part of wide-reaching interrelated social and economic changes between the seventh and sixth millennia B.C.

Exploiting Varied Microenvironments as a Form of Risk Management

The possibility of some cereals growing at wetter edges (as well as in dryer areas of land) was suggested by the presence of phytoliths extracted from several excavation units, with silica-skeletons consisting of between 100–327 cells. Additionally, a number of examples of entirely whole silicified cereal husks (with many hundreds of conjoining cells) were found. The presence of silica skeletons with greater than 100 conjoined cells is suggested by Rosen and Weiner (1994) as an indicator of high water availability, and poorly drained clay-rich soils. Although silica skeletons are broken down by laboratory processing (Shillito 2011), the survival of examples with greater than 100 cells from several units shows that at least some large enough examples (to suggest large amounts of water availability) can survive processing. It is perhaps surprising that more examples of wheat husk silica skeletons with large numbers of conjoined cells were not present in phytolith assemblages from a greater number of excavated archaeological units. Phytolith assemblages from the majority of sediment samples contained wheat husk silica skeletons composed of smaller numbers of conjoined cells potentially suggesting dryer environments; however it is also possible that taphonomic factors such as trampling reduced the size of silica skeletons (Rosen 2005; Ryan 2013). Nevertheless, the variation in silica skeleton size fits with the possibility of crops growing in a patchy wet and dry-land environment. To address the potential issue that smaller numbers of conjoined cells may indicate lower water availability or multicell breakdown, a new method was established by Jenkins (2011) to compare ratios of grass single-cells. The application of this new method to samples from storage bins suggested that some assemblages contained cereals suggestive of wetter environments, while others suggested dryer conditions. Together, these two methodologies suggested cereals growing in a range of dryer and wetter environments, which is consistent with the picture of a mixed topographic environment and the likelihood that growing conditions were probably highly variable spatially (Ryan 2013). That crops were not consistently growing in wet environments is also shown by the association between dry-land weedy flora and crops in the charred macrobotanical record (Bogaard et al. 2013). In the mixed dry and wetter areas around the site, water availability to areas used for crop growing may have varied annually in relation to rainfall and flood levels. Using a range of cereals across the varied topography would have helped to buffer against these yearly changes.

Phytolith evidence showed wide-ranging use of grasses, reeds, and sedges from both dry and wetland habitats. Basketry was commonly made from wetland grasses and sedges, while evidence from impressions in mudbricks indicated the secondary use of cereal chaff (husk parts not the straw), the use of *Phragmites* reeds, as well as wild grasses and sedges. Actual management of *Phragmites* stands for brick temper is suggested by the regularity and size of impressions found in mudbricks (Mira Stevanovic, personal communication 2011). The use of wild and domestic plants together within construction and storage contexts suggests a multitiered subsistence strategy. The relatively infrequent presence of cereal straw in sediment samples in most of the Çatalhöyük East mound sequence emphasizes the economic importance of nearby wetland wild plant leaves and stems, and perhaps that cereals were harvested high on the stem, with straw left behind in the fields.

Wetland plants, including sedges, *Phragmites* reeds, and wild grasses such as *Bromus*, were also a common component of phytolith assemblages from ashy fuel deposits. Sedges and grasses (and notably not reeds) were found in phytoliths extracted directly from dung pellets. Phytoliths from C_4 wild grasses were associated with penning deposits and dung-fuel rich fire-spots. These may be either from *Aeluropus* (a wetland grass genera) or *Sporobolus*, since these were found associated with dung in the macrobotanical record, along with various wetland taxa (Bogaard et al. 2013, 2014). A smaller number of phytolith samples analyzed from the West Mound have shown the continued presence of wetland plants, including the use of sedges for matting and ashy deposits containing *Phragmites* reeds.

Phragmites Reeds: Changing Patterns of Other Economically Important Plants

Given the climate data for an 8.2 ka event we could expect wetland plants to decrease at Çatalhöyük over time. Surprisingly, this is not the case. Wetland plants were present throughout all levels at the East Mound, and quantities of *Phragmites* exploited increased in the later phases. This trend was seen in the analysis of phytoliths from midden sediments analyzed throughout the East Mound sequence from the South P Area onward, and particularly from South Q (Figures 5.4 and 5.5). The trend was also recognized in analyses from fuel deposits sampled from ovens, hearths and fire-spots. The increase in *Phragmites* reeds occurred at the same time as other site-wide alterations, such as new configurations of spatial use including large external open activity areas dating from this time (Regan et al. 2008). This increase over time is in part related to cultural selection, but it is also possible that *Phragmites* was more abundant due to its well-known ability to outcompete many other wetland plants, reducing biodiversity (Ailstock 2001; Cronk and Fennessy 2001; Silliman 2004). *Phragmites* is often accompanied by other co-dominant invasive species such as *Scirpus sensu latu* and *Typha*, and *Bolboschoenus syn. Scirpus* is one of the most prominent taxa in the macrobotanical record (Fairbairn et al. 2005).

One competitive trait of *Phragmites* is its tolerance of wide-ranging soil moisture conditions (Chambers 2003), and once dominant, *Phragmites* is also known to lower water tables (Shay and Shay 1986). It is possible that the human use of wetland areas at Çatalhöyük created the right conditions for *Phragmites* invasions, with exploited wetland taxa creating room for invasive plant expansion. Pits excavated around the site by inhabitants may have created additionally watery areas and bare sediments for *Phragmites* invasion, and in general, *Phragmites* reed invasions may be one example of human and thing "entanglements" at the site (Hodder 2011). The changing biodiversity and physical form of wetland areas would have fundamentally altered plant and land-use opportunities,

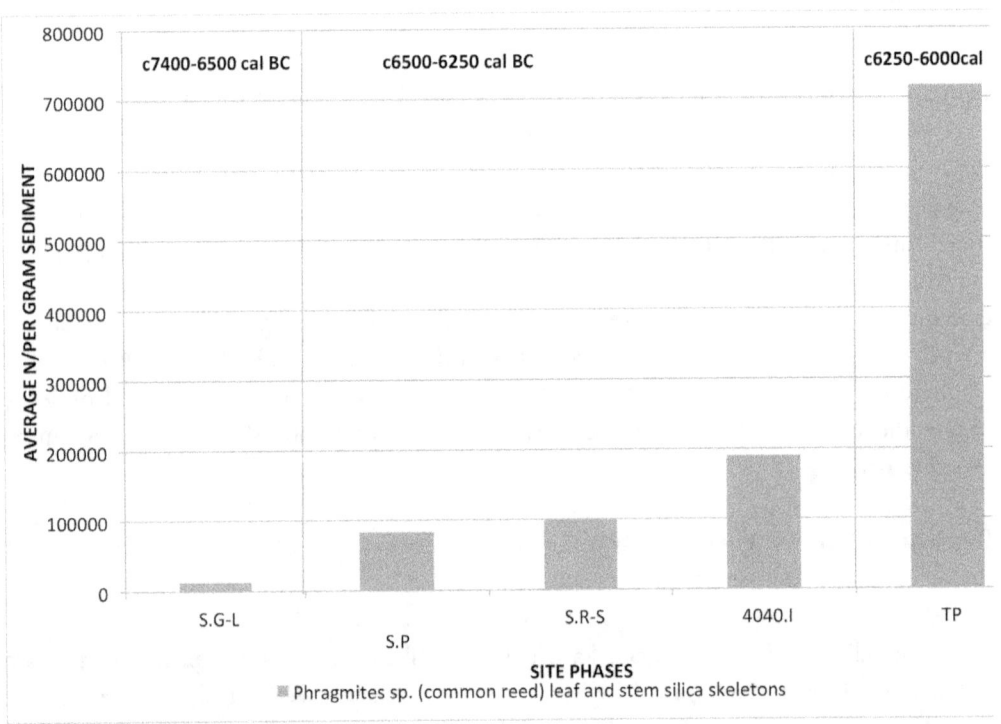

FIGURE 5.4. Histogram showing *Phragmites* sp. (common reed) phytoliths over time (East Mound).

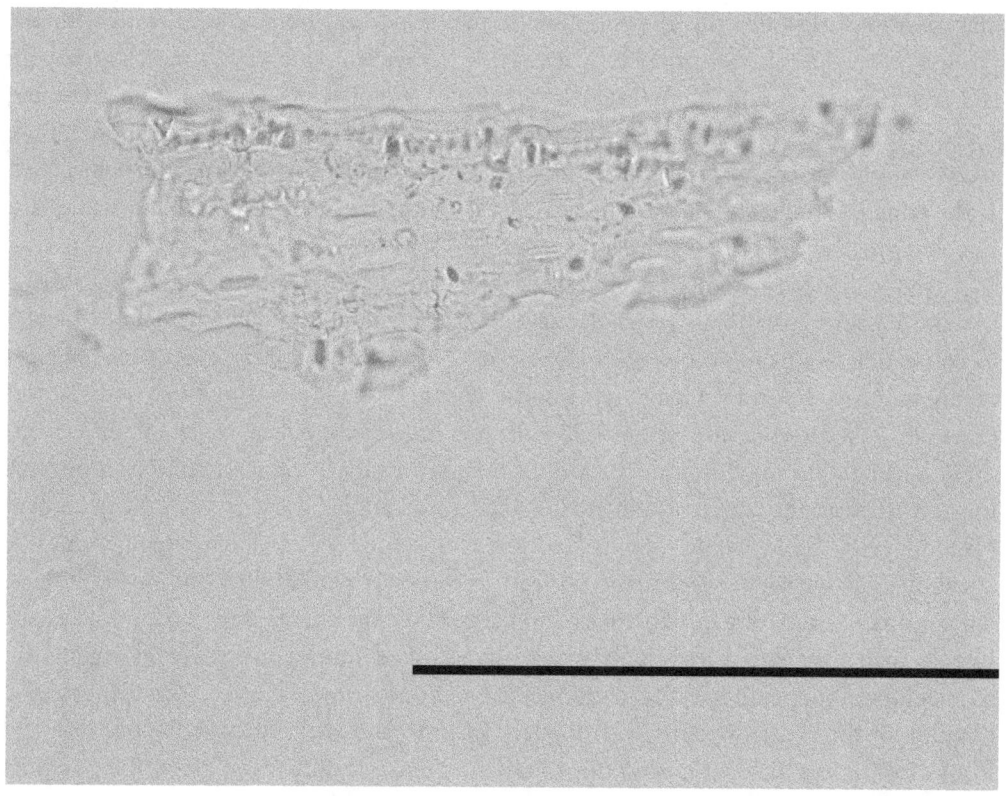

FIGURE 5.5. Image of *Phragmites* sp. (common reed) multicell phytolith, scale 100 microns (East Mound).

including the distribution of microenvironments used for cereal growing. It is difficult to interpret the role of climate change from changing proportions of wetland plant taxa, because increases in invasive taxa and changing biodiversity may be connected to anthropogenic factors. Furthermore, the dramatic increase in reeds seems to begin within the 300 years prior to the 8.2 ka event. It is also possible that cooler temperatures during the 8.2 ka event timeframe may have reduced evaporation and actually led to an expansion of watery areas during spring flooding.

Conclusions

The phytolith data suggest that both wetland and dry-land plants were routinely exploited at Çatalhöyük throughout the Neolithic occupation, and that overall there was a very consistent pattern of plant and land use. The introduction of hulled barley roughly coincided with the 8.2 ka event, however, it is difficult to assess how and why this cereal fits within crop choice decision making. Whether or not the cereal was used as human food or animal fodder, it is interesting that two-row barley is more drought tolerant than six-row barley. Equally, the adoption of the new cereal may be connected with the various other changes occurring at Çatalhöyük. However, whichever the exact purpose and cause, the adoption of the new crop suggests adaptive flexibility within buffering systems and Marston (2011) describes such increased diversification within crop species as the "most efficient form of risk buffering." It is possible that buffering systems, such as the targeting of diverse environmental microhabitats for cereal growing and wild plant exploitation, played a role in the long-term success of the settlement.

The evidence for temporal trends in plants present (barley and reeds) is supported by co-occurring changes in other datasets. However, this also highlights the difficulty in correlating the plant record with climate data since any shifts in plant use were also deeply embedded within wide-ranging social changes. We also suggest that increases or decreases in the record for wetland taxa do not necessarily reflect wetter or dryer conditions, but rather may indicate alterations in biodiversity in response to anthropogenic impacts. Increasing paleoenvironmental evidence for an 8.2 ka event does suggest that Neolithic populations would have had to cope with some level of climate change, and it is likely that any changes in seasonality of frost-free days or rainfall distribution would have impacted cereal agriculture, particularly in marginal environments. However, imprecisions such as the varied signals for the 8.2 ka event recorded from across different regions (in terms of amplitude and seasonal patterns), the debates surrounding the exact timing and duration of the event, together with the lack of independent information about how the Konya region was affected during this timeframe, make it difficult to assess the potential impact of climate on plant-based subsistence systems as distinct from other social and ecological factors. Nevertheless, it seems possible that flexible and stable diversification systems played a role in the durability of the settlement.

Acknowledgments

We would like to thank the Çatalhöyük team for their intellectual support and the many stimulating discussion about Neolithic life on the Konya Plain. Special thanks are due to Ian Hodder and Shahina Farid for their help and support in the field. Many thanks also go to Amy Bogaard, Mike Charles, Dorian Fuller, Katheryn Twiss, and Karen Wright.

References Cited

Ailstock, M. S. 2000 Adaptive Strategies of Common Reed Phragmites australis. In *The Role of Phragmites in the Mid-Atlantic Region*, edited by J. Rooth April 17, Chesapeake Bay National Estuarine Research Reserve in Maryland. Maryland Department of Natural Resources.

Alley, R. B., P. A. Mayewski, T. Sowers, M. Stuiver, K. C. Taylor, and P. U. Clark 1997 Holocene Climate Instability: A Prominent, Widespread Event 8200 yr ago. *Geology* 25(6):483–486.

Asouti, E. 2009 The Relationship between Early Holocene Climate Change and Neolithic Settlement in Central Anatolia, Turkey: Current Issues and Prospects for Future Research. *Documenta Praehistorica* 36:1–5.

Asouti, E. 2013 Woodland Vegetation, Firewood Management and Woodcrafts at Neolithic Çatalhöyük. In *Humans and Landscapes of Çatalhöyük: Reports from the 2000–2008 Seasons*, edited by I. Hodder, pp. 129–160 Monographs of the Cotsen Institute of Archaeology [volume 8]. University of California at Los Angeles, Los Angeles.

Asouti, E., and A. Fairbairn 2002 Subsistence Economy in Central Anatolia during the Neolithic. In *The Neolithic of Central Anatolia. Internal Developments and External Relations during the 9th–6th Millennia Cal. BC. Proceedings of the International CANeW Table Ronde Istanbul, 23–24 November 2001*, edited by F. Gerard and L. Thissen, pp. 181–192. Ege Yayinlari, Istanbul.

Baird, D. 2002 Early Holocene Settlement in Central Anatolia: Problems and Prospects as Seen from the Konya Plain. In *The Neolithic of Central Anatolia*, edited by F. Gerard and L. Thissen, pp. 139–159. Ege Yayinlari, Istanbul.

Bar-Matthews, M., A. Ayalon, and A. Kaufman 1997 Late Quaternary Paleoclimate in the Eastern Mediterranean Region from Stable Isotope Analysis of Speleothems at Soreq Cave, Israel. *Quaternary Research* 47:155–168.

Biehl, P. F., I. Franz, S. Ostaptchouk, D. Orton, J. Rogasch, and E. Rosenstock 2012 One Community and Two Tells: The Phenomenon of Relocating Tell Settlements at the Turn of the 7th and 6th Millennia in Central Anatolia. In *Socio-Environmental Dynamics over the Last 12,000 Years: The Creation of Landscapes*, edited by J. Müller, pp. 53–66. Offa, Kiel.

Bogaard, A., M. Charles, K. C. Twiss, A. Fairbairn, N. Yalman, D. Filipovic, G. A. Demirergi, F. Ertug, N. Russell, and J. Henecke 2009 Private Pantries and Celebrated Surplus: Storing and Sharing Food at Neolithic Çatalhöyük, Central Anatolia. *Antiquity* 83:649–668.

Bogaard, A., M. Charles, A. Livarda, M. Ergun, D. Filipovic, and G. Jones 2013 The Archaeobotany of Mid-Later Neolithic Çatalhöyük. In *Humans and Landscapes of Çatalhöyük: Reports from the 2000–2008 Seasons*, edited by I. Hodder. Monographs of the Cotsen Institute of Archaeology [volume 8]. University of California at Los Angeles, Los Angeles.

Bogaard, A., and M. Charles 2010 Archaeobotany Preliminary Report on West Mound Trenches 5, 6 and 7. In *Çatalhöyük Archive Report 2010*, pp. 66-69, Electronic document http://www.catalhoyuk.com/archive_reports.

Bogaard, A., P. Ryan, N. Yalman, E. Asouti, K. C. Twiss, C. Mazzucato and S. Farid 2014 Assessing Outdoor Activities and Their Social Implications at Çatalhöyük. In *Integrating Çatalhöyük: Themes from the 2000–2008 Seasons*, edited by Ian Hodder. Cotsen Institute of Archaeology University of California at Los Angeles [volume 10], Los Angeles.

Boyer, P., N. Roberts, and D. Baird 2006 Holocene Environment and Settlement on the Carsamba Alluvial Fan, South-Central Turkey: Integrating Geoarchaeology and Archaeological Field Survey. *Geoarchaeology* 21:675–698.

Cappers, T., S. Bottema, H. Woldring, H. van der Plicht, and H. J. Streurman 2002 Modeling the Emergence of Farming: Implications of the Vegetation Development in the Near East during the Pleistocene-Holocene Boundary. In *The Dawn of Farming in the Near East*, edited by R. Cappers and S. Bottema, pp. 3–14. Studies in Early Near Eastern Production, Subsistence, and Environment 6. Ex Oriente, Berlin.

Cessford, C. 2005 Absolute Dating at Çatalhöyük. In *Changing Materialities at Çatalhöyük: Reports from the 1995–1999 Seasons*, edited by I. Hodder, pp 65–100. McDonald Institute for Archaeological Research, Cambridge.

Chambers, R., D. Osgood, D. Bart, and F. Montalto 2003 *Phragmites australis* Invasion and Expansion in Tidal Wetlands: Interactions among Salinity, Sulfide, and Hydrology. *Estuaries and Coasts* (26):398–406.

Charles, M., C. Doherty, E. Asouti, A. Bogaard, E. Henton, C. Spencer Larsen, C. B. Ruff, P. Ryan, J. W. Sadvari, and K. C. Twiss 2014 Landscape and Taskscape at Çatalhöyük:

An Integrated Perspective. In *Integrating Çatalhöyük: Themes from the 2000–2008 Seasons*, edited by I. Hodder. Monographs of the Cotsen Institute of Archaeology [volume 10]. University of California at Los Angeles, Los Angeles.

Cronk, J., and M. Fennessy 2001 *Wetland Plants: Biology and Ecology*. Lewis Publishers, London.

Demirergi, G. A., K. Twiss, A. Bogaard, L. Green, P. Ryan, and S. Farid 2014 Of Bins, Basins, and Banquets: Storing, Handling, and Sharing Food at Neolithic Çatalhöyük. In *Integrating Çatalhöyük: Themes from the 2000–2008 Seasons*, edited by I. Hodder. Monographs of the Cotsen Institute of Archaeology [volume 10]. University of California at Los Angeles, Los Angeles.

Doherty, C., M. Charles, and A. Bogaard 2007 Preliminary Sediment Coring to Clarify "Clay" Sources and Potential Land Use Around Çatalhöyük. In *Çatalhöyük Archive Report 2007*, pp. 381–390. Electronic document, http://www.catalhoyuk.com/archive_reports.

Doherty, C., M. Charles, and A. Bogaard 2008 Landscape Coring. In *Çatalhöyük Archive Report* pp. 263–272. Electronic document, http://www.catalhoyuk.com/archive_reports.

Düring, B. S. 2006 Constructing Communities: Clustered Neighbourhood Settlements of the Central Anatolian Neolithic, ca. 8500–5500 Cal. BC. Nederlands Instituut voor het Nabije Oosten, Leiden.

Fairbairn, A., D. Martinoli, A. Butler, and G. Hillman 2007 Wild Plant Seed Storage at Neolithic Çatalhöyük East, Turkey. *Vegetation History and Archaeobotany* 16:467–479.

Fairbairn, A., J. Near, and D. Martinoli 2005 Macrobotanical Investigation of the North, South, and Kopal Area Excavations. In *Inhabiting Çatalhöyük, Reports from the 1995–1999 Seasons*, edited by I. Hodder, pp. 137–202. Mcdonald Institute for Archaeological Research, London.

Farid, S. 2004 Excavations Overview. In *Çatalhöyük Archive Report 2004*. Electronic document, http://www.catalhoyuk.com/archive_reports/2004/ar04_04.html.

Fontugne, M., C. Kuzucuoglu, M. Karabiyikoglu, C. Hatté, and J. F. Pastre 1999 From Pleniglacial to Holocene: A ^{14}C Chronostratigraphy of Environmental Changes in the Konya Plain, Turkey. *Quaternary Science Reviews* 18:573–591.

Hajar, L., C. Khater, and R. Cheddadi 2008 Vegetation Changes during the Late Pleistocene and Holocene in Lebanon: A Pollen Record from the Bekaa Valley. *The Holocene* 18:1089–1099.

Halstead, P. 1989 The Economy Has a Normal Surplus: Economic Stability and Social Change among Early Farming Communities of Thessaly, Greece. In *Bad Year Economics*, edited by P. Halstead and J. O'Shea, pp. 68–80. Cambridge University Press, Cambridge.

Halstead, P., and G. Jones 1989 Agrarian Ecology in the Greek Islands: Time Stress, Scale, and Risk. *Journal of Hellenic Studies* 109:41–55.

Halstead, P., and J. O'Shea 1989 Introduction: Cultural Response to Risk and Uncertainty. In *Bad Year Economics*, edited by P. Halstead and J. O'Shea, pp. 1–7. Cambridge University Press, Cambridge.

Hillman, G., A. J. Legge, and P. A. Rowley-Conwy 1997 On the Charred Seeds from Epipalaeolithic Abu Hureya: Food or Fuel? *Current Anthropology* 38:651–655.

Hodder, I. 2011 Human-Thing Entanglement: Towards an Integrated Archaeological Perspective. *Journal of the Royal Anthropological Institute (N.S.)* 17:154–177.

Jenkins, E. L., K. Jamjoum, and S. Nuimat 2011 Irrigation and Phytolith Formation: An Experimental Study. In *Water, Life, and Civilization: Climate, Environment, and Society in the Jordan Valley*, edited by S. J. Mithen, and E. Black, pp. 347–372. Cambridge University Press, Cambridge.

Jones, G., and P. Halstead 1995 Maslins, Mixtures, and Monocrops: On the Interpretation of Archaeobotanical Crop Samples of Heterogeneous Composition. *Journal of Archaeological Science* 22:103–114.

Kuijt, I. 2000 People and Space in Early Agricultural Villages: Exploring Daily Lives, Community Size, and Architecture in the Late Pre-Pottery Neolithic. *Journal of Anthropological Archaeology* 19:75–102.

Kuzucuoğlu, C. 2002 The Environmental Frame in Central Anatolia from the 9th to the 6th millennia cal. BC. In *The Neolithic of Central Anatolia. Internal Developments and External Relations During the 9th–6th Millenia Cal BC. Proceedings of the International CANeW Table Ronde Istanbul, 23–24 November 2001*, edited by F. Gérard and L. Thissen, pp. 33–58. Ege Yayinlari, Istanbul.

Marciniak, A., and L. Czerniak 2008 The Excavations of the TP (Team Poznań) Area in the 2008 Season. *Çatalhöyük Archive Report* 2008, pp. 73–82. Electronic document, http://www.catalhoyuk.com/archive_reports.

Marston, J. M. 2011 Archaeological Markers of Agricultural Risk Management. *Journal of Anthropological Archaeology* 30:190–205.

Martinoli, D., and M. Nesbitt 2003 Plant Stores at Pottery Neolithic Höyücek, Southwest Turkey. *Anatolian Studies* 53:17–32.

Migowski, C., M. Stein, S. Prasad, J. F. W. Negendank, and A. Agnon 2006 Holocene Climate Variability and Cultural Evolution in the Near East from the Dead Sea Sedimentary Record. *Quaternary Research* 66:421–431.

Moore, A. M. T., and G. Hillman 1992 The Pleistocene to Holocene Transition and Human Economy in Southwest Asia: The Impact of the Younger Dryas. *American Antiquity* 57:482–494.

O'Shea, J. 1989 The Role of Wild Resources in Small Scale Agricultural Systems: Tales from the Lakes and the Plains. In *Bad Year Economics*, edited by P. Halstead and J. O'Shea, pp. 57–67. Cambridge University Press, Cambridge.

Peyron, O., S. Goring, I. Dormoy, U. Kotthoff, J. Pross, J.-L. de Beaulieu, R. Drescher-Schneider, B. Vannière, and M. Magny 2011 Holocene Seasonality Changes in the Central Mediterranean Region Reconstructed from the Pollen Sequences of Lake Accesa (Italy) and Tenaghi Philippon (Greece). *The Holocene* 21:131–146.

Piperno, D. R. 2006 *Phytoliths: A Comprehensive Guide for Archaeologists and Paleoecologists*. Altamira Press, Lanham.

Pross, J., U. Kotthoff, U. C. Müller, O. Peyron, I. Dormoy, G. Schmiedl, S. Kalaitzidis, and A. M. Smith 2009 Massive Perturbation in Terrestrial Ecosystems of the Eastern Mediterranean Region Associated with the 8.2 kyr B.P. Climatic Event. *Geology* 37:887–890.

Riehl, S. 2009 Archaeobotanical Evidence for the Interrelationship of Agricultural Decision-Making and Climate Change in the Ancient Near East *Quaternary International* 197:93–114.

Regan, R., F. Sadarangani, and J. Taylor 2008 Building 75, Middens and Buildings 80 & 70. In *Çatalhöyük Archive Report* 2008, pp. 62–73. Electronic document, http://www.catalhoyuk.com/archive_reports.

Roberts, N., P. Boyer, and J. Merrick 2007 The KOPAL On-Site and Off-Site Excavations and Sampling. In *Excavating Çatalhöyük: Reports from the 1995–1999 Seasons*, edited by I. Hodder, pp. 553–570. McDonald Institute of Archaeological Research, Cambridge.

Roberts, N., J. M. Reed, M. J. Leng, C. Kuzucuoglu, M. Fontugne, J. Bertaux, H. Woldring, S. Bottema, S. Black, E. Hunt, and M. Karabiyikoglu 2001 The Tempo of Holocene

Climatic Change in the Eastern Mediterranean Region: New High-Resolution Crater-Lake Sediment Data from Central Turkey. *The Holocene* 11:721–736.

Roberts, N., and A. Rosen 2009 Diversity and Complexity in Early Farming Communities of Southwest Asia: New Insights into the Economic and Environmental Basis of Neolithic Çatalhöyük. *Current Anthropology* 50:393–402.

Rohling, E. J., and H. Palike 2005 Centennial-scale Climate Cooling with a Sudden Cold Event around 8,200 Years Ago. *Nature* 434:975–979.

Rollefson, G. O., and I. Köhler-Rollefson 1992 Early Neolithic Exploitation Patterns in the Levant: Cultural Impact on the Environment. *Population & Environment* 13:243–254.

Rosen, A., and S. Weiner 1994 Identifying Ancient Irrigation: A New Method Using Opaline Phytoliths from Emmer Wheat. *Journal of Archaeological Science* 21:132–135.

Rosen, A. M. 1992 Preliminary Identification of Silica Skeletons from Near Eastern Archaeological Sites: An Anatomical Approach. In *Phytolith Systematics*, edited by G. Rapp and S. Mulholland, pp. 129–148. Plenum Press, New York and London.

Rosen, A. M. 2005 Phytolith Indicators of Plant and Land Use at Çatalhöyük. In *Inhabiting Çatalhöyük, Reports from the 1995–1999 Seasons*, edited by I. Hodder, pp. 203–212. McDonald Institute for Archaeological Research, Cambridge.

Rossignol-Strick, M. 1993 Late Quaternary Climate in the Eastern Mediterranean Region. *Paléorient* 19:135–152.

Rossignol-Strick, M. 1995 Sea-Land Correlation of Pollen Records in the Eastern Mediterranean for the Glacial-Interglacial Transition: Biostratigraphy versus Radiometric Time-Scale. *Quaternary Science Reviews* 14:893–915.

Rossignol-Strick, M. 1999 The Holocene Climatic Optimum and Pollen Records of Sapropel 1 in the Eastern Mediterranean, 9000–6000 BP. *Quaternary Science Reviews* 18:515–530.

Ryan, P. 2010 Diversity of Plant and Land Use during the Near Eastern Neolithic: Phytolith Perspectives from Çatalhöyük. Unpublished PhD thesis, Institute of Archaeology, University College, London.

Ryan, P. 2011 Plants as Material Culture in the Near Eastern Neolithic: Perspectives from the Silica Skeleton Artifactual Remains at Çatalhöyük. *Journal of Anthropological Archaeology* 30:292–305.

Ryan, P. 2013 Plant Exploitation from Household and Landscape Perspectives: The Phytolith Evidence. In *Humans and Landscapes of Çatalhöyük: Reports from the 2000–2008 Seasons*, edited by I. Hodder, pp. 161–188. Monographs of the Cotsen Institute of Archaeology. University of California at Los Angeles, Los Angeles.

Russell, N., K. C. Twiss, D. C. Orton, and A. G. Demirergi 2013 More on the Catalhoyuk Mammal Remains. In *Humans and Landscapes of Çatalhöyük: Reports from the 2000–2008 Seasons*, edited by I. Hodder, pp. 209–254. Monographs of the Cotsen Institute of Archaeology [volume 8]. University of California at Los Angeles, Los Angeles.

Russell, N. and K. McGowan 2005 The Çatalhöyük Bird Bones. In *Inhabiting Çatalhöyük: Reports from the 1995–1999 Seasons*, edited by I. Hodder, pp. 99–110. McDonald Institute for Archaeological Research, Cambridge.

Silliman, B. R., and M. D. Bertness 2004 Shoreline Development Drives Invasion of *Phragmites australis* and the Loss of Plant Diversity on New England Salt Marshes. *Conservation Biology* 18(5):1424–1434.

Sadori, L., S. Jahns, and O. Peyron 2011 Mid-Holocene Vegetation History of the Central Mediterranean. *The Holocene* 21:117–129.

Shay, J. and C. Shay 1986 Prairie Marshes in Western Canada, with Specific Reference to the Ecology of Five Emergent Macrophytes. *Canadian Journal of Botany* 64:443–454.

Shillito, L. M. 2011 Taphonomic Observations of Archaeological Wheat Phytoliths from Neolithic Çatalhöyük, Turkey, and the Use of Conjoined Phytolith Size as an Indicator of Water Availability. *Archaeometry* 53(3):631–641.

Wiersma, A. P., and H. Renssen 2006 Model-Data Comparison for the 8.2 ka BP Event: Confirmation of a Forcing Mechanism by Catastrophic Drainage of Laurentide Lakes. *Quaternary Science Reviews* 25:63–88.

Wick, L., G. Lemcke, and M. Sturm 2003 Evidence of Lateglacial and Holocene Climatic Change and Human Impact in Eastern Anatolia: High-Resolution Pollen, Charcoal, Isotopic, and Geochemical Records from the Laminated Sediments of Lake Van, Turkey. *The Holocene* 13:665–675.

Weninger, B., E. Alram-Stern, E. Bauer, L. Clare, U. Danzeglocke, O. Jöris, C. Kubatzki, G. Rollefson, H. Todorova, and T. Van Andel 2006 Climate Forcing Due to the 8200 Cal Yr BP Event Observed at Early Neolithic Sites in the Eastern Mediterranean. *Quaternary Research* 66:401–420.

Woldring, H. 2002a Climate Change and the Onset of Sedentism in Cappadocia. In *The Neolithic of Central Anatolia: Internal Developments and External Relations during the 9th–6th Millenia CAL BC*, edited by F. Gerard and L. Thissen, pp. 59–66. Ege Yayinlari, Istanbul.

Woldring, H. 2002b The Early-Holocene Vegetation of Central Anatolia and the Impact of Farming. In *The Dawn of Farming in the Near East. Studies in Early Near Eastern Production, Subsistance, and Environment 6, 1999*, edited by R. Cappers and S. Bottema, pp. 39–48. Ex Oriente, Berlin.

Wollstonecroft, M., P. Ellis, G. Hillman, and D. Fuller 2008 Advances in Plant Food Processing in the Near Eastern Epipalaeolithic and Implications for Improved Edibility and Nutrient Bioaccessibility: An Experimental Assessment of *Bolboschoenus maritimus* (L.) Palla (Sea Club-Rush). *Vegetation History and Archaeobotany* 17:19–27.

Wright, H., and J. Thorpe 2003 Climatic Change and the Origin of Agriculture in the Near East. In *Global Change in the Holocene*, edited by A. Mackay, R. Battarbee, J. Birks, and F. Oldfield, pp. 49–62. Oxford University Press, London.

Van Zeist, W., and S. Bottema 1991 Late Quaternary Vegetation of the Near East. In *Late Quaternary Vegetation of the Near East*. L. Reichert, Wiesbaden.

CHAPTER SIX

The 8.2 Event and the Neolithic Expansion in Western Anatolia

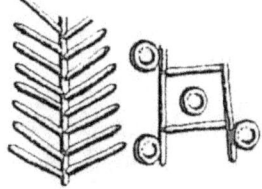

Bleda S. Düring

Abstract *Over the past few years the claim that the 8.2 event triggered the expansion of farming toward Europe has been put forward by various archaeologists and climate researchers. Paradoxically, the archaeological evidence from western Anatolia, a region of key significance in this Neolithic expansion episode, has not featured prominently in these hypotheses. This neglect may partly stem from the circumstances that relevant data, mostly published exclusively in Turkish, have only become available in recent years. Here, new data from western Asia Minor, in particular the Lake District, Aegean Anatolia, and the Marmara Region, will be considered.*

It will further be argued that synchronicity in ecology and archaeology has often been erroneously equated with causality, and that synchronicity in itself does not prove anything. Instead, it is necessary to reconstruct ecological changes in particular regions and to explain why particular developments in the archaeological sequence would have been related to ecological changes, rather than other factors. In order to evaluate the role of the 8.2 event in relation to the Neolithic expansion, I will discuss the chronology of the Neolithic expansion that occurred in Asia Minor during the seventh millennium, proxy records of ecological changes, and, finally, the archaeology of the early Neolithic in western Turkey and what that can tell us about the mechanisms that made this expansion possible. On this basis we can evaluate whether or not the 8.2 event might have played a significant role in this particular Neolithic expansion episode.

Introduction

The central issue addressed in this paper is whether the climatic fluctuation known as the "8.2 event" was a factor in the Neolithic expansion episode that occurred in the seventh millennium B.C. in western Anatolia.

In the past few years we have witnessed a resurgence of deterministic explanations in studies dealing with Near Eastern Prehistory, in which climate changes triggered substantial shifts in how human societies developed. One case in particular that can illustrate this broader development, undoubtedly linked to anxieties in the modern world about climate change and how this may affect us, is that a large number of scholars have claimed that the 8.2 event triggered the expansion of farming toward Europe (Bar-Yosef 2001; Berger and Guilaine 2009; Budja 2007; Clare et al. 2008; Turney and Brown 2007; Wagner et al. 2002; Weiss and Bradley 2001; Weninger et al. 2006).

In these studies, a perceived chronological fit between the 8.2 event climatic oscillation and the expansion of farming has been taken as proof for the link between climate change and the Neolithic expansion. The arguments put forward in these papers vary somewhat from one study to the next—for instance, some authors link the expansion of farming with the so-called PPN-B collapse in the Levant, whereas others argue that the westward spread of farmers started in Central Anatolia—but they have three implicit assumptions in common. First, it is argued that the 8.2 event significantly affected the climate in the Near East. Second, it is postulated that adverse climatic effects would have impacted farming societies of the Near East, resulting in a series of crop failures and famines. Third, it is claimed that, faced with these conditions, farmers massively decided to migrate toward more temperate climatic regions such as western Anatolia, Greece, and the Balkans.

The approach taken in this paper will be to investigate the validity of these three assumptions by critically evaluating the ecological and archaeological data of Asia Minor in the seventh millennium B.C. The aim is to assess possible relations between the expansion of farming in western Asia Minor in the seventh millennium B.C. on the one hand and climatic changes on the other. Before proceeding to this discussion, however, it is necessary to make some more general comments on the manner in which the relation between ecological and culture change has often been conceptualized, and the chasm between archaeologists and climate researchers.

Relating Archaeology and Climate Studies

At first sight, the relation between archaeology and climate studies may appear like another instance of the typical encounter between the "sciences" and the "humanities." For example, there are a large number of publications linking climate change and cultural change in "science" journals, such as *Science*, *Quarternary Research*, and *Quarternary Science Reviews*. By contrast, papers of similar content are almost completely absent from the core journals of archaeology, such as *Antiquity*, *Current Anthropology*, or the *Journal of World Prehistory*.

However, I will argue that it would be a mistake to construe these differences as an opposition between the sciences and humanities. The discipline of archaeology is saturated with scientific methodologies, and archaeologists discussing matters such as agricultural changes and stratigraphic sequences and their chronology base themselves on a large corpus of systematically collected scientific data (Jones 2002). Instead, I will argue that archaeologists and climate researchers typically espouse fundamentally different understandings of the way in which cultural systems function.

The implicit view one often encounters in studies by climate researchers is that cultural systems are essentially stable over long periods of time and changes are triggered by factors upsetting the balance of the ecological-cultural equilibrium. Thus, time and again one encounters the idea that if we can establish synchronicity between ecological changes and cultural changes, it is reasonable to assume that ecological changes triggered cultural changes (Berger and Guilaine 2009; Clare et al. 2008; Turney and Brown 2007; Weiss and Bradley 2001).

This perception of culture is not dissimilar to the "systems theory" model of culture that was popular in the New Archaeology of the 1960s and 1970s, in which cultural systems were perceived as a relatively stable constellation of relations between cultural and ecological subsystems. This model of cultural systems came under attack in the 1980s, one of the main critiques being that this model offered no scope for agency and change (Hodder 1982). From that time onward, archaeologists have regarded cultural systems as dynamic; that is, changing on a more or less constant basis.

The dynamic culture model has significant implications for how archaeologists perceive studies relating archaeology and climate. A possible synchronicity of ecological and cultural changes is not intrinsically interesting given that change is a constant factor in archaeology in any case, and it is always possible to find cultural changes at any moment in time that may or may not be linked with ecological changes (Nieuwenhuyse et al., this volume). This is why the technique of highlighting a specific wiggle in a climate reconstruction and relating it to cultural changes—a popular technique in publications of climate researchers—is almost completely meaningless to archaeologists. One wonders why that particular climatic wiggle is highlighted and why it would have any bearing on a set of cultural changes likewise selected more or less randomly; for example, a posited link between the 8.2 event and the shift from Çatalhöyük East to Çatalhöyük West (Weninger et al. 2006:410). Instead, a typical archaeologist would want to know more about ecological changes in a particular region, how these changes would have affected local communities, and why particular cultural changes are best explained as resulting from ecological changes rather than other factors.

I argue that possible evidence for synchronicity in ecology and archaeology has often been erroneously equated with causality, and that synchronicity in itself does not prove anything. If we focus on the discussion on the 8.2 event and attendant farming expansion, the following questions need to be addressed: first, what were the actual effects of the 8.2 event in the heartlands of farming and elsewhere; second, why would these climatic changes have stimulated migrations rather than other strategies such as diversification of the economy; third, are the radiocarbon data from the regions to which the Neolithic expansion took place synchronous with the 8.2 event; and fourth, do the archaeological data from Central Anatolia and western Asia Minor support the model of large-scale migration of farmers out of their former heartland?

The 8.2 Event in Asia Minor: Regional Perspectives

The 8.2 event, which can be traced in climatic data from many parts of the globe, was caused by the release of a massive amount of cold water from Lake Agassiz in Canada into the Atlantic after the ice dams behind which it was trapped burst (Alley and Agustsdottir 2005; Wagner et al. 2002). The 8.2 event was marked by a drop in temperatures and reduced precipitation in many regions of the earth. On the basis of ice cores, the 8.2 event can be dated to between 6300 and 6100 B.C. (Akkermans et al. 2010).

What evidence is there for climatic changes in Asia Minor in the seventh millennium B.C. in general, and the 8.2 event in particular? First, I would like to argue against a monolithic perspective on climate change. There is often a tendency to model climatic development as blanket events, whereas in fact climates are complicated systems and the local effects of climatic changes might vary greatly (Bottema 1995; Van Andel 2005). For example, it is clear that the effects of the 8.2 event were not homogeneous across the globe. Many Near Eastern climate proxy records show no or little effects of the 8.2 event (Alley and Agustsdottir 2005; Berger and Guilaine 2009:38–40; Morrill and Jacobsen 2005; Van Andel 2005). It follows that we need to work with regional proxy data for modeling the effects of this climatic oscillation and its possible effect on local societies. This point is especially relevant in Asia Minor because of its ecological diversity, which means that large climatic changes might have had little impact in some regions, whereas in other areas even small climate changes could have had large effects on subsistence economies of prehistoric societies.

A substantial number of dated sequences with climate proxy data, consisting of pollen, diatoms, and isotopes, are available for Asia Minor and surrounding regions (Eastwood et al. 1999; Fleitmann et al. 2009; Van Zeist and Bottema 1991; Woldring and Bottema 2002). Many of these studies have been published after the "discovery" of the 8.2 event, but as far as I know no one has ever published any evidence for an 8.2 event in their proxy records in Central Anatolia. It has been suggested that this reflects sampling resolution factors; that the 160 years of the 8.2 event were too short to leave a mark in lake pollen records and speoleothems (Roberts et al. 2011:150), but one wonders why it does show up in other lake pollen samples, such as at Tenaghi Philippon (Kothoff et al. 2008). In my mind, we should therefore be careful with reconstructing dramatic ecological changes that drove people out of Central Anatolia. This is relevant because arguments for a possible link between the 8.2 event and demographic movements have revolved around the idea that it was in the steppe region of Central Anatolia, where early farmers had been settled for millennia, that the climatic oscillation resulted in droughts and famines and it was in these circumstances that people decided to migrate to western Asia Minor. It seems that at the moment this argument cannot be substantiated.

By contrast, in climate proxy records in the Aegean and the Marmara Region, much clearer evidence of the 8.2 event has been found. For example, at Tenaghi Philippon, in northeastern Greece, a pronounced change in vegetation probably related to the 8.2 event has been recognized, and this has been interpreted as a mesoclimatic effect, limited to a specific region (Pross et al. 2009). The 8.2 event also seems to be present in other northern Aegean Sea proxy datatets, and possibly in the Yenişehir pollen sequence in the Marmara Region (Bottema et al. 2001:339; Kotthoff et al. 2008:1028).

A somewhat different climate reconstruction is provided by the fine-grained climate proxy data from deep-sea cores in the southern Aegean (Clare et al. 2008; Rohling and Pälike 2005). In these data it not so much the 8.2 event but a broader climatic oscillation which can be observed. Between about 6600–6000 B.C., the southern Aegean Sea was 2–3 degrees colder in winter. Effects were probably more severe on land, and the regular occurrence of severely cold and dry winters in the north of the Aegean and in the Marmara Region is postulated. Recent stalactite evidence from Sofular Cave on the western Turkish Black Sea seems to provide similar data (Fleitmann et al. 2009). Thus, there is some evidence in the Aegean and Black Sea region for the 8.2 event and a broader, but milder climatic oscillation, which I will label the *mega 8.2 event* for lack of a better designation.

In summary, ecological effects of both the 8.2 event and the mega 8.2 event remain to be established for the Central Anatolian heartland of farmers, whereas possibly significant effects of these climatic oscillations have been documented for the Aegean and the Marmara Region, with seemingly greater changes occurring in the north than to the south. Thus, it is possible that the (mega) 8.2 event might have significantly impacted local hunter-gatherer-fisher groups in the Aegean and in the Marmara region.

The precise effects of these climate and ecological changes on prehistoric subsistence strategies in the Aegean and Marmara regions are unknown. More research is required to establish, for example, what the effects were on, for example, fish and mollusks that were of importance to Mesolithic communities along the coast.

The Chronology of the Neolithic Expansion in Western Asia Minor

Apart from the regional ecological effects of climatic oscillations such as the 8.2 event, any consideration of the effects of climate changes on cultural systems has to include a discussion of the chronology of archaeological developments. For the 8.2 event/Neolithic expansion model evaluated here, the chronology of western Asia Minor is of key significance, because it is here that the earliest Neolithic expansion occurred. Paradoxically, the archaeological evidence from Asia Minor has not featured prominently in studies that link the Neolithic expansion with the 8.2 event. No doubt this can be explained in part because much of the relevant data has been published in recent years and in some cases only in Turkish (Özdoğan and Başgelen 2007). However, it is also the case that in some of the articles linking the Neolithic expansion with the 8.2 event there is a selective use of data that fit with the postulated link.

At first sight, the proposed link between the Neolithic expansion and the 8.2 event seems attractive. After the uptake of farming in southern Central Anatolia around 8500 B.C. at sites such as Aşıklı Höyük and Boncuklu Höyük (Figure 6.1) (Esin and Harmankya 2007), it is only in the mid-seventh millennium B.C., nearly 2,000 years later, that farming seems to expand farther westward (Düring 2011; Özdoğan 2010). In theory, at least this Neolithic expansion could have been triggered by a climatic change.

Here, I will briefly summarize the chronological evidence for the spread of farming in western Asia Minor. To facilitate this summary I will distinguish three subregions: the Lake District, Aegean Anatolia, and the Marmara Region. These regions can be distinguished in part in terms of ecology. Central Anatolia, the region where the earliest

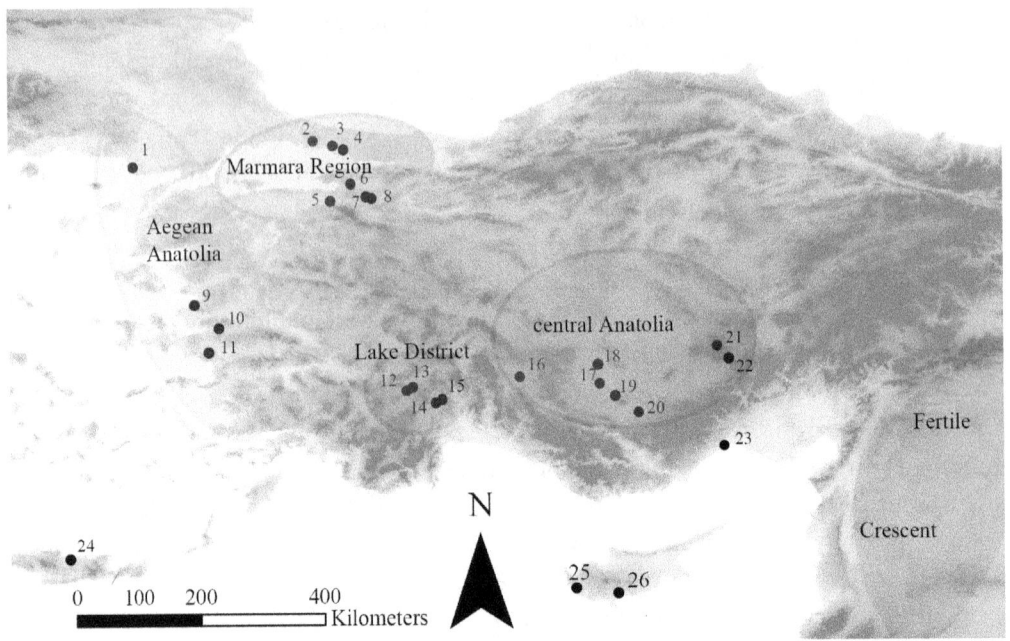

FIGURE 6.1. Neolithic sites of Anatolia. 1—Hoca Çesme; 2—Yarımburgaz and Yenikapı; 3—Fikirtepe; 4—Pendik; 5—Aktopraklik; 6—Ilipnar; 7—Mentese; 8—Barcin Höyük; 9—Ege Gübre; 10—Ulucak; 11—Dedecik-Heybelitepe; 12—Hacılar; 13—Kuruçay; 14—Bademağacı; 15—Höyücek; 16—Erbaba; 17—Çatalhöyük East and West; 18—Boncuklu Höyük; 19—Pınarbaşı; 20—Canhasan; 21—Aşıklı Höyük; 22—Kaletepe; 23—Mersin-Yumuktepe; 24—Knossos; 25—Mylouthkia; 26—Shillourakambos.

Neolithic of Asia Minor is documented, has a steppe climate similar to much of the Fertile Crescent. The Lake District and Aegean Anatolia have a Mediterranean climate, with dry summers and mild winters. Finally, the Marmara Region has a more temperate climate with summer rains and frequent frost in winter.

It is not entirely clear when the Neolithic sequence in the Lake District starts. A few years ago Weninger et al. (2006), in a paper linking the 8.2 event and the Neolithic expansion, claimed that the Neolithic sequence in this region started at around 6200 B.C. This chronology was based on the Hacılar radiocarbon dates, of which four dates were considered too early and discarded. However, Hacılar is only one of a number of investigated Neolithic sites in the Lake District, which also include the sites of Kuruçay, Höyücek, and Bademağacı. The collective evidence from the Lake District sites clearly demonstrates that the region was settled by sedentary farmers much earlier than 6200 B.C., and possibly even as early as 7000 B.C.[1]

In particular, there is some evidence from Bademağacı that its occupation seems to have begun around 7000 cal B.C. (Duru 2004). Unfortunately, the "Early Neolithic I" exposure at this site was very small, and we know very little about the early seventh millennium in this region. However, recent data from the site of Ulucak, which will be discussed later, adds credence to the possible existence of this early horizon.

In contrast with the presently elusive data for the first half of the seventh millennium B.C., there is strong chronological data for Neolithic strata immediately after 6500 B.C. at all excavated sites in the Lake District. This 6500 B.C. date is one that recurs in other areas of western Asia Minor also (Thissen 2005).

For Aegean Anatolia, where evidence for early Neolithic strata has been obtained in recent years, the data are similar to those of the Lake District. Two recently obtained radiocarbon dates on charcoal from the oldest level (Vg) at Ulucak suggest the site was inhabited already in the first half of the seventh millennium B.C. (Çilingiroğlu 2009a:12, 2009b:47). More information concerning these levels and radiocarbon dates on short-lived samples rather than charcoal would be useful, because the present samples have wide ranges and could suffer from the old wood problem. Occupation at Ulucak is continuous into the later seventh and early sixth millennia B.C. (Çilingiroğlu and Çilingiroğlu 2007). From around 6500 B.C. we also have radiocarbon-dated sequences at the sites of Ege Gübre, Yeşilova, and Hoca Çeşme (Derin 2007; Özdoğan 2007; Sağlamtimur 2007).

For the Marmara region, Özdoğan and Gatsov (1998) have argued for the existence of an Aceramic Neolithic phase, to be dated from about 7000 B.C. onward. The evidence upon which they base this consists of two survey sites, Çalca and Musluçesme, for which we lack absolute dates.

The excavated and dated Neolithic strata in the Marmara Region date from around 6500 B.C. Strata dating to the latter half of the seventh millennium B.C. have been excavated at Menteşe, Barcın Höyük, and Aktopraklık (Karul 2007; Roodenberg and Alpaslan-Roodenberg 2007; Roodenberg et al. 2008). These earliest Neolithic strata have been investigated in small trenches at sites such as Menteşe and Barcın Höyük, and we await further details from the recently excavated site of Aktopraklık.

Summarizing the chronological evidence for the Neolithic expansion into western Asia Minor as a whole (Figure 6.2), two conclusions can be drawn. First, there is some

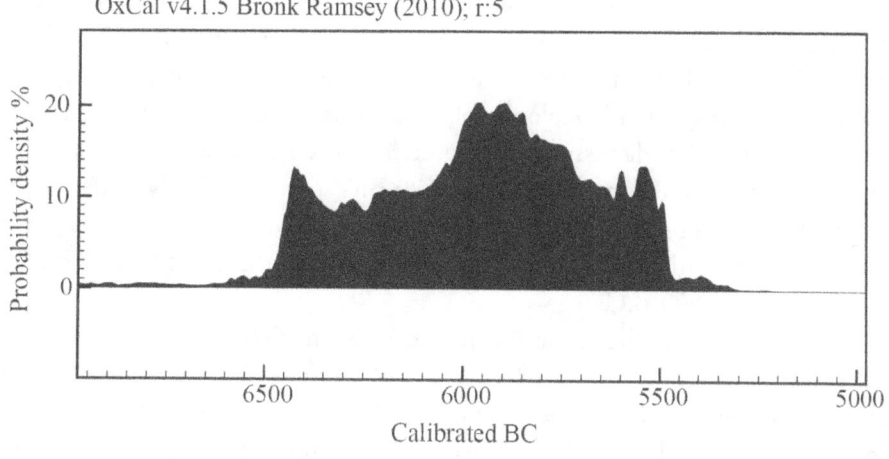

FIGURE 6.2. Cumulative radiocarbon plot for the Lake District, Aegean Anatolia, and the Marmara Region (n=135, data obtained from CONTEXT (http://context-database.uni-koeln.de/) database [Böhner and Schyle 2006] augmented with data from Özdoğan and Başgelen 2007).

evidence for a Neolithic expansion into western Asia Minor in the first half of the seventh millennium B.C. This phase has been tentatively identified with survey data in the Marmara Region, has been excavated in a small sounding at Bademağacı in the Lake District, and has also been excavated in a small exposure at Ulucak. Much remains to be learned about this, as yet elusive, earliest Neolithic horizon in western Asia Minor. Second, there seems to be a substantial increase in Neolithic settlements in all three areas from about 6500 B.C. onward, mirroring the emergence of Neolithic settlements in Greece (Pérles 2001; Reingruber 2005; Thissen 2005), a circumstance that could suggest that the two developments were not entirely unrelated.

Evaluating the Synchronicity of the Neolithic Expansion in Western Asia Minor and the 8.2 Event

On the basis of the chronological data for the Neolithic expansion in western Asia Minor, we can now evaluate the question: To what degree is this expansion synchronous with the 8.2 event? However, there are a number of caveats surrounding this synchronicity discussion. First, although the 8.2 event is dated to about 6300–6120 B.C. in calendar years in the Greenland ice cores (Akkermans et al. 2010), one frequently encounters significantly earlier dates for the 8.2 event: up to 6500 B.C. (Berger and Guilaine 2009; Turney and Brown 2007; Weninger et al. 2006). Second, there is the broader but milder climatic oscillation of the mega 8.2 event, dated between about 6400 and 5900 B.C. in calendar years (Maher et al. 2011:8; Rohling and Pälike 2005).

The fact that different scholars have used different dates for the 8.2 event has resulted in a group of scholars opting for an "early 8.2 event" from about 6500 B.C. to explain the expansion of farming toward western Turkey and the Aegean (Bar-Yosef 2001; Budja 2007; Clare et al. 2008; Turney and Brown 2007; Weninger et al. 2006), whereas another group has used the more robust dates for the 8.2 event to explain transformations in Mesolithic societies in the Balkans and their shift toward a Neolithic way of life (Berger and Guilaine 2008; Bonsall et al. 2003) or changes in the later Neolithic of Central Anatolia (Biehl and Rosenstock 2009; Roberts and Rosen 2009). Clearly, there is a need for accurate chronologies of both climatic oscillations and archaeological developments; otherwise, discussing synchronicities becomes rather pointless.

It is clear that the 8.2 event of 6300–6120 B.C. postdates the Neolithic expansion by a considerable margin, even if we exclude the recent evidence for sites dating to the first half of the seventh millennium B.C. and focus on the rapid spread of farming settlements around 6500 B.C.

The 8.2 event/Neolithic expansion model rests on, firstly, a flawed chronology of the 8.2 event; and secondly, on a selective use of chronological data from archaeology (for example, by discarding dates preceding the 8.2 event as "unreliable"). Thus, from a chronological point of view the 8.2 event occurs too late to explain the westward Neolithic expansion in Asia Minor (see also Berger and Guilaine 2009; Bonsall et al. 2003).

What about the "mega" 8.2 event and the farming expansion? As is the case for the 8.2 event, one can encounter earlier date ranges for this climatic oscillation, such as 6600–6000 B.C. (Clare et al. 2008), while the more accurate range is probably between about 6400 and 5900 B.C. (Maher et al. 2011:8; Rohling and Pälike 2005). This would put the start of this oscillation about a century after the main agricultural expansion in western Turkey around 6500 B.C. However, we have to admit that this chronology rests on relatively few dates, and we cannot exclude the possibility that the mega 8.2 event and farming expansion were synchronous. While the mega 8.2 event would, chronologically, better fit the acceleration of the spread of farming that occurred around 6500 B.C.— recently, this link has been proposed by Weninger and Clare (personal communication, 5 April 2010), abandoning an earlier link between the 8.2 event and farming expansion—its effects on the climate of the Near East are much less pronounced, and it does not show up at all in many proxy records (Berger and Guilaine 2009).

The crucial points that emerge out of the presented discussion are: firstly, for neither the 8.2 event nor the mega 8.2 event do we have proxy data for ecological changes in Central Anatolia, though both climatic oscillations can be documented in proxy record from the (northern) Aegean and the Marmara Region; and secondly, the earliest Neolithic strata in the Lake District and Aegean Turkey predate both climatic oscillations. The conclusion, at least for me, is that if climate changes played a role in the transformation of Mesolithic/Neolithic groups in Asia Minor, we have to focus on what happened in western Turkey rather than Central Anatolia.

The Archaeology of the Neolithic Expansion in Western Asia Minor

From the data that have been presented so far it has already become clear that both in the Lake District and in Aegean Anatolia, at the sites of Bademağacı and Ulucak, there is evidence for Neolithic settlements that predate the climatic oscillations of the seventh millennium B.C. Further, there is a sudden boom in of Neolithic settlements in the Lake District, Aegean Anatolia, and the Marmara Region occurring around 6500 B.C. It is this acceleration of a Neolithization process already underway in which climatic oscillations might have played a role. If so, it would be the mega 8.2 event that would fit in terms of chronology, and it might have been predominantly local groups in the Aegean and Marmara Region that shifted into a farming way of life around 6500 B.C., given that the climatic effects were most prominent in these regions.

Such a perspective, in which one could postulate an indigenous uptake of farming by local hunter-gatherer groups in western Asia Minor, has not been popular among scholars investigating Turkish prehistory (who have tended to opt for migrations out of Central Anatolia [Özdoğan 2010]), but it can be further substantiated along two lines of evidence. The first line of evidence concerns a cultural continuity between Neolithic traditions and those of preexisting Mesolithic groups, and the second argument concerns the heterogeneity of the Neolithic of western Asia Minor.

The Mesolithic in Asia Minor is poorly investigated. There are a number of cave sequences in the Antalya region with excavated Mesolithic strata, at sites such as Öküzini, Beldibi, and Belbaşı, but these are, unfortunately, poorly published and understood (Yalçınkaya et al. 2002). For Aegean Anatolia we know even less, though it is conceivable that the Latmos rock paintings date to the Mesolithic (Peschlow-Bindokat 1996). Finally, in the Marmara Region we have the so-called "Ağaçlı group," a group of Mesolithic sites known only from surveys (Gatsov and Özdoğan 1994).

Interestingly, there is a marked cultural continuity from these Ağaçlı sites to the earliest Neolithic sites in the Marmara Region, known as the Fikirtepe horizon (Gatsov and Özdoğan 1994; Özdoğan 2007). The chipped stone industries of Ağaçlı sites and Fikirtepe sites are almost identical, an example being the type of bullet cores and endscrapers used in both complexes. In the Fikirtepe horizon we can distinguish between coastal sites and interior sites. The coastal sites are similar in location to those from the Ağaçlı group. In the Fikirtepe coastal sites we find simple round sunken huts, and fishing constituted an important component to subsistence. It is plausible that fishing and round huts were also common in Ağaçlı sites. In contrast to the coastal Fikirtepe sites, those in the interior have large rectangular wattle and daub houses. Here, the economy, surprisingly, is almost completely dominated by animal husbandry and crop cultivation. The cultural continuity between the Ağaçlı sites and the Fikirtepe sites suggest that local Mesolithic hunter-gatherer groups played an important role in the Neolithization of the Marmara Region. This argument does of course not exclude the possibility that there was also some westward migration from Central Anatolia.

At present, similar continuities between the Mesolithic and Neolithic cannot be established for the Lake District and Aegean Anatolia, but much research remains to be done on documenting the earliest Neolithic and Mesolithic horizons in both regions before we can establish the ancestry of the Neolithic complexes known at present. One interesting issue in this regard is the cultural affiliations of the earliest Neolithic groups at sites such as Bademağacı and Ulucak.

A striking characteristic of the Neolithic in western Asia Minor is its regional diversity (Düring 2011:122–199, 2013). Our current data suggest that there are at least three cultural facies in western Asia Minor: the Lake District, Aegean Anatolia, and the Marmara Region. Within each of these regional horizons there are more or less interchangeable artifacts, iconographical styles, settlements, and burial traditions, but these regions clearly differ both from each other and from the Neolithic of Central Anatolia (Düring 2011; Özdoğan 2010). Thus, for example, Fikirtepe ceramics are only found in the Marmara Region (Özdoğan 2007; Thissen 2001), while in Aegean Anatolia we find red slipped burnished and impressed wares that can be clearly distinguished from those in the Lake District and the Marmara Region (Çilingiroğlu and Çilingiroğlu 2007; Herling et al. 2008; Sağlamtimur 2007), and in the Lake District we find monochrome ceramics with features that are absent in the seventh millennium in Central Anatolia, such as tubular lugs, s-profiled bowls, and globular jars (Duru 2007; Last 2005).

Similarly, if we focus on the types of settlements occurring in these three regions, clear differences are again apparent. The settlements of the earliest Fikirtepe horizon (Özdoğan 2007; Roodenberg and Alpaslan-Roodenberg 2007) consist of groups of sunken huts along the coasts and large rectangular wattle and daub buildings in the interior. In both cases, buildings are freestanding without a clear alignment to streets or the like. In Aegean Anatolia, settlements have a more structured format. At Ulucak (Çilingiroğlu and Çilingiroğlu 2007), rectangular structures of about six by six meters, constructed of wattle and daub, pise, and mud bricks, are arranged along streets. At nearby Ege Gübre (Sağlamtimur 2007), stone foundations of similarly sized buildings were found, arranged around a central court. Finally, in the Lake District (Düring 2011:160–174; Umurtak 2000), the earliest settlements consist of rectangular buildings found in small house clusters and built of mud in various techniques, with a door in the long wall and a hearth on the wall across. All these settlements are distinct from each other and from the settlement type prevailing in Central Anatolia, that of the "clustered neighborhood settlements," in which there were no streets in the blocks of houses that made up a neighborhood and structures were accessed from the roof (Düring 2006).

Other differences between the assemblages found in Aegean Anatolia, the Lake District, the Marmara Region, and Central Anatolia could be added, concerning other artifacts, iconographical styles, burial traditions, and agricultural practices, but this would go too far in the context of this paper (for further discussion, see Düring 2011). The idea here is simply to sketch the degree to which western Turkey was divided into a number of cultural horizons during the seventh millennium B.C. Future research will have to demonstrate whether these are, in part, a product of the archaeological research that has taken place in specific regions, and whether there are intermediate assemblages that combine various elements.

What is pertinent in the context of the issue of how farming expanded in the seventh millennium B.C. in Turkey is how we can explain this regionalization of the Neolithic of western Asia Minor, a pattern that contrasts with large spreads of more or less homogenous Neolithic horizons emerging a few centuries later in Europe, such as the LBK and the Cardial (Barnett 2000; Bogucki 2000). The cultural diversity of Neolithic horizons in western Asia Minor could point to a development in which local groups played an important role in the articulation of distinctive Neolithic horizons in their respective regions. This would also explain the strong continuities between Mesolithic and Neolithic cultural traditions in the Marmara Region. On the other hand, it is not unlikely that at least some of the actors in this process were migrant farmers from Central Anatolia.

Indeed, it is possible to explain the Neolithic expansion of 6500 B.C. in Asia Minor as a combination of small-scale migration from Central Anatolia and local hunter-gatherer groups opting into farming. In this scenario, the migrants would have contributed farming expertise and the hunter-gatherers contributed knowledge of the local environment and its resources. The (mega) 8.2 event could have accelerated this process, which appears to have been set in motion earlier, although this is a hypothesis that needs to be investigated further rather than a firm conclusion.

Conclusions

In this paper, I have argued that proponents of the 8.2 event/Neolithic expansion hypothesis have mistakenly taken possible evidence for synchronicity as proof of climatic causation. Further, the 8.2 event in its traditional sense was shown to chronologically postdate the Neolithic expansion. A better fit in terms of chronology is the "mega" 8.2 event, dating to between ca. 6400 and 5900 B.C., which might have accelerated the Neolithization of western Asia Minor already under way.

The ecological effects of climatic oscillations need to be investigated in local climate and vegetation proxy records, because local climates might have been affected quite differently. At present, the best evidence for ecological changes in the seventh millennium B.C. can be found in western Asia Minor rather than in Central Anatolia. Thus, if climate change played a role in the Neolithization of western Asia Minor, it is most likely that it triggered local Mesolithic groups into taking up farming.

This scenario finds some circumstantial support in the Mesolithic-Neolithic continuity established in the Marmara Region, and in the regional facies of the Neolithic of western Asia Minor, which suggest local Neolithization processes rather than the arrival of large groups of migrants from Central Anatolia.

Acknowledgments

I would like to thank Peter Biehl and Olivier Nieuwenhuyse for inviting me to contribute to this volume. The paper was originally presented in a symposium at Leiden University on the 8.2 event and Near Eastern archaeology. I would like to thank Peter Akkermans, Olivier Nieuwenhuyse, Anna Russell, and Hans van der Plicht for the possibility to present at that occasion. This paper was written in the course of postdoctoral research funded by the Netherlands Organization for Scientific Research (N.W.O. grant 275-62-002), and the Faculty of Archaeology at Leiden University.

Note

1. An even earlier date at around 8000 B.C. has been suggested for Aceramic Hacılar (Mellaart 1970), but this is based on a single radiocarbon date taken from a small sounding that has no chronological parallels and is rejected by most researchers working on Anatolian prehistory (Duru 1989; Schoop 2005:178–179; Thissen 2002).

References Cited

Akkermans, P. M. M. G., J. van der Plicht, O. P. Nieuwenhuyse, A. Russel, A. Kaneda, and H. Buitenhuis 2010 Weathering Climate Change in the Near East: Dating and Neolithic Adaptations 8200 Years Ago. *Antiquity* 84(325): project gallery.

Alley, R. B., and A. M. Agustsdottir 2005 The 8k Event: Cause and Consequences of a Major Holocene Abrupt Climate Change. *Quaternary Science Reviews* 24:1123–1149.

Barnett, W. K. 2000 Cardial Pottery and the Agricultural Transition in Europe. In *Europe's First Farmers*, edited by T. Douglas Price, pp. 93–116. Cambridge University Press, Cambridge.

Bar-Yosef, O. 2001 From Sedentary Foragers to Village Hierarchies: The Emergence of Social Institutions. In *The Origin of Human Social Institutions,* edited by W. Garry Runciman, pp. 1–38. The British Academy, Oxford.

Berger, J.-F. and J. Guilaine 2009 The 8200 cal BP Abrupt Environmental Change and the Neolithic Transition: A Mediterranean Perspective. *Quarternary International* 200:31–49.

Biehl, P. F., and E. Rosenstock 2009 Von Çatalhöyük Ost nach Çatalhöyük West: Kulturelle Umbrüche an der Schwelle vom 7. zum 6. Jt. v. Chr. in Zentralanatolien. In *Zurück zum Gegenstand, Festschrift für Andreas E. Furtwängler*, edited by R. Einicke, S. Lehman, H. Löhr, G. Mehnert, A. Mehnert, and A. Slawisch, pp 471–482. Langenweißbach, Beier & Beran.

Böhner, U., and D. Schyle 2006 Radiocarbon CONTEXT Database. Electronic document, http://context-database.uni-koeln.de/, accessed 25 September 2013.

Bogucki, P. 2000 How Agriculture Came to North-Central Europe. In *Europe's First Farmers*, edited by T. Douglas Price, pp. 197–218. Cambridge University Press, Cambridge.

Bonsall, C., M. G. Macklin, R. W. Payton, and A. Boroneanţ 2003 Climate, Floods, and River Gods: Environmental Change and the Meso-Neolithic Transition in Southeast Europe. *Before Farming* 4(2):12.

Bottema, S. 1995 The Younger Dryas in the Eastern Mediterranean. *Quarternary Science Review* 14:883–891.

Bottema, S., H. Woldring, and İ. Kayan 2001 The Late Quaternary Vegetation History of Western Turkey. In *The Ilıpınar Excavations II*, edited by J. J. Roodenberg and L. Thissen, pp. 327–354. Nederlands Instituut voor het Nabije Oosten, Istanbul.

Budja, M. 2007 The 8200 Cal BP "Climate Event" and the Process of Neolithisation in South-Eastern Europe. *Documenta Praehistorica* 34:191–201.

Çilingiroğlu, Ç. 2009a Of Stamps, Loom Weights, and Spindle Whorls: Contextual Evidence on the Function(s) of Neolithic Stamps from Ulucak, Izmir, Turkey. *Journal of Mediterranean Archaeology* 22:3–27.

Çilingiroğlu, Ç. 2009b Central-West Anatolia at the End of the 7th and Beginning of the 6th Millenium BCE in the Light of Pottery from Ulucak (Izmir). Tübingen, Unpublished PhD Thesis.

Çilingiroğlu, A. and Ç. Çilingiroğlu 2007 Ulucak. In *Anadolu'da Uygarlığın Doğuşu Avrupaya Yayılımı Türkiye'de Neolitik Dönem*, edited by Mehmet Özdoğan and Nezih Başgelen, pp. 361–372. Arkeoloji ve Sanat Yayınları, Istanbul.

Clare, L., E. J. Rohling, B. Weninger, and J. Hilpert 2008 Warfare in Late Neolithic / Early Chalcolithic Pisidia, Southwestern Turkey. Climate Induced Social Unrest in the Late 7th Millenium cal BC. *Documenta Praehistorica* 35:65–92.

Derin, Z. 2007 Yeşilova Höyüğü. In *Anadolu'da Uygarlığın Doğuşu Avrupaya Yayılımı Türkiye'de Neolitik Dönem*, edited by M. Özdoğan and N. Başgelen, pp. 377–384. Arkeoloji ve Sanat Yayınları, Istanbul.

Düring, B. S. 2006 *Constructing Communities, Clustered Neighbourhood Settlements of the Central Anatolian Neolithic, ca. 8500–5500 Cal. BC*. Instituut voor het Nabije Oosten, Leiden, Nederlands.

Düring, B. S. 2011 *The Prehistory of Asia Minor: From Complex Hunter-Gatherers to Early Urban Societies*. Cambridge University Press, Cambridge.

Düring, B. S. 2013 Apples and Oranges? Comparing Neolithic horizons of Upper Mesopotamia and Asia Minor. In *Interpreting the Late Neolithic of Upper Mesopotamia*, edited by O. Nieuwenhuyse, R. Bernbeck, P. Akkermans and J. Rogasch, pp. 367–376. Turnhout, Brepols.

Duru, R. 1989 Were the Earliest Cultures at Hacılar Really Aceramic? In *Anatolia and the Ancient Near East, Studies in Honor of Tahsın Özgüç*, edited by K. Emre, M. Mellink, B. Hrouda, and N. Özgüç, pp. 99–106. Türk Tarih Kurumu, Ankara.

Duru, R. 2004 Excavations at Bademağacı: Preliminary Report, 2002 and 2003. *Belleten* 68:540–560.

Duru, R. 2007 Göller Bölgesi Neolitiği: Hacılar—Kuruçay Höyüğü—Höyücek—Bademağacı Höyüğü. In *Anadolu'da Uygarlığın Doğuşu Avrupaya Yayılımı Türkiye'de Neolitik Dönem, Yeni Kazılar, Yeni Bulgular*, edited by M. Özdoğan and N. Başgelen, pp. 331–360. Istanbul, Arkeoloji ve Sanat Yayınları.

Eastwood, W. J., N. Roberts, H. F. Lamb, and J. C. Tibby 1999 Holocene Environmental Change in Southwest Turkey: A Palaeoecological Record of Lake and Catchment-Related Changes. *Quarternary Science Reviews* 18:671–695.

Esin, U., and S. Harmankaya 2007 Aşıklı Höyük. In *Anadolu'da Uygarlığın Doğuşu Avrupaya Yayılımı Türkiye'de Neolitik Dönem*, edited by M. Özdoğan and N. Başgelen, pp. 255–272. Arkeoloji ve Sanat Yayınları, Istanbul.

Fleitmann, D., H. Cheng, S. Badertscher, R. L. Edwards, M., Mudelsee, O. M. Göktürk, A. Fankhauser, R. Pickering, C. C. Raible, A. Matter, J. Kramers, and O. Tüysüz 2009 Timing and Climatic Impact of Greenland Interstadials Recorded in Stalagmites from Northern Turkey. *Geophysical Research Letters* 36:L19707.

Gatsov, I., and M. Özdoğan 1994 Some Epi-Palaeolithic Sites from the NW Turkey, Ağaçli, Domali, and Gümüsdere. *Anatolica* 20:97–120.

Herling, L., K. Kasper, C. Lichter, and R. Meriç 2008 Im Westen nichts Neues? Ergebnisse der Grabungen 2003 und 2004 in Dedecik-Heybelitepe. *Istanbuler Mitteilungen* 58: 13–65.

Hodder, I. 1982 Theoretical Archaeology: A Reactionary View. In *Symbolic and Structural Archaeology*, edited by I. Hodder, pp. 1–16. Cambridge, Cambridge University Press.

Jones, A. 2002 *Archaeological Theory and Scientific Practice*. Cambridge University Press, Cambridge.

Karul, N. 2007 Aktopraklik: Küzeybatı Anadolu'da Gelişkin bir Köy. In *Anadolu'da Uygarlığın Doğuşu Avrupaya Yayılımı Türkiye'de Neolitik Dönem*, edited by M. Özdoğan and N. Başgelen, pp. 387–392. Arkeoloji ve Sanat Yayınları, Istanbul.

Kotthoff, U., U. C. Müller, J. Pross, G. Schmiedl, I. T. Lawson, B. van de Schootbrugge, and H. Schulz 2008 Lateglacial and Holocene Vegetation Dynamics in the Aegean Region: An Integrated View Based on Pollen Data from Marine and Terrestrial Archives. *The Holocene* 18:1019–1032.

Last, J. 2005 Pottery from the East Mound. In *Changing Materialities at Çatalhöyük, Reports from the 1995–1999 Seasons*, edited by I. Hodder, pp. 101–138. McDonald Institute / British Institute of Archaeology at Ankara, Cambridge.

Maher, L. A., E. B. Banning, and M. Chazan 2011 Oasis or Mirage? Assessing the Role of Abrupt Climate Change in the Prehistory of the Southern Levant. *Cambridge Archaeological Journal* 21:1–29.

Mellaart, J. (editor) 1970 *Excavations at Hacılar*. Edinburgh University Press, Edinburgh.

Morrill, C., and R. M. Jacobsen 2005 How Widespread Were Climate Anomalies 8200 Years Ago? *Geophysical Research Letters* 32:L19701.

Özdoğan, M. 2007 Marmara Bölgesi Neolitik Çağ Kültürleri. In *Anadolu'da Uygarlığın Doğuşu Avrupaya Yayılımı Türkiye'de Neolitik Dönem*, edited by M. Özdoğan and N. Başgelen, pp. 401–426. Arkeoloji ve Sanat Yayınları, Istanbul.

Özdoğan, M. 2010 Westward Expansion of the Neolithic Way of Life: Sorting the Neolithic Package into Distinct Packages. In *Proceedings of the 6th International Congress on the Archaeology of the Ancient Near East*, edited by P. Matthiae, F. Pinnock, L. Nigro, and N. Marchetti, pp. 883–897. Harrasowitz, Rome.

Özdoğan, M., and I. Gatsov 1998 The Aceramic Neolithic Period in Western Turkey and in the Aegean. *Anatolica* 24:209–232.

Özdoğan, M., and N. Başgelen (editors) 2007 *Anadolu'da Uygarlığın Doğuşu Avrupaya Yayılımı Türkiye'de Neolitik Dönem*. Arkeoloji ve Sanat Yayınları, Istanbul.

Perlès, C. 2001 *The Early Neolithic in Greece*. Cambridge University Press, Cambridge.

Peschlow-Bindokat, A. 1996 *Der Latmos: Eine unbekannte Gebirgslandschaft an der türkischen Westküste*. Phillip von Zabern, Mainz am Rhein.

Pross, J., U. Kotthoff, U. C. Müller, O. Peyron, I. Dormoy, G. Schmiedl, S. Kalaitzidis, and A. M. Smith 2009 Massive Perturbation in Terrestrial Ecosystems of the Eastern Mediterranean Region Associated with the 8.2 kyr B.P. Climatic Event. *Geology* 37:887–890.

Reingruber, A. 2005 The Argissa Magoula and the Beginning of the Neolithic in Thessaly. In *How Did Farming Reach Europe?* edited by C. Lichter. Ege Yayınları, Istanbul.

Roberts, N., and A. Rosen 2009 Diversity and Complexity in Early Farming Communities of Southwest Asia: New Insights into the Economic and Environmental Basis of Neolithic Çatalhöyük. *Current Anthropology* 50:393–402.

Roberts, N., W. J. Eastwood, C. Kuzucuoglu, G. Fiorentino, and V. Caracuta 2011 Climatic, Vegetation, and Cultural Change in the Eastern Mediterranean during the Mid-Holocene Environmental Transition. *Holocene* 21:147–162.

Roodenberg, J., and S. Alpaslan-Roodenberg 2007 Ilıpınar ve Menteşe: Doğu Marmara'da Neolitik döneme ait iki yerleşme. In *Anadolu'da Uygarlığın Doğuşu Avrupaya Yayılımı Türkiye'de Neolitik Dönem*, edited by M. Özdoğan and N. Başgelen, pp. 393–400. Arkeoloji ve Sanat Yayınları, Istanbul.

Roodenberg, J., A. van As, and S. Alpaslan-Roodenberg 2008 Barcın Hüyük in the Plain of Yenişehir (2005–2006): A Preliminary Note on the Fieldwork, Pottery, and Human Remains of the Prehistoric Levels. *Anatolica* 34:53–60.

Rohling, E. J., and H. Pälike 2005 Centennial-Scale Climate Cooling with a Sudden Cold Event around 8,200 Years Ago. *Nature* 434:975–979.

Sağlamtimur, H. 2007 Ege Gübre Neolitik Yerlişimi. In *Anadolu'da Uygarlığın Doğuşu Avrupaya Yayılımı Türkiye'de Neolitik Dönem*, edited by M. Özdoğan and N. Başgelen, pp. 373–376. Arkeoloji ve Sanat Yayınları, Istanbul.

Schoop, U. D. 2005 *Das anatolische Chalkolithicum*. Albert Greiner, Remshalden.

Thissen, L. 2001 The Pottery of Ilıpınar, Phases X to VA. In *The Ilıpınar Excavations II*, edited by J. Roodenberg and L. Thissen, pp. 3–154. Nederlands Historisch Archaeologisch Istituut te Istanbul, Istanbul.

Thissen, L. 2002 Appendix I: The CANeW 14C Database, Anatolia 10,000–5000 cal BC. In *The Neolithic of Central Anatolia*, edited by F. Gérard and L. Thissen, pp. 299–337. Ege Yayınları, Istanbul.

Thissen, L. 2005 Coming to Grips with the Aegean in Prehistory: An Outline of the Temporal Framework, 10.000 to 5500 Cal BC. In *How Did Farming Reach Europe?*, edited by C. Lichter. Ege Yayınları, Istanbul.

Turney, C. S. M., and H. Brown 2007 Catastrophic Early Holocene Sea Level Rise: Human Migration and the Neolithic Transition in Europe. *Quarternary Science Reviews* 26:2036–2041.

Umurtak, G. 2000 A Building Type of the Burdur Region from the Neolithic Period. *Belleten* LXIV(241):683–716.

Van Andel, T. H. 2005 Coastal Migrants in a Changing World?, An Essay on the Mesolithic in the Eastern Mediterranean. *Journal of the Israel Prehistoric Society* 35:381–397.

Van Zeist, W., and S. Bottema 1991 *Late Quarternary Vegetation of the Near East*. Dr. Ludwig Reichert Verlag, Wiesbaden.

Wagner, F., B. Aaby, and H. Visscher 2002 Rapid Atmospheric CO_2 Changes Associated with the 8,200-years-B.P. Cooling Event. *Proceedings of the National Academy of Sciences* 99:12011–12014.

Weiss, H., and R. S. Bradley 2001 Archaeology: What Drives Societal Collapse? *Science* 291(5504):609–610.

Weninger, B., E. Alram-Stern, E. Bauer, L. Clare, U. Danzeglocke, O. Jöris, C. Kubatzki, G. Rollefson, H. Todorova, and T. van Andel 2006 Climate Forcing Due to the 8200 cal yr BP Event Observed at Early Neolithic Sites in the Eastern Mediterranean. *Quarternary Research* 66:401–420.

Woldring, H., and S. Bottema 2002 The Vegetation History of East-Central Anatolia in Relation to Archaeology: The Eski Acıgöl Pollen Evidence Compared with the Near Eastern Environment. *Palaeohistoria* 43/44:1–34.

Yalçınkaya, I., M. Otte, J. Kozlowski, and O. Bar-Yosef (editors) 2002 *La Grotte d'Öküzini: Evolution du Paleolithique Final du Sud-Ouest de l'Anatolie*. Universite de Liege, Liege.

PART II

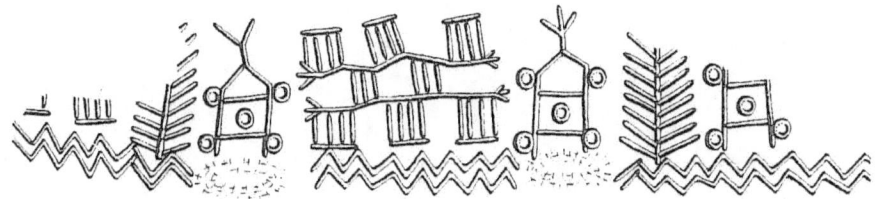

Europe

CHAPTER SEVEN

"Singing in the Rain"

Khirokitia (Cyprus) in the Second Half
of the Seventh Millennium cal B.C.

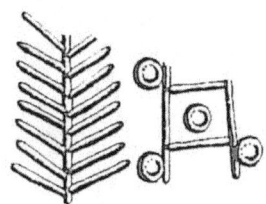

Odile Daune-Le Brun and Alain Le Brun

Abstract *The history of Khirokitia, an Aceramic Neolithic village in Cyprus, is marked by a series of events clearly evidenced by variations in the spatial extent and organization of the village. The most significant shift of the built area happened around the end of the seventh millennium B.C. As for climatic changes, the study of the hydromorphological evolution of the nearby riverbed has revealed the existence of torrential flows and violent erosion, which seems to indicate the foundation of an erratic and concentrated pluviosity during the occupation of the village. The paper will focus on the main events that affected the village and investigate changes and continuities that can be observed in the ancient environment as well as in the village's relations to the environment, in the form of subsistence strategies, craft techniques and activity organization, architectural practices, social organization, and rituals.*

Introduction

The second half of the seventh millennium cal B.C. is marked in Cyprus by the emergence of the so-called Khirokitia Culture, the peak of the Cypriot Aceramic Neolithic. This culture, the most distinctive features of which are the persistence of a circular in plan architecture as well as a brilliant stone craft industry, is illustrated by several sites scattered throughout the island: Cap Andreas-Kastros (Le Brun 1981), a fishing hamlet in the east at the tip of the Karpas peninsula; Troulli and Petra tou Limniti in the north; Kholetria Ortos in the west; Kalavasos-Tenta[1] and Khirokitia in the south, a few kilometers from the actual coastline; and Dhali-Agridhi and Kataliondas, both

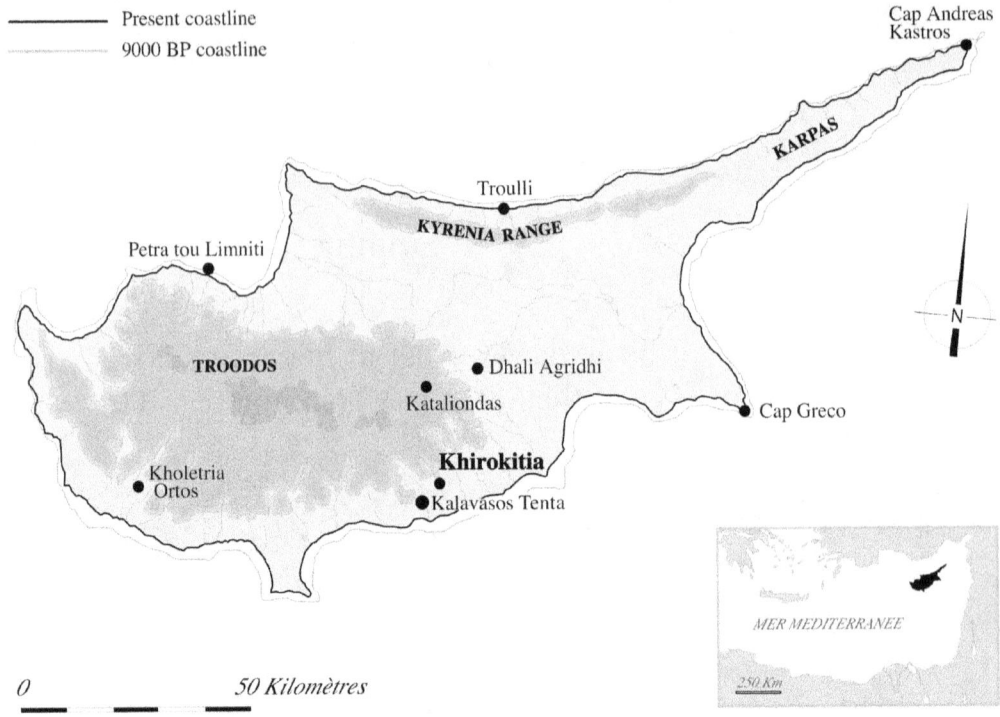

FIGURE 7.1. Map of Cyprus showing the locations of the Late Aceramic Neolithic sites belonging to the "Khirokitia Culture" (7th–6th millenium B.C.).

located inland. However, only Khirokitia will be considered here, because of its long and consistent stratigraphical sequence and the richness of the cultural and environmental data that have been collected.

This single point of view obviously presents a double risk: one is to take a blinkered view of events and the other is, on the contrary, to draw general conclusions from a particular case that could in fact have been just an exception. This should not be forgotten.

Climate Change and Village Historiography

Khirokitia village spreads over the sides of a hill surrounded by a meander of the Maroni River. It is a closed space encircled by an enclosure wall, the outline of which has been modified several times. The surface of the built-up area at the time of its maximum extent may be estimated at 3 ha (Figures 7.2–7.4, pages 155, 156 157).

Research conducted on this site by the French Archaeological Mission (Le Brun 1984, 1989, 1994) since 1976 has followed on from excavations carried out by P. Dikaios, the discoverer of this settlement, which he explored between 1936 and 1946 (Dikaios

FIGURE 7.2. Khirokitia. General view of the southern hillside (photo Th. Saggory) with the two successive walls enclosing the Neolithic village (Photos Mission Archéologique Française).

1953). Excavations concentrated mainly on two areas: the first on top of the hill and on the saddle linking it with the neighboring western relief, which was recently extended to the northern slope; and the second down the eastern hillside, where archaeological remains are in contact with the river. The excavations were supplemented with geoarchaeological and paleogeographical investigations in order to reconstruct the paleoenvironmental conditions, to precisely determine the behavior of the river and its impact on the settlement life (Hourani 2008). Results from this research demonstrate several modifications in the hydro-geomorphologic configuration of the river that occurred before, during, and after the occupation on the site. As a matter of fact, the stratigraphic sequence observed next to the river is marked by torrential flow episodes (hereafter *Potamos*, Figures 7.5, page 158 and 7.6, page 159).

Running at first very close to the foothill, the river moved away and thus exposed an area liable to flooding which is covered by significant deposits resulting from slope erosion and from river floods, in the form of a one meter–thick succession of horizontal layers of sand and coarse gravels containing archaeological artifacts, levels P10 and P9. These deposits represent the first major natural event recognized on the section before

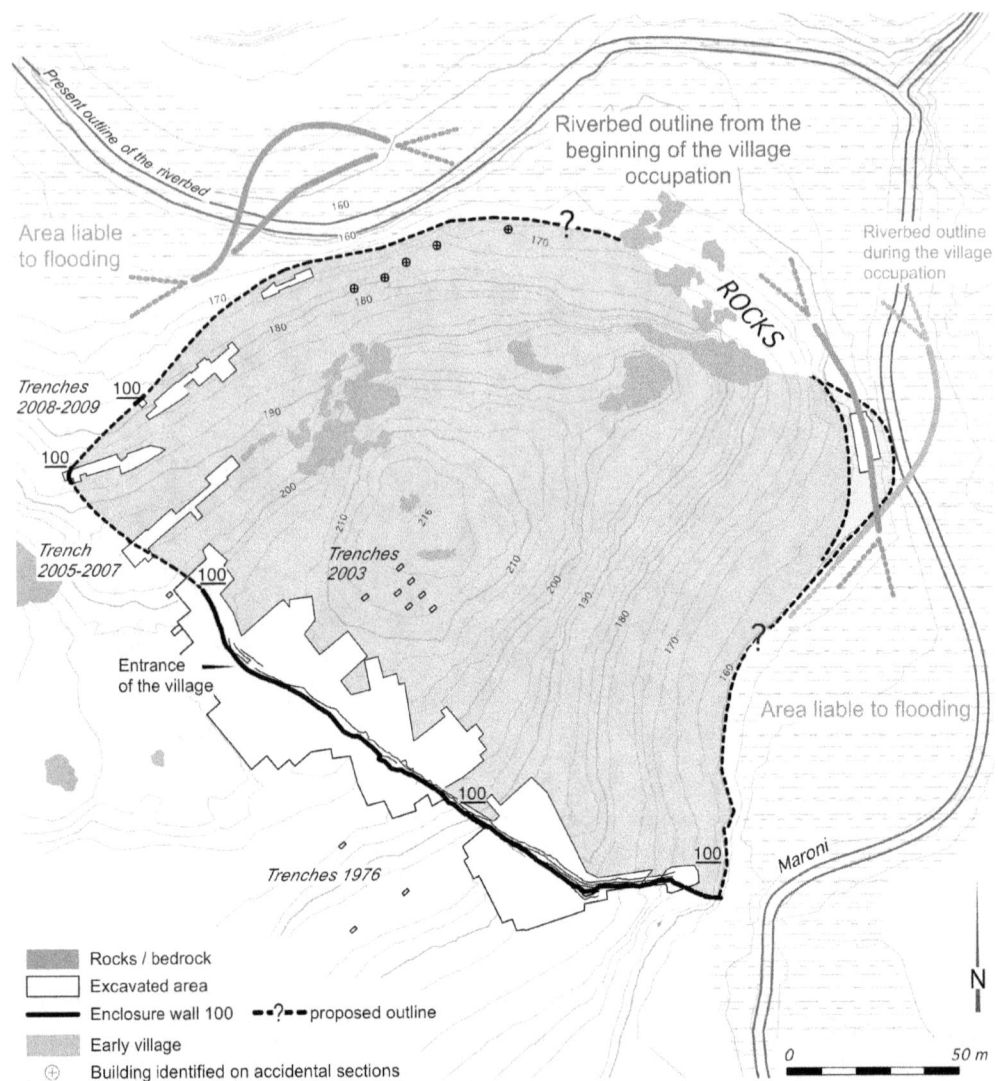

FIGURE 7.3. Khirokitia. The early village (levels J-B).

the occupation of this area. Taking advantage of the shift of the riverbed, the inhabitants extended the village there; the first constructions in this area (level P8b-structure S.75) are built on these deposits.

For all that, the activity of the river doesn't slow down. From this level until level P4, a succession of river floods can be observed, due to torrential rains as evidenced by pebble flows brought by water and sandy deposits left as the water level dropped. These violent floods also caused repeated heavy destruction, to which villagers replied with new architectural works.

This repeated fight, a true Sisyphean labor, was abruptly stopped by a catastrophic event combining both an attack of floodwaters and a sediment flow coming from the hill

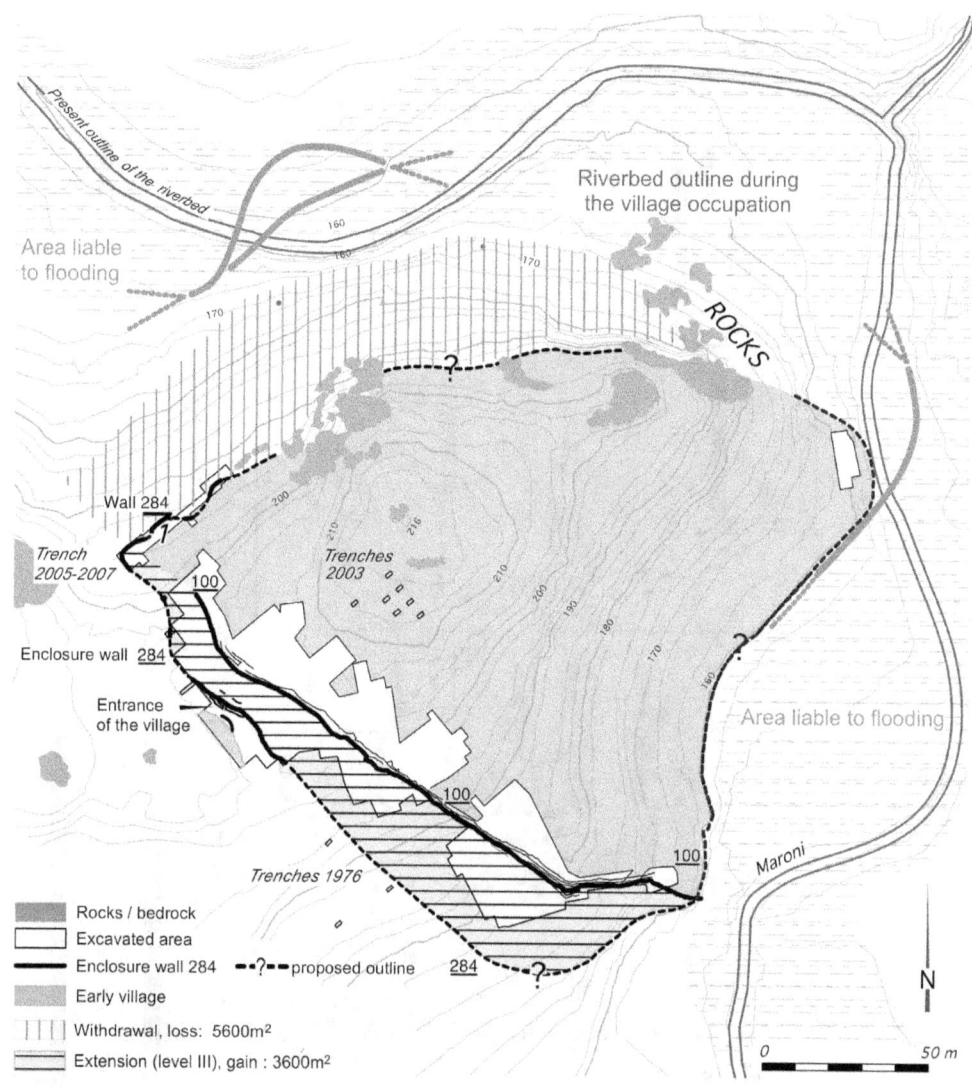

FIGURE 7.4. Khirokitia. The recent village after the shift (levels III and A).

slope, level P4a. The motion of these sediments most likely expressed a deep embankment of the river, probably together with a lateral shift of its course, concurrently with violent torrential rains falling on a bare and vegetation-less slope.

Yet life returned to normal and the area was again occupied during level P3d-b, but only for a short time before a new erosive episode occurred at level P3a. The destruction resulting from a new river flood was again combined with those resulting from new slope erosion deposits, ravaging the entire built area.

From this episode onward, environmental conditions appear to be more geomorphologically stable. Its course having moved away and being deeply embanked, the river seems to have come to a stable equilibrium, causing no more hillside erosion.

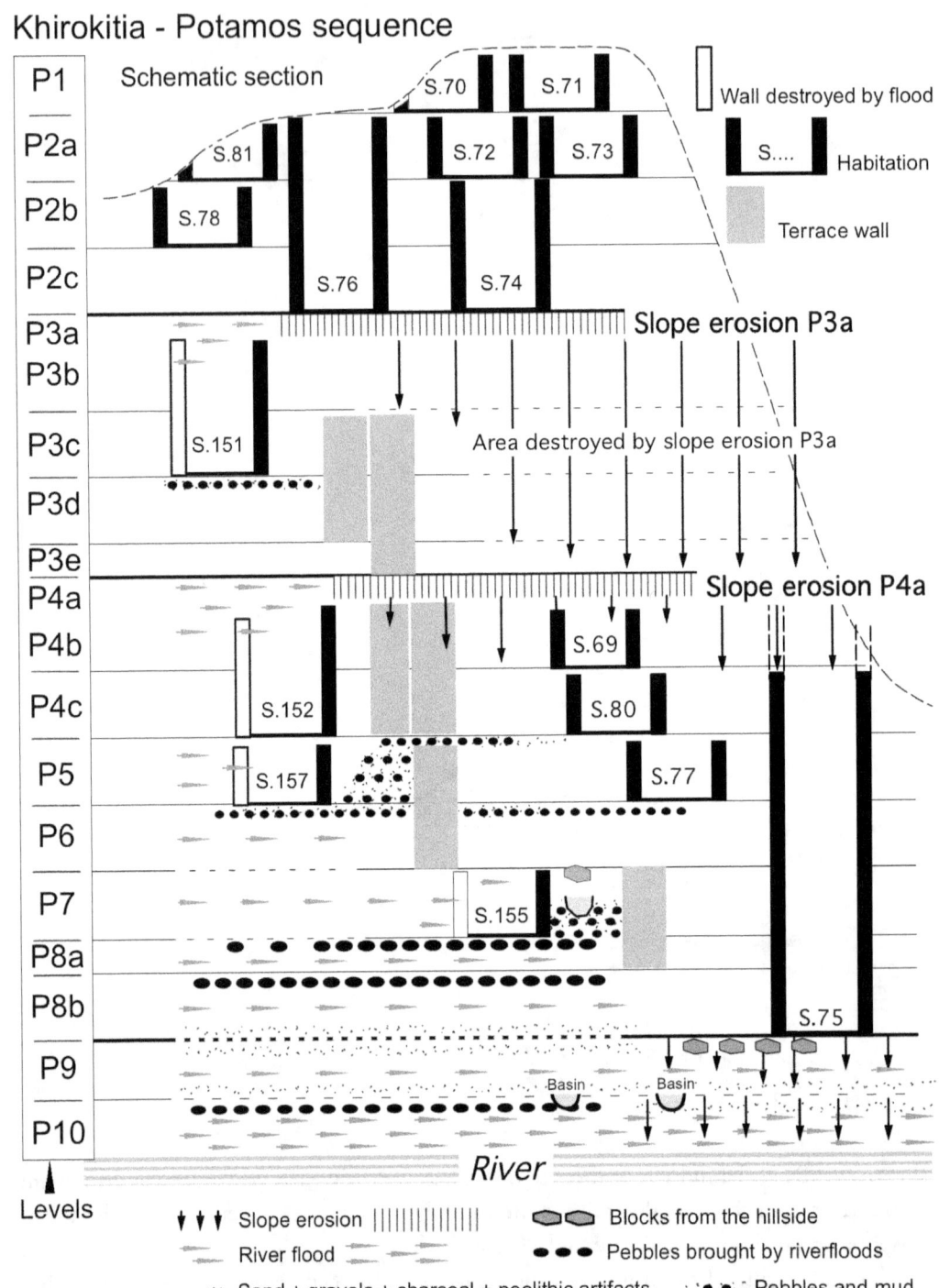

FIGURE 7.5. Khirokitia. The Potamos sequence.

FIGURE 7.6. Khirokitia. View on the Potamos sequence. Levels P10–P4 (Photo Mission Archéologique Française).

Correlating Climate and Culture Change

Considering these different sediment episodes that express a period of climatic instability, one may naturally wonder if this has to do with a simple local event or, on the contrary, with a regional transcription of the climatic changes observed elsewhere (such as in the Jordan Valley, Hourani and Courty 1998), generally called the 8.2 event. A significant change in the vegetal cover at this time has been revealed by anthracological analysis (Thiébault 2003) of samples collected on top of the hill. But before going farther, it is necessary to present the sequence (hereafter, reference sequence) and see if it is possible to adjust these two sequences based on observations made in two points located 150 m apart which cannot be strictly related for topographic reasons.

This so-called reference sequence consists of two series of architectural levels (Figure 7.7), the first comprising levels J to B and the second levels III to I. From J to B, the levels steadily follow one another within the same space enclosed by an imposing wall (100) which is repaired and modified several times, its outline remaining the same (Figure 7.3). This continuity ceases at the end of level B. The village then undergoes both a shift and a contraction of its built space (Le Brun and Daune-Le Brun 2010). Half of the northern slope is abandoned, and the loss in terms of surface is more significant than the gain obtained by extending on the western slope; there, the land, which had previously remained unoccupied, is built up during level III, and the new installation is accompanied by the simultaneous construction of a new enclosure wall (284), which again separates the inhabited territory from the outside world (Figure 7.4).

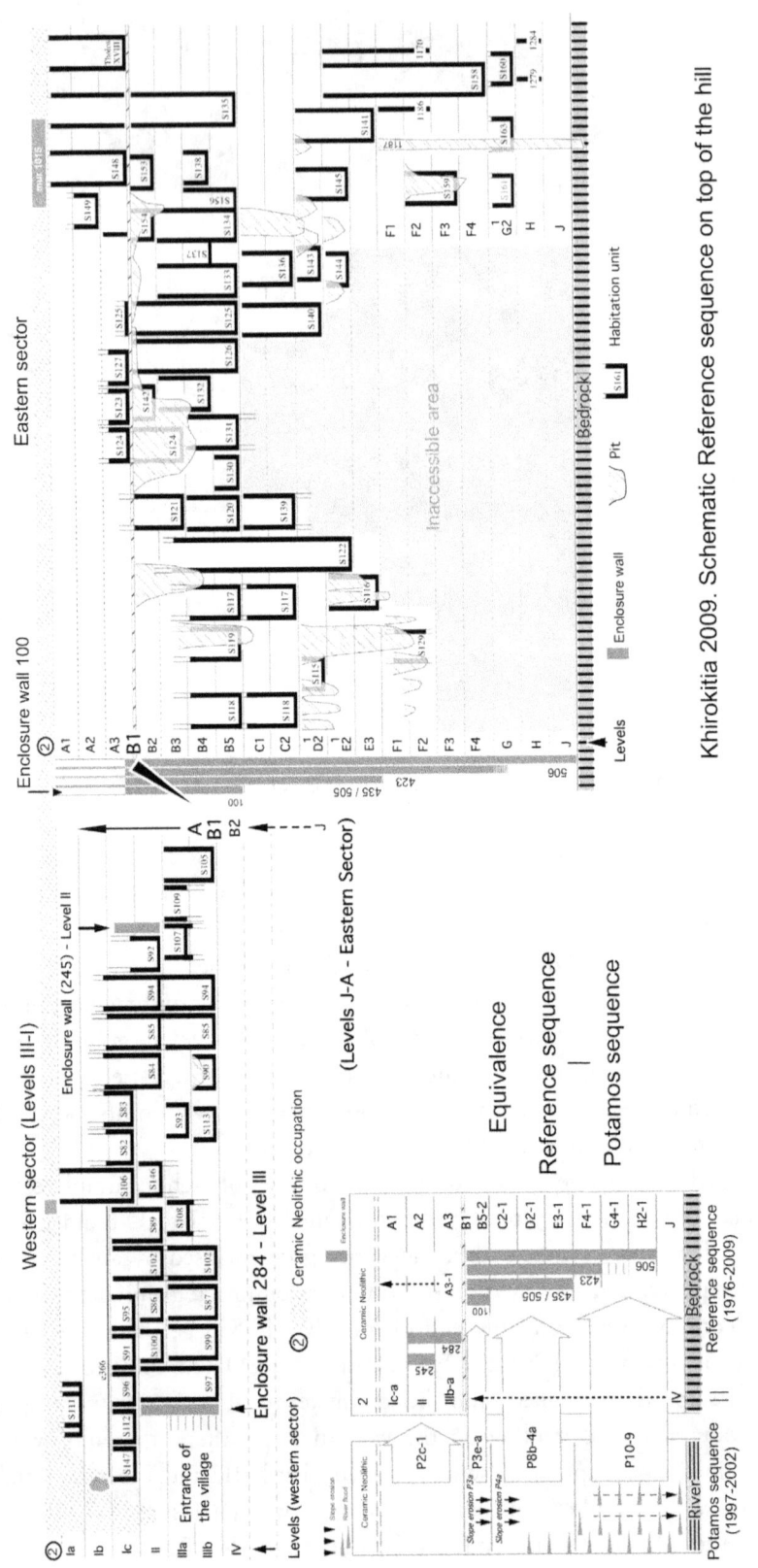

FIGURE 7.7. Khirokitia. Temporal correlations between the excavated Reference sequence and the Potamos sequence.

Although no strict equivalence in terms of level can be established between the reference and Potamos sequences, several convergent observations give the opportunity to propose a reliable adjustment (Figure 7.7). The earliest evidence of architectural activity in the Potamos sequence rests on deposits attesting to a human occupation: Aceramic Neolithic artifacts as well as a basin and trampling marks. Therefore, this first architectural level (P8) cannot equal the earliest levels of the reference sequence, levels J or H, which rest directly on the bedrock. Moreover, the comparison of material from levels P8–P4—that is, during the torrential flows marking this sequence—with material from the top of the hill, the evolution of which can be traced level by level, allows for more precision.

In the course of time, from level J to level B of the reference sequence, the incidence of limestone vessels, largely predominant in early levels, declines in relation to diabase stone vessels, the inversion of the trend happening between levels C and B. Yet limestone vessels that prevail at Potamos, their ratio (64 percent of the assemblage) being comparable to levels F-E of the reference sequence. As nothing in the data indicates a divergent evolution of both areas, one may at least suggest that Potamos deposits are earlier than those of level B.

The faunal assemblage provides other clues. The proportion of each large mammal species within the Potamos faunal assemblage is of the same order as in levels E to C of the reference sequence (Davis, personal communication 2011). Considering the percentage of ovines compared to goats, or the evolution in size of ovines, leads to the same conclusion. Although these observations do not constitute indisputable evidence, their convergence nevertheless leads one to consider the probable equivalence of early architectural Potamos levels P8–P4 with levels E-C of the reference sequence. In this case, it becomes possible to draw a parallel between the violent erosive episodes marking the Potamos sequence and the redistribution of the village and the break in the anthacological sequence, both events occurring during the transition of levels B and III.

The changes in vegetation—a decrease of Rosaceae (*Prunus sp.* and *Pomoïdeae*) as well as pistachio and olive trees, and an increase of pine and the appearance of juniper, which require less humid weather conditions—may have several explanations. It may result from a change in wood supply sources, or from a climatic change. But it could just as well result from an intensification of anthropic pressure on the environment, due to an increase of agriculture and/or ovine husbandry. The development of agriculture would have led to the use of new techniques for clearing vegetation—by using fire, for instance—resulting in an important expansion of pyrophyte plants such as pine. Regarding the increase in sheep husbandry, it would have led to overgrazing, vegetal cover deterioration, and the reconquest of the environment by species with fewer needs, such as pine and juniper (Thiébault 2003:228).

Whatever the reasons, a new relation between the environment and the landscape comes to light, evidenced in several fields. This results in the assertion of already existing tendencies as well as the emergence of new economic and cultural characteristics.

At the same time, that is, after the redistribution of the village at the end of level B, another change occurs which concerns faunal resources (Davis 2003, personal communication 2011). The faunal assemblage of large mammals—fallow deer, sheep, goat, and pig—remains the same, but each species ratio, rather steady until level B, changes in level III. Fallow deer, which are woodland animals, decrease from 38 percent in level B to 22 percent in level III, while sheep and goats increase from 41 percent to 65 percent (reaching 80 percent in level II), and pigs, animals that tend to prefer more humid environments, decrease from 21 percent to 14 percent and further to 10 percent in level II.

The prevalence of goats in comparison to sheep also shows an evolution. The ratio of 1 goat bone to 4 sheep bones identified in level B becomes 1 to 10 in level III, then 1 to 28 in level II. Moreover, while sheep do not change in size throughout the sequence, it is also after Level B that a change has been observed, with sheep gradually increasing in size.

The concept of the built space, considered as a closed space as well as the implantation model, remains unchanged. The village entrance keeps the same axial location, not far from a place in the village fabric that can be considered as a privileged place. It is here that, throughout the village occupation—including in level A (after the village spatial redistribution)—large buildings follow one after another, accompanied by open spaces equipped with oval-shaped platforms.

This continuity shows up despite the fact that the modification of the enclosure wall's outline resulted in a modification of the daily landscape, too—the everyday life landmarks were no longer the same. Moreover, the new enclosure wall is not built the same way as the previous one either in material use or technical assembly. Unlike the first enclosure, which is made of a thick earthen massif with an outer stone facing, the new one is only made of stones pointed with earth: limestone blocks available nearby on the hill but also large diabase pebbles collected in the riverbed. This change implies a different approach to environment as well as a different labor organization, which concerned the whole village; as a matter of fact, considering the enormous investment in time and labor required for the construction and the maintaining of these walls, building was a technical activity shared by the community.

The village pulse does not pound at the same rate. The steady rhythm of early levels, characterized by construction phases alternating with temporary withdrawal episodes, is followed from level III on by a rapid and almost jerky rhythm. In level II, which is soon after the village shift, the outline of the village limit is again modified, and this time the whole northern slope is abandoned (Figure 7.8).

If the constructions are still circular in plan, they are not built the same way, and their internal arrangement has changed. The "modeled" walls made of small stones embedded in trodden earth give way to a new type of walls combining an outer ring of stones and an inner ring of mudbricks, which come in general use. The wall thickening results in increasing the size of the buildings, however, without modifying their inner diameter. But the inner domestic space is not treated the same way. Its fragmentation, very loose in early levels, is amplified by the apparition and the development of partition walls. On the contrary, the small platform supporting a fireplace first located apart from the usual trapezoidal floor platform tends to be linked to it in the recent levels.

Regarding the massive pillars located inside some buildings, depending on their interpretation, as supporting a loft, or element of symbolic language, or both functions,

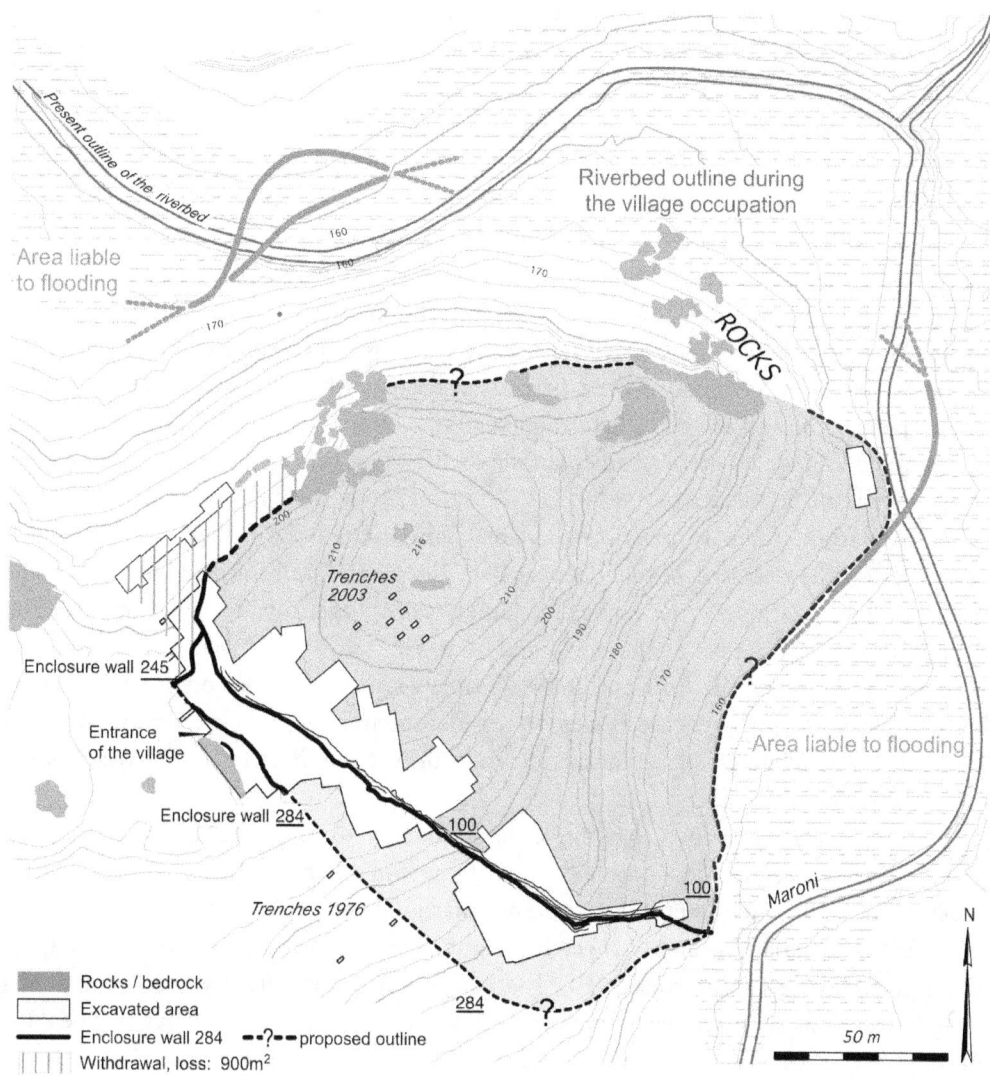

FIGURE 7.8. Khirokitia. Level II. Abandonment of the northern slope.

their decrease may be interpreted differently. In one case it may express a different conception of domestic space; in the other, it would mean at least a significant modification of symbolic practices.

Another striking thing as well, if one may consider together two different types of objects, is the diverging diachronic distribution of what are called "mace-heads," usually considered as prestige objects that are found in the early levels of the reference sequence, and the engraved pebbles, objects of enigmatic function, which are found in one other Cypriot site only (Simmons 1996) and which proliferate in the recent levels.

Craft activities reflect also this different approach to the environment and resources. Limestone was mainly employed in early levels for stone vessel manufacturing, as in earlier Cypriot sites dated from the ninth and eighth millennia, but this predominance decreased gradually during the site occupation until diabase became the preferred stone just before the village shift.

Regarding the bone industry, variations observed in the percentage of the faunal assemblage of different species seem not to effect the choice of raw material used, fallow deer bones, metapodes mainly, remaining the preferred bones. But the bone manufacturing shows innovations (Legrand 2007:126–130). In recent levels, the use of longitudinal segments with epiphysis replaces the use of segments without epiphysis. Regarding the tool manufacturing technique, direct percussion decreases for the benefit of a combined use of percussion and sawing. These modifications imply both a higher degree of technicity and a change of the instrumentation. At the same time, a preference emerges for needles with an ogival head.

In contrast, funerary practices did not change. From the earliest until the most recent levels, the deceased people uncovered in the village are buried in pits dug inside buildings that continue to be inhabited. Funerary language, too, remains the same: no change as to the position of the body, and as to the use of objects that in some cases accompany the deceased, such as knobbed stones or scapula placed on the body, stone vessels, or necklaces. These objects were sometimes intentionally broken during the burial ceremony, or deliberately placed upside down in the grave, which in a sense symbolically "destroyed" them.

This continuity is such that the same vocabulary—stone vessels, knobbed stones, scapula, fallow deer antlers—was still used at the end of the sequence, in level A, for sanctioning the abandonment of a building, that is, its end (Le Brun 2003).

All these variations and changes happened without the concept of the built space framework having been reconsidered. The village space was considered over time as a closed space, closed by an enclosure wall. This concept is revised in level I, when the northern slope is reoccupied (Figure 7.9). The village spreads out then freely, without any visible limit concretely marked by a wall for separating the inhabited from the outside world. On the contrary, constructions are built that overlap and obliterate the eroded remains of the ancient limit, blotting out even its souvenir.

Regarding constructions, if the circular plan continues to be the rule, expressed by buildings of larger dimensions, it is reappraised (Figure 7.9). It is the object of experimentations such as a model made of a central element surrounded by radiating compartments. It is also challenged by a construction subrectangular in plan with round corners that prefigures by several centuries the regular plan of Sotira Ceramic culture buildings (Dikaios 1961).

After ca. 5500 cal B.C., the Khirokitia culture disappears from the archaeological record of Cyprus, and there is a lengthy gap of about a millennium for evidence of any occupation of the island. During this time, the populations on Cyprus likely reverted to mobile small communities, far less visible archaeologically. However, the Aceramic sites

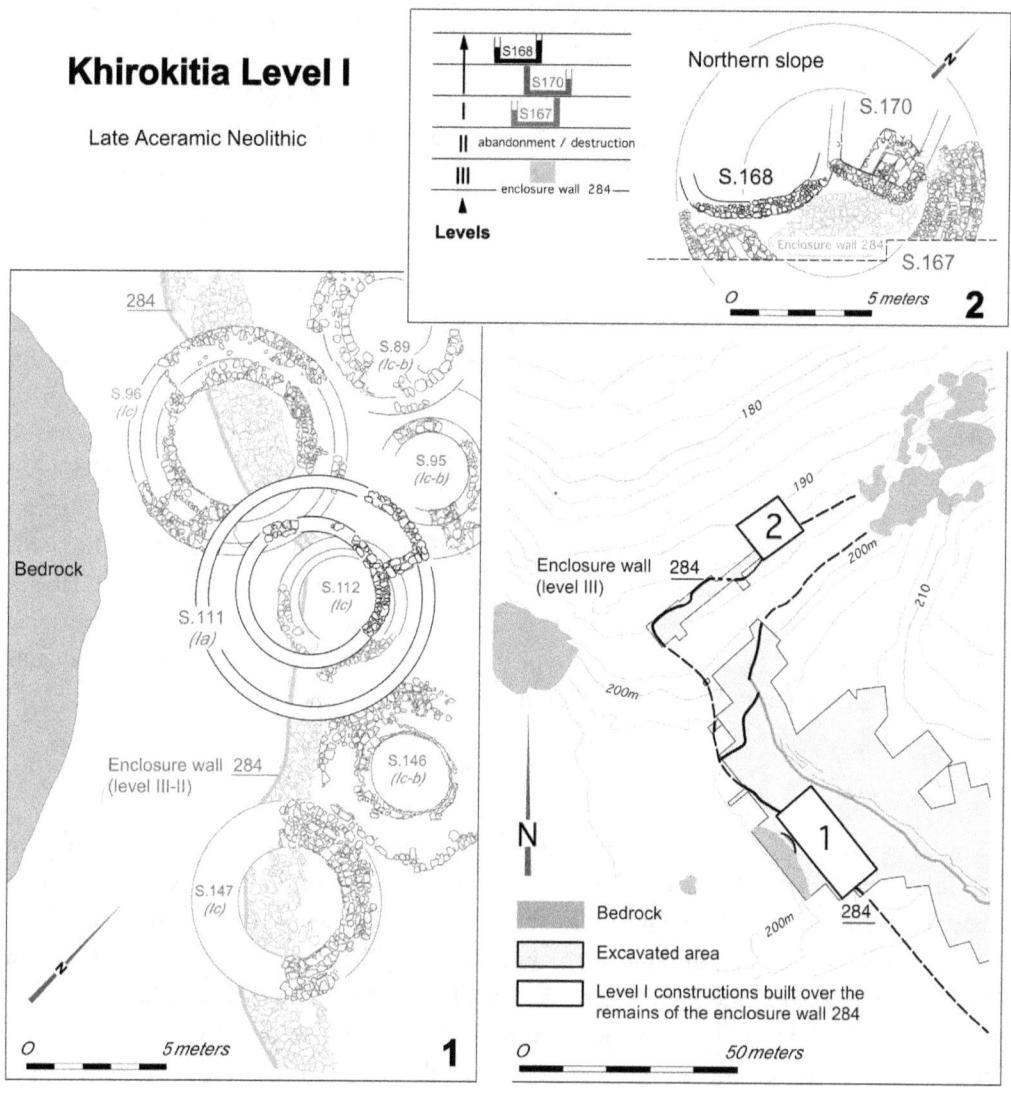

FIGURE 7.9. Khirokitia. Level I. New constructions built over the remains of the enclosure wall 284.

were not destroyed violently, and neither human agency nor natural causes by themselves are evident to explain this large-scale settlement crisis.

A likely explanation for this collapse may have been the vast size of the Khirokitia settlement, which would have exceeded the supposed socioeconomic basis of the traditional Neolithic society on the island (Peltenburg 2003). Some would even suggest the term *mega-site* for describing sites from this period (McCartney 2007:77), after a model inspired by the southern Levant. The term has been criticized; what does the term *mega-site* (Verhoeven 2006) exactly mean? In any case, the society would have been unable to cope with a scalar stress inherent in such a large community. The collapse of Khirokitia, the island's major site, would have then affected the whole of Cyprus.

The changes observed in level I, when the concept of the village as a closed space was abandoned, might certainly reflect a social upheaval. But the response to it, the renewed vigor sensible in the architectural activity and even in the improvement of life conditions and diet, as evidenced by the result of dental pathologies analysis, does not fit with such a catastrophic scenario.

The diminution over time of both dental diseases—2.5 carries per individual in early levels compared to 1.01 in recent levels—and periodontal diseases—50 percent affected individual in early levels compared to 35 percent in recent levels—suggests a change in the immune system, but probably also an improvement of life conditions and a more balanced diet (F. Le Mort and M. Sansilbano-Collilieux, personal communication 2004). Moreover, one may reverse the argument proposed of a "subsistence stress" (Peltenburg 2003:85) resulting from a decrease of fallow deer hardly compensated by the increase of sheep/goats, husbandry being qualified as a "more laborious means of production." The increase of sheep/goats may as well be interpreted as indicating an improvement in breeding techniques, also suggested by the increase of sheep size (Simon Davis, personal communication 2011), without mentioning also that the human death rate remains unchanged throughout the sequence (Le Mort 2003).

Conclusions

Supposing that the correlation proposed between the sedimentary episodes observed at Khirokitia and the 8.2 event observed elsewhere is somehow reliable, we must admit that these events don't have the same abrupt effects in Cyprus and in the Mainland: Khirokitia certainly goes through them but survives. The persistence, on one hand, of some characters along the sequence as observed in circular architecture, in funerary practices, or else in lithic industry, which is constantly marked by local raw material supply, poor technical investment, and limited retouched flaking (Astruc 2002), and, on another hand, the progressive assessment level after level of some other characters as illustrated, for instance, by stone vessel manufacture, show an overall cultural continuity. There is no doubt that the progressive dissolution of the links between Cyprus and the mainland and the renewed assertion of the population to maintain its identity may not be a complete stranger to this overall continuity. This last factor adds to the complexity of the interaction between human communities and their environments. It does not lessen the reality of the changes due to the redistribution of the village space.

How do the manifold cultural innovations and continuities observed at Khirokitia relate to climate shifts? What role did climate play as a causal agent? For the moment we are obliged to leave these questions unanswered because all these observations of these changes rely on one site only and require confirmation on other sites elsewhere in the island. However, the parallel that could be observed between weather vagaries and cultural changes appears most provocative.

Note

1. "Period 1" (Todd 2003).

References Cited

Astruc, L. 2002 *L'outillage lithique taillé de Khirokitia. Analyse fonctionnelle et spatiale*. CRA Monographies 25. CNRS Editions, Paris.

Davis, S. J. M. 2003 The Zooarchaeology of Khirokitia (Neolithic Cyprus), Including a View from the Mainland. In *Le Néolithique de Chypre*, edited by J. Guilaine, and A. Le Brun, pp. 253–268. Bulletin de Correspondance Hellénique Supplément 43. Ecole française d'Athènes, Athènes.

Dikaios, P. 1953 *Khirokitia*. Oxford University Press, Oxford.

Dikaios, P. 1961 *Sotira*. University of Pennsylvania, Philadelphia,

Hourani, F. 2008 Khirokitia (Chypre), un village néolithique les pieds dans l'eau. In *L'Eau. Enjeux, usages et représentations*, edited by A. M. Guimer-Sorbets, pp. 159–169. De Boccard, Paris.

Hourani, F., and M.-A. Courty 1998 L'évolution morpho-climatique de 10500 à 5 500 BP dans la vallée du Jourdain. *Paléorient* 23(2):95–105.

Le Brun, A. 1981 *Un site néolithique précéramique en Chypre: Cap Andreas-Kastros*. Association pour la Diffusion de la Pensée Française, Paris.

Le Brun, A. 1984 *Fouilles récentes à Khirokitia (Chypre), 1977–1981*. Editions Recherche sur les Civilisations, Paris.

Le Brun, A. 1989 *Fouilles récentes à Khirokitia (Chypre), 1983–1986*. Editions Recherche sur les Civilisations, Paris.

Le Brun, A. 1994 *Fouilles récentes à Khirokitia (Chypre), 1988–1991*. Editions Recherche sur les Civilisations, Paris.

Le Brun, A. 2003 Idéologie et symboles à Khirokitia: la "fermeture" d'un bâtiment et sa mise en scène. In *Le Néolithique de Chypre*, edited by J. Guilaine and A. Le Brun, pp. 341–349. Bulletin de Correspondance Hellénique, Supplément 43. Ecole française d'Athènes, Athènes.

Le Brun, A., and O. Daune-Le Brun 2010 Khirokitia (Chypre): la taille et les pulsations de l'établissement néolithique précéramique, nouvelles données. *Paléorient* 35(2):67–76.

Legrand, A. 2007 *Fabrication et utilisation de l'outillage en matières osseuses du Néolithique de Chypre: Khirokitia et Cap Andreas-Kastros*. BAR International Series 1678. Archeopress, Oxford.

Le Mort, F. 2003 Les restes humains de Khirokitia: particularités et interprétations. In *Le Néolithique de Chypre*, edited by J. Guilaine and A. Le Brun, pp. 313–325. Bulletin de Correspondance Hellénique, Supplément 43. Ecole française d'Athènes, Athènes.

McCartney, C. 2007 Lithics. In *On the Margins of Southwest Asia. Cyprus during the 6th to 4th Millennia BC*, edited by J. Clarke, pp. 72–90. Oxbow Books, Oxford.

Peltenburg, E. J. 2004 Social Space in Early Sedentary Communities of Southwest Asia and Cyprus. In *Neolithic Revolution. New Perspectives on Southwest Asia in Light of Recent Discoveries on Cyprus*, edited by E. Peltenburg and A. Wasse, pp. 71–89. Levant Supplementary Series 1. Oxbow Books, Oxford.

Simmons, A. H. 1996 Preliminary Report on Multidisciplinary Investigation at Neolithic Kholetria-Ortos. *Report of the Department of Antiquities, Cyprus* 1996:29–41. Nicosia.

Thiébault, S. 2003 Les paysages végétaux de Chypre au Néolithique: premières données anthracologiques. In *Le Néolithique de Chypre*, edited by J. Guilaine and A. Le Brun, pp. 221–230. Bulletin de Correspondance Hellénique Supplément 43. Ecole française d'Athènes, Athènes.

Todd, I. A. 2003 Kalavasos-Tenta: A Reappraisal. In *Le Néolithique de Chypre*, edited by J. Guilaine and A. Le Brun, pp. 35–44. Bulletin de Correspondance Hellénique Supplément 43. Ecole française d'Athènes, Athènes.

Verhoeven, M. P. F. 2006 Megasites in the Pre-Pottery Neolithic B. Evidence for "Proto-Urbanism"? In *Domesticating Space: Landscapes and Site Structures in the Prehistoric Near East*, edited by E. B. Banning and M. Chazan, pp. 75–79. SENEPSE 6. Ex oriente, Berlin.

Chapter Eight

Early Holocene Climatic Fluctuations and Human Responses in Greece

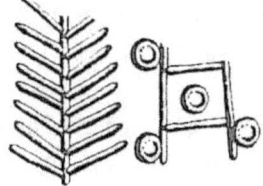

Catherine Perlès

Abstract *This paper investigates the effects of climatic fluctuations in two different socioeconomic contexts: hunter-gatherers of the Late Pleistocene and Early Holocene, sedentary farmers of the mid-Holocene. Two different approaches are used in order to search for correlation between climatic events, settlements patterns, site use, subsistence economy, etc. A fine-grained approach, focused on the sequence of Franchthi Cave, will be applied for the early part of the period and foraging economies. A broad-based approach on a large geographical scope will look for disruptions linked to 8.2 event among Early/Middle Neolithic communities. Neither approach allowed us to find unambiguous synchronicities and causal relations between climatic and socioeconomic change. Besides methodological problems inherent to each approach, such as problems of equifinality for the first, of chronological resolution for the second, this probably also reflects the resilience of these ancient socioeconomic systems to environmental change. In the case of the 8.2 event, whose environmental impact is well documented in Greece, one can also question our own mental templates and the very notion of "climatic deterioration."*

Introduction

All climatic records from Greece and the Aegean Sea concur to demonstrate important climatic and environmental transformations in the Early Holocene (ca. 12,000–8000 cal B.P.), characterized by a sharp rise in rainfall and the expansion of the forest, subsequently interrupted by the dry and cool episode of the 8.2 ka cal B.P. climatic event (henceforth the 8.2 event). These climatic transformations, however, were only one aspect of more wide-ranging transformations that included the arrival of new populations, sedentism, and a complete economic shift. As a consequence, the investigation of human responses to climatic change involves disentangling the effects of independent but potentially interrelated dynamics: an environmental dynamic and various socioeconomic ones. The Early Holocene in Greece indeed corresponds to two very different phases in terms of human adaptation and way of life. Until the beginning of the ninth millennium B.P., Greece was still occupied by groups of mobile hunter-gatherers. Around 8700/8600 cal B.P., sedentary farmers started to settle in the region and introduced a mixed farming economy that led to the rapid disappearance of traditional ways of life. We can thus theoretically expect very different answers to climatic and environmental transformations among the mobile Mesolithic groups, exploiting a wide range of natural resources, and among the sedentary Neolithic groups who relied almost exclusively on domestic resources, brought along from the Near East.

In addition to its anthropological relevance, the period studied here offers the interest of diametrically opposed situations in terms of quality and quantity of the data. This will lead us to investigate the heuristic potential of two very different methodological approaches: a narrow-focused approach with high resolution for the Mesolithic, and a very broad focus for the Neolithic, relying on a much larger data base albeit with lower resolution. The discussion on the effects of the Younger Dryas/Preboreal transition on Mesolithic hunter-gatherers' adaptations will concentrate on the Franchthi Cave, one of the very few sites in Greece to offer a well-dated sequence. Despite abundant, varied, and detailed data, differing interpretations in terms of environmental impact have been recently published and will serve as a basis of discussion. In a second section, we shall investigate the effects of the following major climatic event, the 8.2 event, this time on early farming communities. Contrary to what obtains with the preceding climatic change, the chronological resolution of available ^{14}C dates does not allow an investigation of this short-term event on individual sites. We shall thus address the question from a large geographical scope, and search for transformations in settlement patterns and subsistence economy that could be related to this climatic event. In both cases, however, the procedure that will be followed is the same, a classical procedure for archaeologists or historians investigating the impact of climatic change: a search for temporal correlations between climatic events, environmental transformations, and changes in site location, site use, type of occupation, food acquisition or production (Berger and Guilaine 2009:32). While not enough to infer causality, establishing synchronicity is an essential first step.

It will be shown, however, that in both cases strict correlation between climatic or environmental transformations and shifts in the socioeconomic bases are difficult to bring to light. This is in part, but only in part, related to important discrepancies between purely environmental data and site-specific data, located in different topographic environments and of widely differing resolution. The importance of surface finds in our Neolithic database, whose relative dating is imprecise in relation to the brevity of the 8.2 event, hampers the search for tight chronological correlations. In addition, as already argued, even when the latter can be established, a correlation is not necessarily a cause, and other potential factors of change must be taken into account. Socioeconomic systems result from an interplay between environmental constraints and the groups' choices and traditions, so that social and ideological factors must be considered as equally relevant in the interpretation of economic transformations. Potential changes in the status of the site will thus interfere with climatic and environmental factors in the interpretation of economic changes at Franchthi. However, the major obstacle may well be, in the end, the fact that the economic strategies of both hunter-gatherers and farmers were sufficiently resilient to withstand the environmental transformations of the Early and Middle Holocene in Greece.

Mesolithic Hunter-Gatherers and the Advent of the Holocene

Although of a general Mediterranean character, with dry hot summers and a concentration of precipitation in the cooler, winter months, the climate and natural environments in Greece vary according to three gradients: a west-east gradient of decreasing rainfall, a north-south gradient of decreasing rainfall and increasing temperatures, and an altitudinal gradient of increasing rainfall and decreasing temperatures (Bottema 2003; Polunin 1987). Since these gradients are linked with topography and latitude, they were necessarily operative during the Holocene and must be kept in mind when one has, by necessity, to transfer the results of environmental analyses from one region to another.

Despite intensive investigations, several uncertainties remain concerning the Late Glacial and Early Holocene climate in Greece (Tzedakis 2007; Wilson et al. 2008). The Early Holocene witnessed the gradual expansion of forests to the detriment of the steppe vegetation that prevailed during the Late Glacial. This phase of forest development corresponds chronologically to the deposition of the Sapropel S1 in the Aegean Sea (Figure 8.1), between ca. 10,000–6000 cal B.P. (Rossignol-Strick et al. 1982; Schmiedl et al. 2010), which is generally considered to mark a relatively mild and wet climate of definite Mediterranean character (Peyron et al. 2011; Roberts et al. 2011).

In Aegean core SL 152 (Figure 8.2, page 173), the sapropel deposition is coeval to an increase in terrestrial *Quercus* pollens, with estimated winter temperatures of ca. 4–6° and winter precipitations of ca. 270 mm (Kothhoff et al. 2008). Other climatic proxies support the hypothesis of a wetter climate in the Early Holocene: the $\delta^{18}O$ from biogenic and endogenic lake carbonates (Roberts et al. 2008) shows minimal $\delta^{18}O$ values between 11 and 8 ka in Lake Ioannina in northwest Greece (Figure 8.2). Similarly, the species composition and $\delta^{18}O$ of marine dinoflagellates and planktonic foraminifera from several cores in the northern, central, and southeastern Aegean Sea (Figure 8.2), indicate an increase in precipitations estimated to about 20 percent for the eastern Mediterranean as a whole (Marino et al. 2009; Schmiedl et al. 2010).

However, this hypothesis is contradicted, according to Tzedakis (2007), by a peak in Mediterranean schlerophyls during the first half of the sapropel S1 deposition, such as *Pistacia, Phyllyrea,* or evergreen *Quercus* (Rossignol-Strick 1997; Magri and Tzedakis 2000). He thus suggests that summer aridity was more pronounced than today, and that, "if an increase of humidity did occur at all, it must have been confined to autumn/winter" (Tzedakis 2007:2054; see also Peyron et al. 2011). Regional differences must also be taken into account: for instance, the depletion of $\delta^{18}O$ in the carbonates of Lake Kopais in Central Greece is much less pronounced than in Lake Ioannina (Griffiths et al. 2002;

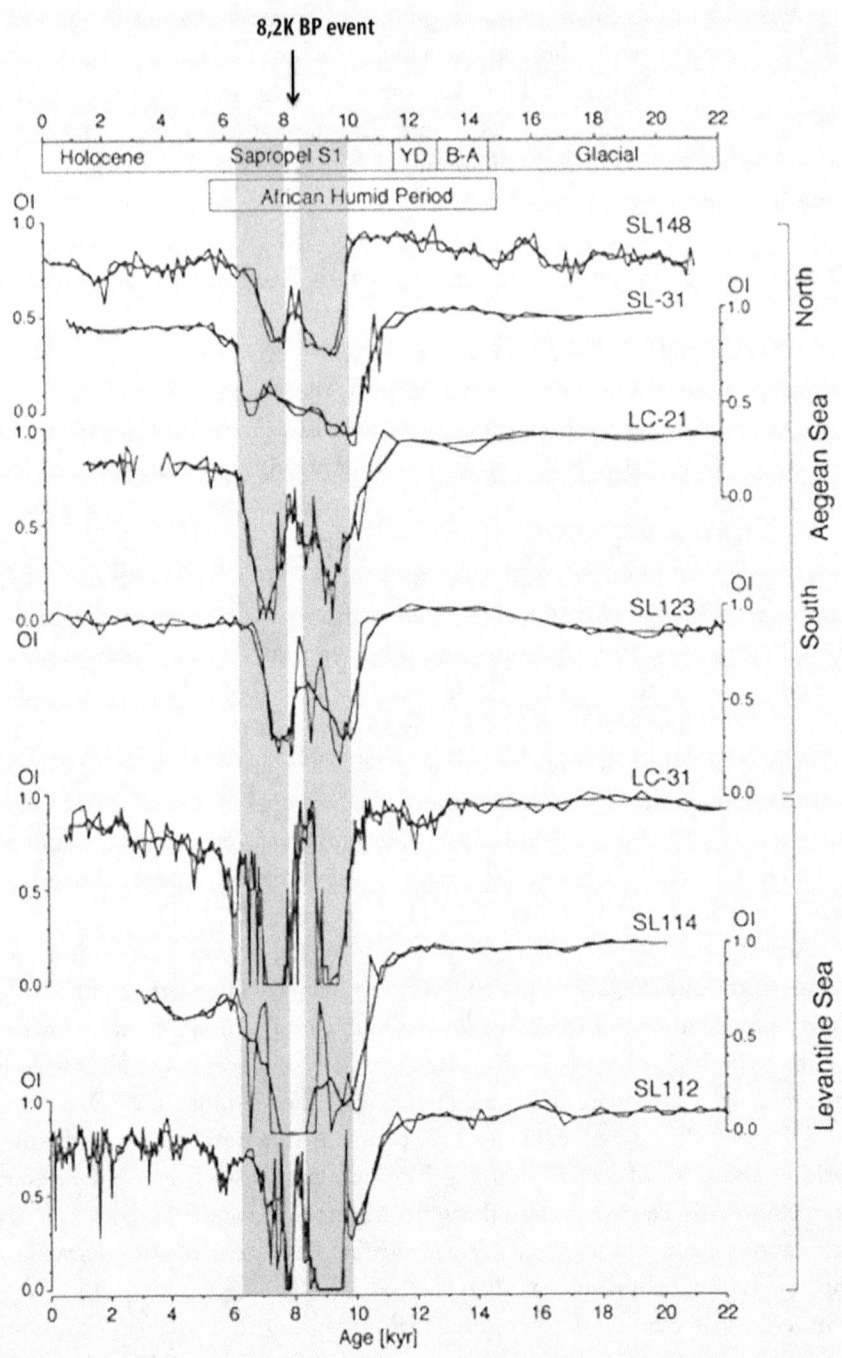

FIGURE 8.1. Benthic foraminiferal Oxygen Index (OI) for eastern Mediterranean sediment cores along an NW-SE transect. The sapropel S1 is characterized by depleted OI values, interrupted by a sharp rise during the 8.2 event (after Schmiedl et al. 2010). Copyright Elsevier; used with permission.

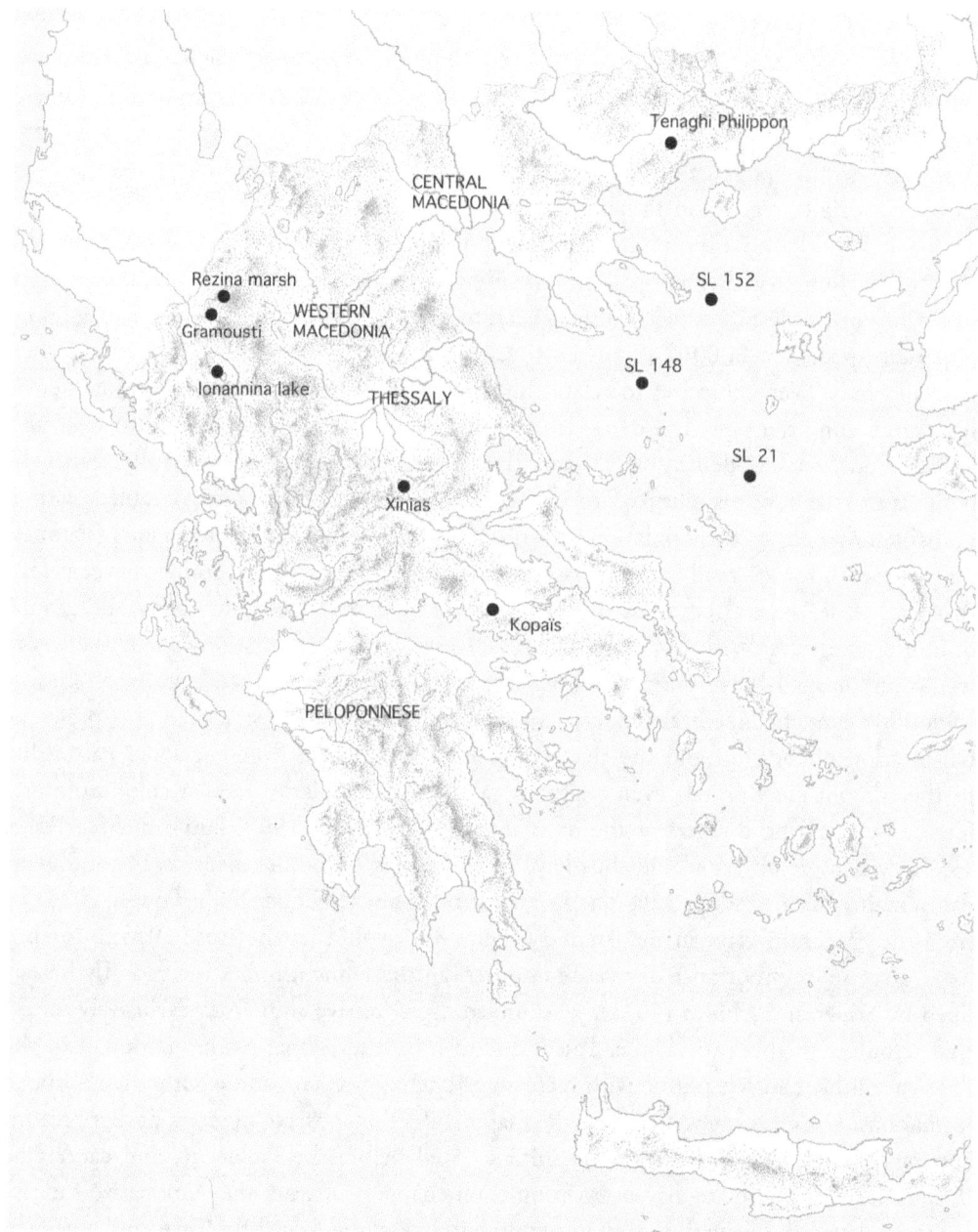

FIGURE 8.2. Location of the cores discussed in the text (background map by G. Monthel).

Roberts et al. 2008). These regional differences are also reflected in pollen records. At the beginning of the Holocene, an open deciduous oak forest gradually developed at slightly different tempi according to the location. It became prevalent around 9.4 ka, at least at elevations of 400 m or more, and then gave way, in the moister western Pindus, to a more diversified woodland including *Fraxinus, Tilia, Ulmus, Corylus, Pistacia,* etc. (Bottema 1991, 1994, 2003; Kotthoff et al. 2008; Lawson et al. 2004; van Zeist and Bottema 1982; Willis 1994). To the contrary, the open oak woodland remained dominant in Macedonia and Thessaly (Bottema 2003), while farther south, at Franchthi Cave in the Argolid, where no pollen record is available for this period, the carbonized seeds suggest an open garrigue vegetation rather than an oak forest (Hansen 1991).

We should thus expect rather different responses to the Early Holocene climatic changes by hunter-gatherers groups, depending on the region. Mesolithic sites remain scarce in Greece despite a recent intensification of research (Galanidou and Perlès 2003; Kyparissi-Apostolika 2000; Laskaris et al. 2011; Runnels 2004, 2009; Sampson 2011). It would have been tempting to relate the scarcity of Mesolithic sites to the new environmental conditions, with a dense tree or bush cover, large game limited to deer and available only in low density, if Upper Paleolithic sites had not been equally scarce. In both cases, the low visibility of the short-term open-air settlements, coupled with a probable low density of population and the selection of specific environments (Runnels 2009), explains the small number of sites recorded. There is no doubt, nevertheless, that the denser vegetal cover in the Holocene created new problems to mobile groups of hunter-gatherers. With few rivers to provide easy inland penetration routes, the coasts would undoubtedly have provided the easiest routes of circulation, and almost all known Mesolithic sites in Greece are indeed coastal or close to the coast. Given this preferential location of the sites and the changes in lithic production from the Final Paleolithic to the Mesolithic, Runnels even suggested a shift in population and a colonization of Greece by seafaring foragers at the dawn of the Mesolithic (1995, 2004; but see Perlès 1990). Marine resources would thus have been a major economic focus, as the abundant fish remains from Kyklops cave on the island of Youra (Mylona 2003; Powell 2003) or from the Franchthi Cave in the Argolid (Figure 8.3) would testify (Rose 1995, in prep.).

The rise in importance of marine resources in the Franchthi diet was recently underlined by Stiner and Munro (2011), who linked a progressive shift from exclusively terrestrial resources in the Early Upper Paleolithic to mixed terrestrial-marine resources in the Late Paleolithic and Mesolithic with increasing foraging pressure and environmental transformations. However, they stress that the major shifts they observe in terms of diet breadth and range of habitats exploited were initiated well before the Holocene, and cannot be directly related to climate-driven environmental changes. Instead, they emphasize the rise in sea level and the reduction of the coastal plain below the Franchthi Cave, and a general decline in high-ranked faunal resources. Although my own interpretation of the sequence differs (Perlès 2010a), I concur that the relations between economic transformations and climatic fluctuations in the Franchthi sequence are all but straightforward. A more detailed examination of the Final Pleistocene/Early Holocene transition will illustrate the point.

Despite the presumed increase in humidity during the Early Holocene, especially in comparison with the Younger Dryas, the vegetation around Franchthi, as reconstructed from the carbonized seeds, appears to have remained very stable. A park-woodland or woodland-steppe vegetation was established at least by the fifteenth millennium cal B.P. and it seems to have been resilient to the climatic fluctuations of the Final Pleistocene

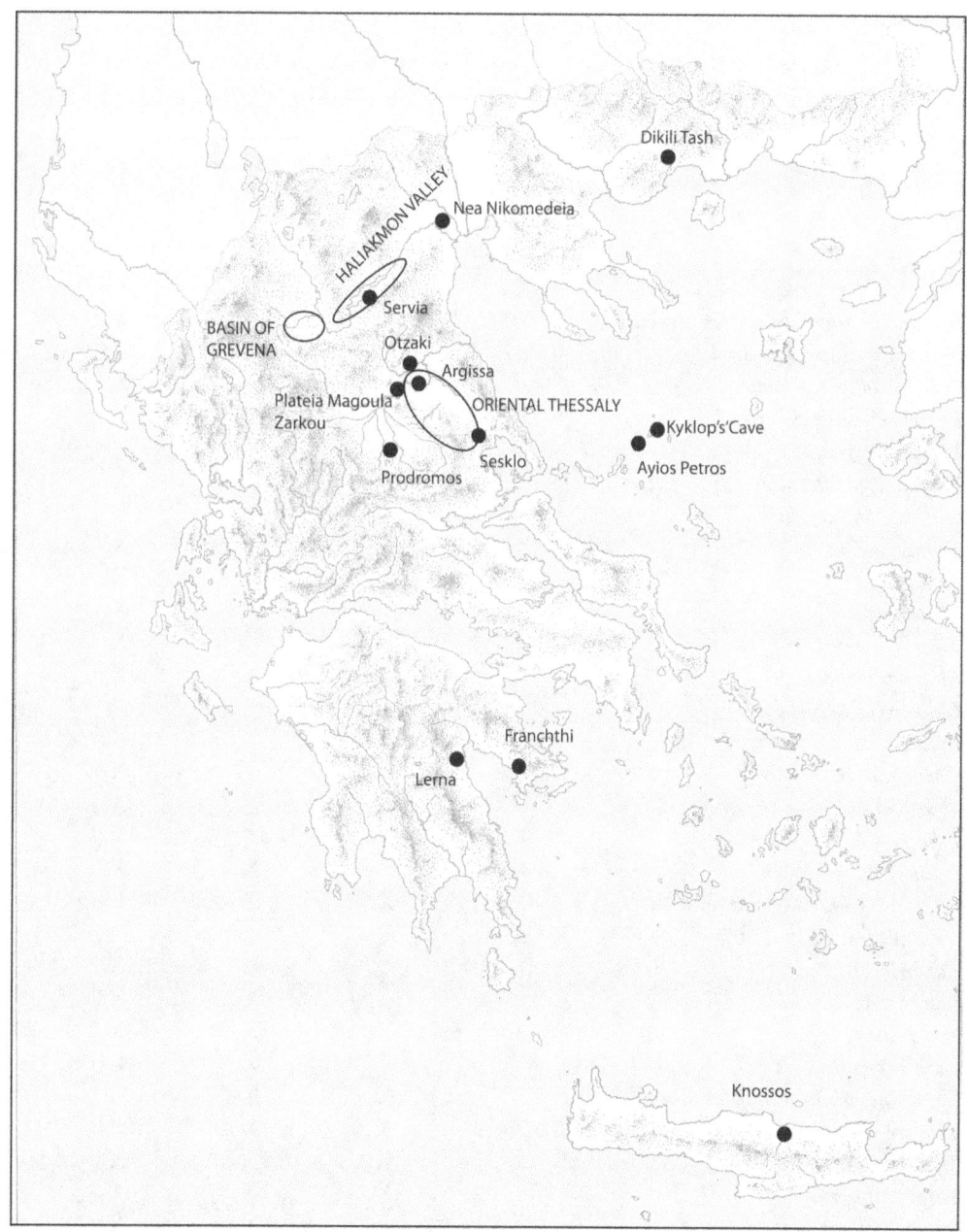

FIGURE 8.3. Location of the sites and archaeological regions discussed in the text (background map by G. Monthel).

and Early Holocene. There are no significant changes in the main taxa exploited—which comprise *Prunus amygdalus, Pistacia* cf. *lentiscus, Pyrus amygdaliformis, Lens* sp. *Lathyrus* sp., *Vicia ervilia, Avena* sp., *Hordeum vulgare* ssp. *spontaneum*—throughout the Bölling, Alleröd, Younger Dryas, and Preboreal (Table 8.1, after Hansen 1991). During the

TABLE 8.1
COMPOSITION OF THE SEED ASSEMBLAGES AT FRANCHTHI CAVE,
FROM THE BÖLLING TO THE PREBOREAL (DATA AFTER HANSEN 1991)

	Phase 4 Late Palaeo, (Bölling)	Phase 5 Final Palaeo, (Alleröd)	Phase 6 Final Palaeo, (Dryas III)	Phase 7 Early Mesolithic (Preboreal)
Buglossoides arvensis (uncarbonized)	X	X	X	X
Alkanna cf *orientalis* (uncarbonized)	X	X	X	X
Anchusa sp. (uncarbonized)	X	X		X
Lithospermum officinale		X		X
Pistacia cf. *lentiscus*	X	X	X	X
Prunus amygdalus	X	X	X	X
Pyrus amygdaliformis	X	X	X	X
Celtis cf. *tournefortii*,		X		X
Adonis sp.	X		X	X
Fumaria sp.	X		X	X
Phalaris sp.	X		X	X
Medicago sp.	X			X
Malva parviflora			X	X
Capparis cf. *spinosa*				X
Erodium sp.				X
Calendula sp.				X
Crucifera sp.				X
Liliacae	X	X	X	X
Galium sp.	X	X	X	X
Cf. *Colchicum/Polygonatum*		X	X	X
Vinis vitifera ssp. *sylvestris*	X			X
Fumaria sp.	X			X
Cirsium sp.			X	X
Avena sp.	X	X	X	X
culm nodes ind.			X	X
Hordeum vulgare ssp. *spontaneum*	X	X	X	X
Lens sp.	X	X	X	X
Lathyrus sp.	X	X		X
Large *Lathryrus*		X	X	X
Small *Lathryrus*			X	X
Lathyrus cicera/sativa				X
Vicia/Lathyrus sp.		X	X	X
Vivia ervilia	X	X	X	X
Pisum/vicia sp.		X	X	X
Large *Pisum* sp.			X	X
Pisum elatius/humilis				X
Small Leguminosae		X	X	X
Medium Leguminosae		X	X	X
Large Leguminosae				X
TOTAL	422	240	244	Ca. 28,000

Alleröd, intense occupation of the cave is marked by the deposition of thick land-snail middens (*Helix figulina*), while the higher-ranked faunal species previously exploited, *Bos primigenius* and *Equus hydruntinus* decline and disappear. Deer and wild boar are now the dominant hunted species, accompanied by *Capra*, small carnivores, pond-turtles, hare, and mollusks (Stiner and Munro 2011). Edible marine mollusks are collected for the first time, while small-scale fishing is attested.

The intensity of occupation dropped down markedly during the Younger Dryas. While botanical species remain unchanged, the faunal assemblage, still dominated by deer, wild boar, hare, and fox, is drastically reduced. Seashells and land snails were still collected, but in small quantities. The few concentrations of carbonized seeds, land snails, and shells all come from different excavation units, as though different resources were primarily exploited at each visit. This accords with the seasonality indicators, which show that the seashells were variously collected during the summer, the fall, and the winter (Deith and Shackleton 1988). This would accord with sporadic visits to the cave by mobile groups relying on a broad-spectrum economy based on hunting, snaring, collecting, and fishing.

The cave was abandoned during the later part of the Younger Dryas and reoccupied during the Preboreal. This Lower Mesolithic occupation is dated to a few centuries between ca. 10,800 and 10,300 cal B.P. It corresponds to a markedly increased intensity of occupation, to the extent that permanent settlement has been suggested (Runnels 1995). A denser tree cover is inferred from the global rise in rainfall, but, as stated above, the seed assemblages do not reflect any major environmental transformation. A few seeds of capers (*Capparis* cf. *spinosa*) and *Celtis* cf. *tournefortii* may indicate a warmer weather, if their presence is not solely due to a vastly enlarged sample.

Subsistence activities are now clearly focused on the gathering of plants and land snails. Large and small game is scarce and, contrary to various statements quoted above, the exploitation of marine resources, whether fish or edible mollusks, is extremely limited when actual number of specimens is considered (Perles 2016; Stiner and Munro 2011:Appendix 1). This is well exemplified by marine mollusks (Shackleton 1988). When ornamental *C. neritea* and *C. rustica*, which account for up to 80 percent of the assemblages, are excluded, the remaining predominant species by far is *Cerithium vulgatum*, followed by *Hexaplex trunculus* and *Cerastoderma glaucum*. *Cerithium* and *Hexaplex* have been considered non-edible (Colonese et al. 2010) and possibly used as bait (Stiner and Munro 2011), but their state of preservation suggests they were eaten, as they are still nowadays in several Mediterranean regions Their overall density remains low, however, with about one shellfish for three liters of sediment and they never constitute shell middens (Table 8.2).

This dietary shift and the heavy reliance on plants and land snails *can* be interpreted as a consequence of environmental change: the climatic amelioration of the Preboreal would have enhanced the primary productivity of the environment—as well as the availability of land snails—while the denser vegetal cover would have rendered hunting and snaring more difficult. However, was the environment very different from that of the Bölling? One could also consider a potential shift in the status of the cave. From the very beginning of its reoccupation in the Holocene, the cave is used, for the first time, as a burial ground (Cullen 1995). Besides numerous scattered human bones considered to

TABLE 8.2
COMPOSITION OF THE MARINE MOLLUSCAN ASSEMBLAGES FROM THE LOWER MESOLITHIC OF TRENCH FAS (DATA AFTER SHACKLETON 1988)

Unit	Volume of sediment (in liters)	*Cerithium vulgatum*	*Hexaplex trunculus*	*Cerastoderma glaucum*	*Columbella rustica*	*Cyclope cf. neritea*	Others	Total marine mollucs
FAS 198	208	14	0	1	3	81	3	102
FAS 197	226	12	0	2	6	58	4	82
FAS 196	554	13	0	18	3	127	20	181
FAS 195	440	7	7	7	7	16	7	51
FAS 194	28	0	4	2	0	51	3	60
FAS 193	30	2	3	3	0	16	0	24
FAS 192	346	50	21	45	5	240	9	370
FAS 191	381	46	40	41	5	141	3	276
FAS 190	312	14	5	14	1	32	3	69
FAS 189	346	38	10	25	0	56	5	134
FAS 188	277	78	13	54	3	112	2	262
FAS 187	277	73	28	16	2	88	3	210
FAS 186	277	84	64	10	2	111	7	278
FAS 185	346	104	37	10	2	125	5	283
FAS 184	450	396	44	4	5	264	16	729
FAS 183	143	189	16	2	1	124	6	338
FAS 182	371	47	120	0	6	239	10	422
FAS 181	151	12	30	0	2	63	3	110
FAS 180	214	119	10	0	6	250	7	392
FAS 179	277	55	2	0	0	236	12	305
FAS 178	14	3	1	0	0	58	2	64
FAS 177	346	23	3	0	7	342	4	379
FAS 176	649	9	3	0	0	243	32	287
FAS 175	112	5	2	0	0	58	8	73
FAS 174	139	28	6	0	9	141	16	200
Total	6924	1421	469	254	75	3272	190	5681

come from disturbed burials, a pit burial of a young man and remains of 10 individuals, collectively buried and, for some, partially cremated, were found in a small area of the cave (trench G1). This may account for some odd features of the Lower Mesolithic assemblages: the relative scarcity of grinding tools, chipped stone tools, and bone tools compared to the overabundance of carbonized seeds—more than 28,000 were recovered from the four water-sieved trenches only, in about 18/20 cubic meters—the abundance of land snails (nearly 13,000 in trench H1B alone, for about 3.4 m^3), and the very large quantities of ornaments. Compared to the Upper and Final Mesolithic, the ratio of land snails per volume of sediment is 2 to 10 times higher, the ratio of seeds is 10 to 30 times higher. The same holds true for the ornaments, which comprise a few perforated pebbles and thousands of ornamental shell species, *Dentalium* sp., *Columbella rustica* and especially *Cyclope neritea,* of which more than 7,000 specimens were recovered from the four water-sieved trenches only. We know they were also extremely abundant in trench G1, where the burials were brought to light, but no precise counts are available (Shackleton 1969). We should thus consider the possibility that the activities that took place in the cave were directly related to the presence of the dead, and that the cave witnessed repeated ceremonies.

It is interesting, in this respect, to compare the sequences from Franchthi and Pupicina Cave, in Istria (Croatia). In both sites, Final Pleistocene occupations during the Younger Dryas appear to have been brief and ephemeral (Miracle 2001). In both cases, sterile layers overlay these levels, whereas the Early Holocene levels witness a dramatic increase in anthropic remains. Through the analysis of the deer remains, Miracle suggests that communal feasting had become more important in the Early Holocene. He further suggests that the large quantities of land snails were also collected in view of these feasts. Significantly, this pattern emerges with the first human remains, suggestive of "celebratory feasts" (Hayden 2001). Franchthi's record is very similar, but the overabundance of food remains and apparent waste here relates to plant food and snails, not ungulates. If these food remains are linked to ceremonial activities in the cave, the economic transformations would then only be fortuitously related to the climatic ones. Social and ideological factors would be predominant.

After a depositional hiatus of three to four hundred years, the occupation of the cave resumes at the beginning of the tenth millennium cal B.P. During the Upper Mesolithic, Franchthi was no longer used as a collective burial ground[1] and the most spectacular economic shifts at that time are the decrease in seed number and the importance of tuna fishing (Rose 1995, in prep.; Stiner and Munro 2011). In trench FAS, fish bones represent 10 to 50 percent of the total weight of animal bones, and the Upper Mesolithic is the only occupational phase when marine resources can indeed be considered as prevalent. There is no indication, however, that this could be related to specific environmental transformations or a further decline in terrestrial resources, even if temperatures had continued to rise (Marino et al. 2009). The temporary accessibility of large schools is a more probable factor, since the rest of the fauna show no conspicuous change (Stiner and Munro 2011: Appendix 1). The seed assemblage is numerically far more restricted, but stable in terms of taxa represented, with the exception of coriander that appears in the Upper Mesolithic. The availability—and probable disappearance—of the tuna fish around the Franchthi shores may thus be related to local marine environmental factors such as upwelling, but the tuna fishing predates the contentious catastrophic "flooding" of the Black Sea (Giosan et al. 2009; Yonko-Hombach et al. 2007) and takes place at a time when no modification of deep-sea ecosystems are recorded in the Mediterranean Sea.

Following another depositional hiatus, the occupational sequence resumes at the very beginning of the ninth millennium. The inhabitants of the cave reverted to a diversified diet based on wild fruit, legumes, cereals, a few land snails, shellfish, and the capture of sardinella and sea-breams. The diet breadth is comparable to that of the Lower Mesolithic in terms of variety, but the proportion of seeds and land snails in relation to the volume of sediments is 20 to 30 times lower, while that of shellfish is four times higher.

FARMING COMMUNITIES AND THE 8.2 EVENT

Soon after, between ca. 8700 and 8500 cal B.P., the first farming communities, originating from the Near East, started to settle in the alluvial basins that local hunter-gatherers had tended to avoid (Halstead 2011; Perlès 2001). The two communities seem to have rapidly and peacefully merged (Perlès 2010b). The arrival of these farmers coincided with a multicentennial slow climatic deterioration (Marino et al. 2009), which did not impede the foundation of hundreds of Early Neolithic villages from the Peloponnese to Macedonia. All relied almost exclusively on domestic resources, cattle, sheep, goat, pig, wheat, barley, lentils, vetches, and peas, with a mainly terrestrial diet even in coastal sites (Papathanasiou 2003).

When the 8.2 event occurred, the Neolithic was thus already well established in Greece. This climatic event actually corresponds to the beginning of the transition from the Early to the Middle Neolithic, primarily defined and characterized by the development of fine, regionally distinctive painted wares (Demoule and Perlès 1992; Gimbutas et al. 1989; Perlès in press). Contrary to recent suggestions (Weninger et al. 2006), there is thus no relation between the 8.2 event and the arrival of farmers in Greece (see also Düring this volume).

In most of the 13 pollen diagrams from Greece analyzed by Bottema (2003), the 8.2 event cannot be formally identified, in large part because of insufficient temporal resolution. This is, unfortunately, the case with the diagrams from the lakes nearest to the regions of dense settlement in the Early Neolithic, such as Kopais in Boeotia (Greig and Turner 1974; Okuda et al. 1999; Turner and Greig 1975) or Xinias near Lamia (Bottema 1979; Digerfeldt et al. 2000). At the Rezina Marsh (Willis 1992b) and at Gramousti (Willis 1992a) in Epirus, the 8.2 event probably corresponds to a brief and well-marked decline in *Pinus* and a peak in *Quercus* and *Abies*. Nearby, in Lake Ioannina, the 8.2 event has not been formally identified in the diatom and pollen record of core I-284, although a wrong age model might be responsible (Lawson et al. 2004; Wilson et al. 2008). In Central Macedonia, the Giannitsa core shows a pronounced decline of *Pinus* and *Ulmus* dated to the mid-seventh millennium B.P., which might in fact correspond to the 8.2 event (Bottema 2003:Figure 2.2).

To the contrary, the 8.2 event is very well marked in the recent marine Aegean records, where it corresponds to a brief interruption of sapropel S1 deposition (Kotthoff et al. 2008). According to the synthesis recently published by Schmiedl and his collaborators (Schmiedl et al. 2010), the 8.2 event shows "the most prominent impact of abrupt high-latitude climate change on EMS [Eastern Mediterranean Sea] deep-water formation and deep-sea ecosystems." The analyses of $\delta^{18}O$ marine dinoflagellates and planktonic foraminiferal data from cores SL21, in central Aegean Sea (Figure 8.4), show that the 8.2 event corresponds to a prominent short-lived winter sea temperatures cooling

EARLY HOLOCENE CLIMATIC FLUCTUATIONS AND HUMAN RESPONSES

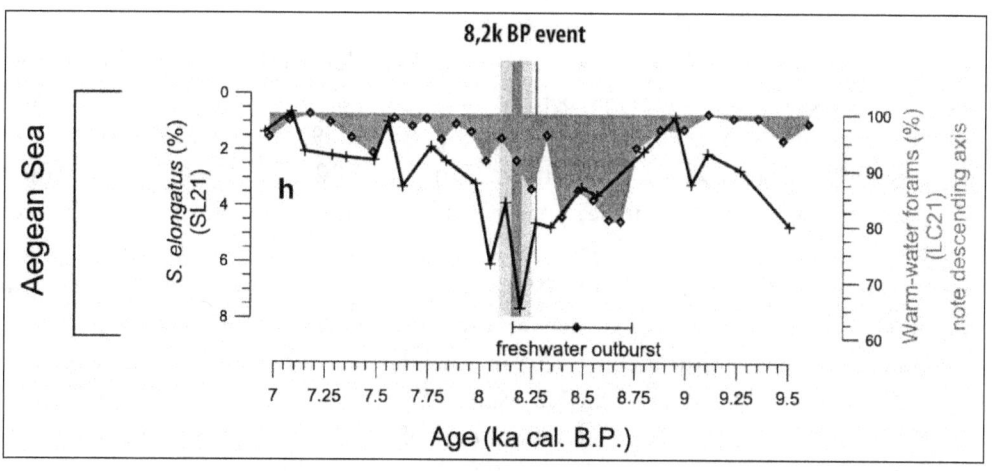

FIGURE 8.4. Dark line: relative abundance (%) of the cool-water dinoflagellate *Spiniferes elongatus* in core SL21 with respect to a gonyaulacoid-only dinocyst sum. Shaded area (with reverse scale): relative abundances (%) of warm-water planktonic foraminifera from core LC21. The light grey bands represent the 8.2 event. After Marino el al. 2009. Copyright Elsevier; used with permission.

event, superimposed on a deteriorating episode between 8.8 and 7.8 ka B.P. (Marino et al. 2009). A strong decrease of broad-leaved tree pollens percentage in core SL 152 in the northern Aegean basin, in particular evergreen *Quercus, Corylus,* and *Ulmus* (Figure 8.5), is considered to reflect a decrease of 2°C of mean winter temperatures (December, January, February). The consistent presence of *Ephedra* indicates a dry climate, but the stability of deciduous oaks, a group sensitive to summer droughts, suggests that droughts may have been especially pronounced during the winter.

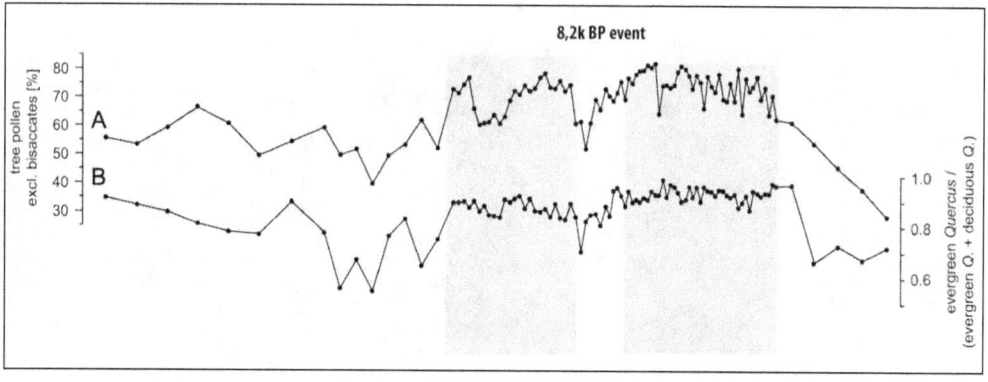

FIGURE 8.5. Broad-leaved tree pollen percentage, (top) and index of evergreen *Quercus*/ evergreen *Quercus* + deciduous *Quercus*) in core SL 152 (after Kotthoff et al. 2008). Copyright Elsevier, used with permission.

An even more pronounced effect is inferred inland, from the new pollen core of Tenaghi Philippon in the Drama plain of Eastern Macedonia (Figure 8.6). Broad-leaved trees drop from 87 percent to 53 percent, mainly at the expense of evergreen oaks, *Corylus, Ulmus,* and *Tilia.* Pross and his collaborators (2009) conclude for a decrease of more than 4°C in mean winter temperatures (DJF) and of 2° C in summer temperatures (JJA), together with a change in seasonality resulting from wetter summers (+ 75 mm) and drier winters (–100 mm) (Peyron et al. 2011). Annual precipitations would have dropped from ca. 800 to ca. 600 mm (Pross et al. 2009). The stronger signal at Tenaghi Philippon than in SL 152, located not far away, is attributed to local topographic conditions and the high (up to 2,200 m) mountains that surround the Drama plain.

There is no doubt, thus, that the 8.2 event had an environmental impact, in particular in northeastern Greece. Elsewhere, the magnitude of the decline in mean annual precipitation must be comprised between the 50–100 mm estimated by predictive models for the Aegean region as a whole (Renssen et al. 2002, quoted by Pross et al. 2009) and the 200 mm decline estimated in extreme conditions at Tenaghi Philippon (Peyron et al. 2011). According to the same sources, winter temperatures (December, January, February) would have dropped from 1–4°C, when summer temperatures would have lost about 1°C (Pross et al. 2009:889). Pross and his collaborators conclude that "the magnitude of these anomalies appear to have been strong enough to have seriously affected Neolithic settlers in the north-

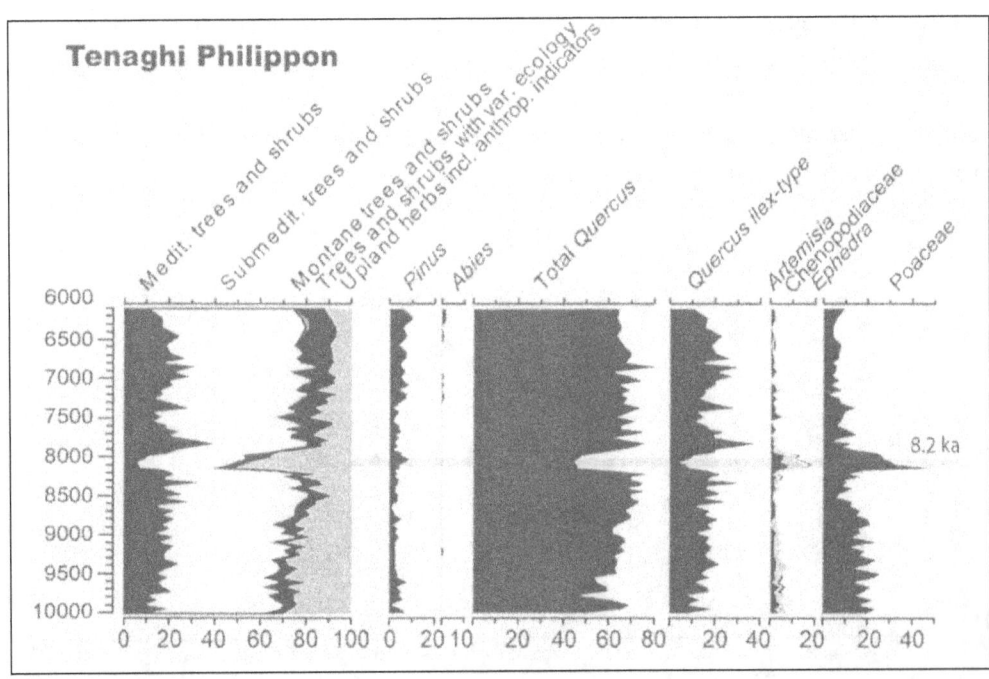

FIGURE 8.6. Simplified pollen diagram of Tenaghi Philippon (after Peyron et al. 2011). Copyright Peyron et al., used with permission.

eastern Mediterranean" (Pross et al. 2009:887). Should this be the case, both settlement patterns and economic practices ought to show important disruptions.

SETTLEMENT PATTERNS

Contrary to these expectations, settlement patterns show an overall continuity between the Early and Middle Neolithic. According to my files, of the ca. 330 sites where Early Neolithic has been recorded from surface surveys and excavations, about 80 percent were still occupied during part or whole of the Middle Neolithic. This does not imply that occupation was always continuous, but, in excavated sites, the stratigraphic gap emphasized by Berger and Guilaine (2009) at Sidari, in northwest Greece, appears to be the exception rather than the rule.

Despite new foundations and an increase in the number of recorded MN sites of ca. 20 percent, the overall geographic distribution of the settlements is stable. The contrast between eastern Greece, drier and densely settled, and western Greece, much wetter and barely occupied, remains unchanged. The contrast between the close networks of villages in Central Greece and Thessaly and the more sparsely populated Peloponnese is maintained, and the Cycladic islands are still not permanently settled.

The nature of the settlements is another element of continuity. In contrast with what will occur in the Late Neolithic, sedentary occupation of permanent villages appears to have remained the norm (Halstead 2000, 2005). Most sites constitute tells or *magoules*, and the "flat sites," now better known in the Late Neolithic thanks to preventive archaeology, remain a very small minority. The use of caves and rock shelters is extremely sparse in both periods, comprising less than 5 percent of the known Early and Middle Neolithic sites. This suggests, in particular, that the cooler and generally drier conditions of the 8.2 event did not lead to the exploitation of more marginal environments, as will be the case later (Halstead 2008).

One-fifth of the Early Neolithic settlements, nevertheless, were abandoned during the Middle Neolithic, and possible causal relations with the 8.2 event must be investigated. These sites, mostly known from surface surveys, tend to concentrate in specific regions: Western Macedonia, the high Pindus, Euboea, and Thessaly (Figure 8.3). Identification of the local variants of MN pottery may have been a problem in some cases, but not in Western Macedonia or Thessaly. In Eastern Thessaly, according to Costas Gallis's atlas of Neolithic settlements (Gallis 1992, updated), 27 of the 118 Early Neolithic sites were abandoned during the Middle Neolithic. This figure is compensated by an equivalent number of creations (29), which shows that the region as a whole did not offer adverse conditions. In addition, 17 out of these 28 sites were abandoned before the end of the Early Neolithic (no EN3 recorded), thus possibly before the 8.2 event. Some of the abandoned sites were located along the Penios River or along the water expanses of the old Lake Voivi (Perlès 2001). Sites along located in the Penios floodplain were occasionally flooded, and a renewed phase of aggradation is—rather roughly—dated to ca. 8.2 cal B.P. (Demitrack 1986; van Andel et al. 1995:Table 1). Rather than leading to a problem of drought, then, the 8.2 event may have locally led to more severe flooding problems, in what Berger and Guilaine consider to be part of a climatic "buffer zone" characterized by "droughts episodes favoring the spreading of fires, alternating with heavy rainfall" (Berger and Guilaine 2009:44). However, even flooded sites could be occupied year-round (Halstead 2005), and no coherent environmental pattern can be distinguished when all sites abandoned during the Middle Neolithic are considered together (Figure 8.7).

North of Thessaly, a survey of the Grevena basin by N. Wilkie and her collaborators brought to light about 20 EN sites, at elevations between 400 and 700 m asl. All these relatively short-lived settlements were abandoned during the MN (Wilkie and Savina 1997). Farther east, in the middle Haliakhmon valley, a survey conducted by A. Hondroyanni-Metoki also revealed that more than half of the EN settlements were abandoned, but this was compensated by an equal number of new foundations (Hondroyianni-Metoki 2009). Whereas the 8.2 event may have had a more drastic effects in these areas, submitted to a more continental and wetter climate than Thessaly, the lack of radiocarbon dates makes it impossible to specify whether these sites were abandoned during or after the 8.2 event. At any rate, the fact that these sites, as well as the whole Grevena basin, were abandoned during the complete duration of Middle Neolithic, when

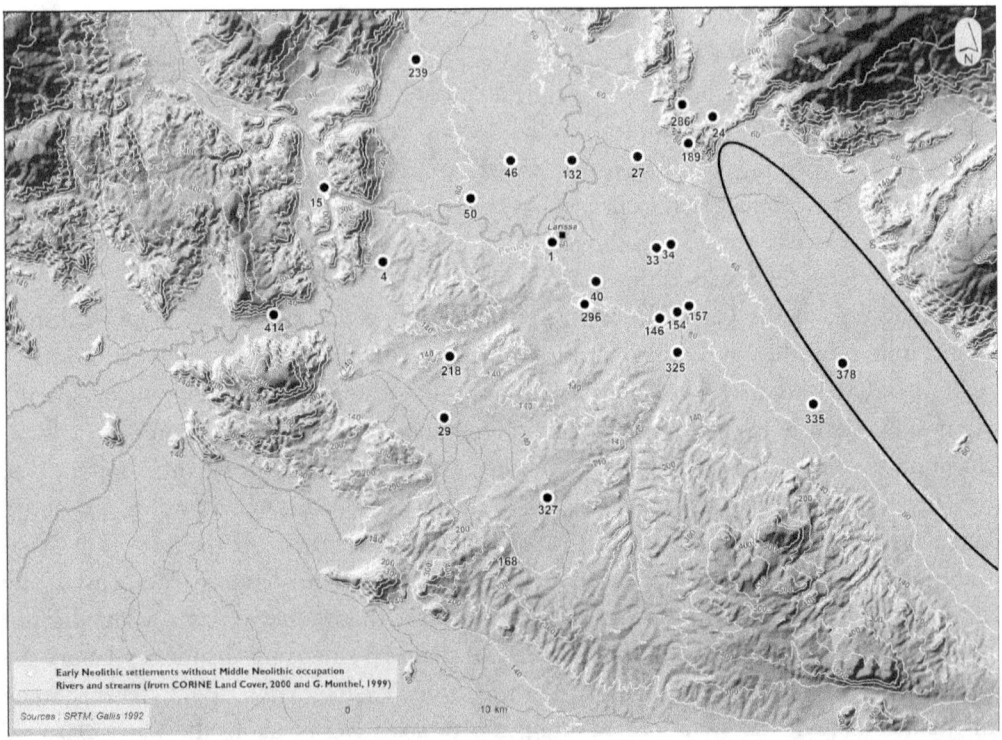

FIGURE 8.7. Early Neolithic sites abandoned during the Middle Neolithic. The black oval circumscribes the old Voivi Lake. Figures for each site refer to Gallis 1992. CAD S. Ménard.

the climate had reestablished itself, suggests that the 8.2 event had little to do with the shift in settlement patterns.

The absence of major disruption in settlement patterns linked to the 8.2 event is further confirmed by the continuous occupation of the Drama plain itself, where the new core of Tenaghi Philippon was extracted. A recent coring campaign at Dikili Tash showed that the site was founded around 8500–8400 B.P. (Lespez et al. 2013) and continuously occupied afterward. Contrary to what the interpretation of the Tenaghi Philippon pollen core suggested (*supra* p. 182; Pross et al. 2009; Peyron et al. 2011), the 8.2 event corresponds, here again, to a period of high water level that provoked a slight lateral shift of the settlement. This is interpreted as the result of the rise in summer rainfall (ca. 75 mm), which, "combined with the decrease of temperatures and the reduction of evapotranspiration, would have maintained groundwater at a high level" (Lespez et al. 2013:42).

To the west of the Drama plain, the 8.2 event corresponds to a period of florescence in Central Macedonia. If this climatic event induced a "massive perturbation in terrestrial ecosystems," it appears to be a favorable one, a point to which we shall return.

Subsistence Economy

The second aspect on which potential effects of the 8.2 event should be most visible is subsistence economy. The EN subsistence basis can be first characterized by a strong focus on domesticated species to the detriment of wild resources. In this respect, the Middle Neolithic is in complete continuity. Wild vertebrates represent between 1 and 6 percent of the Middle Neolithic faunal assemblages, a range almost exactly similar to that of the Early Neolithic[2] (1 to 7 percent). In both cases, red deer and hare predominate in the wild fauna. A later increase in the proportion of wild fauna (von den Driesch 1987) shows this to be a social choice, not the effect of environmental constraints.

Among domesticated animals, ovicaprids predominate numerically in both periods, but *Bos* can be predominant in terms of meat yield. Despite striking fluctuations from assemblage to assemblage which are difficult to explain (Becker 1999), similar faunal compositions can actually be found in both periods (Table 8.3) and suggest small-scale domestic herding in the context of "integrated crop and animal husbandry" (Halstead 2006:49; see also Halstead 2011).

The number of sites with published and quantified data on seed assemblages is very limited for the Middle Neolithic. Nevertheless, a similar continuity seems to obtain, with *Triticum dicoccum* (emmer) and *T. monococcum* (einkorn) predominant (except at Servia where *T. aestivum* would be more common), followed by *Lens culinaris*. All cereals and legumes under cultivation are common to the Early and Middle Neolithic, and no change of agricultural practices are perceived from these—admittedly limited—data (Halstead 2008; Hansen 1988). Despite a relative increase in storage vessels during the Middle Neolithic, there is also no evidence that storing strategies were profoundly modified (Halstead 1995).

The impact of the 8.2 event on agriculture and herding, if any, did not lead to a turnover of the species that were exploited. One may in fact wonder whether this climatic

Table 8.3
Relative Proportion of the Domesticated Faunal Species in Early and Middle Neolithic Settlements

	Early Neolithic			
	Ovicaprids %	Bos %	Sus %	Canis %
Otzaki, EN2	57	33	9	1
Otzaki, EN3	49	30	20	1
Prodromos 1 & 2	56	30	13	0
Prodomos 3	49	41	10	0
Sesklo EN1	64	14	22	0
Sesklo EN2	62	9	24	0,5
Achilleion phase I	88	4	7	1
Achilleion phase II	83	4	11	0,2
Nea Nikomedeia	71	15	15	0,2
Lerna I	63	12	24	1
Knossos EN Ia	75	7	18	0,2

	Middle Neolithic			
	Ovicaprids %	Bos %	Sus %	Canis %
Otzaki MN1	35	18	41	0
Achilleieon III	78	6	14	1
Achilleieon IIIb/IV	74	8	16	1,5
Achilleieon IV	73	6	20	0,3
Plateia Magoula Z.	56	18	23	2
Ayios Petros	84	4	12	1
Franchthi	80	1	18	1

event could actually have adversely affected the most important species exploited. The key issue here is not a decrease of winter rainfall, cooler winters, or cooler summers, which were still within the range to which these animal and plant species were well adapted (see Willcox 2007 for cereals). What matters is late winter and spring rainfall, when the crops would grow, but no direct proxies are available for this season. However, if the estimates from Tenaghi Philippon are right, the decrease in spring rainfall would not have been important. Small and archaeologically imperceptible alterations of the farmers' seasonal calendar were probably enough to adjust to the climatic modifications.

Responses to environmental changes could also be reflected, more indirectly, by transformations in long-distance trade, craft activities, or symbolic activities. I shall not discuss these aspects here, however, since the Early to Middle Neolithic transition in Greece is marked by a strong continuity in most domains (Perlès, in press). The most obvious change, which founded the distinction between the Early and the Middle Neolithic, is the development of regionally distinctive fine pottery. The regionalization of pottery styles, however, is a frequent phenomenon in the dynamics of Neolithic societies. At the turn of the eighth millennium B.P., it affects large regions of the Levant and Anatolia as well as Greece (Biehl and Rosenstock in press), where environmental conditions differ markedly. Considering the stability of settlement patterns, architecture, and subsistence economy, it is best interpreted as a purely sociological phenomenon, without relation to climatic instability.

Long-distance trade was particularly developed in the Early Neolithic of Greece, from the Peloponnese up to Thessaly, where the majority of the lithic tool-kits were made from Milian obsidian and exotic honey-flint blades (Perlès 1992). Despite typological differences between the Early and Middle Neolithic assemblages, no weakening or reorganization of these trade networks is perceptible. Finally, anthropomorphic figurines, which might have echoed a new perception of this changing environment, appear very uniform in the Early and Middle Neolithic (see Gimbutas et al. 1989 as a good illustration).

Conclusions

Although my investigations were undertaken without any theoretical a priori in favor of or against environmental determinism, they failed to demonstrate unambiguous relations between climatic change and transformations in human ways of life. Two radically different approaches have been attempted here, and none proved conclusive. This is not to deny, evidently, that the climatic changes I have considered had no effect whatsoever on human populations in Greece. Before such a conclusion could be reached, however, a number of theoretical and methodological problems would first have to be solved.

The economic changes observed at Franchthi between the Younger Dryas and the Preboreal, for instance, *can* be interpreted as an outcome of climatic and environmental transformations (e.g., Cullen 1995; Runnels 1995; Stiner and Munro 2011). Yet, other factors can be invoked that would have had similar effects. In this specific case, the new status of the cave as a burial ground could also account for the observed differences. We are thus confronted with a classic case of equifinality, with several different causal factors potentially leading to the observed patterns. The analysis of a well-dated sequence with rich environmental and archaeological data, would, a priori, appear to be an optimal situation for the investigation of correlations between climatic and socioeconomic change. However, even with homogeneous conditions of study, a sequence of human occupations cannot compare with laboratory conditions, where single variables can be changed at a time. The climatic changes are but one of the many variables that structure our archaeological record (and hunter-gatherers' ways of life), so that disentangling the role of climatic change versus, for instance, seasonality patterns and status of the site, would require integrated regional studies. This conclusion can hold true for most of the transformations observed during the Late Pleistocene-Early Holocene sequence at Franchthi.

Most transformations in resource exploitation, from phase to phase, are also related to transformations in the status of the cave and how it was inserted in a broader system of territorial exploitation and social construction of space. Whether these transformations are directly related to the effects of climatic fluctuations, rather than to the socioeconomic dynamics of the local groups, remains to be demonstrated.

In order to investigate the effects of the 8.2 event, we have used, to the contrary, a large-scope approach, which should have overcome the above-mentioned problem. Observing concomitant transformations in the late Early Neolithic societies should have been straightforward, since, besides the "massive perturbations in terrestrial ecosystems" inferred by environmentalists (Pross et al. 2009), Berger and Guilaine's survey led them to conclude that: "In our view societies were deeply affected by broad-scale climatic changes transforming both the geomorphology and ecology of the Mediterranean basin and Central Europe, between 8500 and 8000 cal B.P." (Berger and Guilaine 2009:45). The 8.2 k cal B.P. climatic anomaly is indeed well documented in natural records from Northeast Greece and from the Aegean. However, most of the recent coring sites, with a fine enough resolution that the 8.2 event can be identified, are located far away from areas of dense Early or Middle Neolithic settlements. This is all the more problematic in that, given the climatic gradients in Greece and the local topographic effects underlined at Tenaghi Philippon, it is difficult to transfer the results from one region to another. In order to assess the nature and magnitude of these effects in inland alluvial basins, where the vast majority of settlements are located, far more precise local environmental data would be required. The quality of environmental proxies in natural records is, for the moment, completely unmatched in or around archaeological sites. Our terrestrial and archaeological record is, at the time being, far too crude to investigate the possible effects of a sudden and marked climatic change. The need for renewed paleoenvironmental studies in and around inland alluvial basins is pressing, and recently developed techniques, such as the $\delta^{18}O/\delta^{13}O$ on seeds and charcoals (Voltas et al. 2008) would allow a climatic estimate on the very sites we are investigating. Thus far, Dikili Tash, near the pollen core of Tenaghi Philippon, is the only archaeological settlement where the quality of environmental data matches that of natural locations, but the Early and Middle Neolithic levels are only known through coring.

Meanwhile, the consequences of the 8.2 event can only be approached through archaeological data that are themselves not devoid of problems. The relative imprecision of the published ^{14}C dates, usually done decades ago and almost exclusively on charcoal, makes it difficult to decide whether a given phenomenon was contemporaneous or succeeded the climatic change. The necessary reliance on surface surveys to get a broad picture of land use and settlement patterns transformations only adds to the imprecision.

This being said, in the present state of the data, the archaeological record in Greece does not show a drastic reorganization in settlement patterns, nor a profound economic reorganization between the Early and Middle Neolithic. No ideological change, either, is documented. The transition from the Early to the Middle Neolithic appears fundamentally as a "quiet transition," mainly marked by sociological transformations (Perlès, in press). If the 8.2 event had the profoundly disruptive effects that some of our colleagues, both environmentalists and archaeologists, had predicted, it ought to have been reflected, one way or another, in our archaeological data. This is not the case, and despite the methodological problems we have discussed, the stability of the economic and social patterns appears difficult to reconcile with the notion of severe detrimental effects of the 8.2 event. But we should perhaps also question our mental templates: as environmental publications exemplify well, any increase in temperature and/or humidity is considered as an "amelioration," whereas a decrease in temperature and/or rainfall is a "deterioration." Should we not consider to the contrary that, against the background of a typically Mediterranean climate, cooler and wetter summers, associated with drier winters, were a treat that all too briefly benefited human beings, animals, and cultivars alike?

Notes

1. There is one baby burial and a few scattered bones only.
2. The percentage of wild fauna reached 11 percent at Lerna, but on a very small sample.

References Cited

Becker, C. 1999 The Middle Neolithic and the Platia Magoula Zarkou—A Review of Current Archaeozoological Research in Thessaly (Greece). *Anthropozoologica* 30:3–22.

Berger, J.-F., and J. Guilaine 2009 The 8200 Cal BP Abrupt Environmental Change and the Neolithic Transition: A Mediterranean Perspective. *Quaternary International* 200:31–49.

Biehl, P., and E. Rosenstock (editors) In press *Times of Change. The Turn from the 7th to the 6th Millennium BC in the Near East and Southeast Europe*. Cambridge University Press, Cambridge.

Bottema, S. 1979 Pollen Analytical Investigations in Thessaly (Greece). *Palaeohistoria* XXI:19–40.

Bottema, S. 1982 Palynological Investigations in Greece with Special Reference to Pollen as an Indicator of Human Activity. *Palaeohistoria* 24:257–289.

Bottema, S. 1991 Développement de la végétation et du climat dans le bassin méditerranéen oriental à la fin du Pléistocène et pendant l'Holocène. *L'Anthropologie* 95(4):695–728.

Bottema, S. 1994 The Prehistoric Environment of Greece: A Review of the Palynological Record. In *Beyond the Site. Regional Studies in the Aegean Area*, edited by P. N. Kardulias, pp. 45–68. University Press of America, Lanham.

Bottema, S. 2003 The Vegetation History of the Greek Mesolithic. In *The Greek Mesolithic. Problems and Perspectives*, edited by N. Galanidou and C. Perlès, pp. 33–49. BSA Studies 10. British School at Athens, London.

Colonese, A. C., M. A. Mannino, D. E. Bar-Yosef, D. A. Fa, J. C. Finlayson, D. Lubell, and M. C. Stiner 2010 Marine Mollusc Exploitation in Mediterranean Prehistory: An Overview. *Quaternary International* 239:86–103.

Cullen, T. 1995 Mesolithic Mortuary Ritual at Franchthi Cave, Greece. *Antiquity* 69(263):270–289.

Deith, M. R., and N. J. Shackleton 1988 Oxygen Isotope Analysis of Marine Molluscs from Franchthi Cave. In *Marine Molluscan Remains from Franchthi Cave*, pp. 133–156. Excavations at Franchthi Cave, Greece, Fascicle 4. Indiana University Press, Bloomington and Indianapolis.

Demitrack, A. 1986 The Late Quaternary Geologic History of the Larissa Plain, Thessaly, Greece: Tectonic, Climatic, and Human Impact on the Landscape. Unpublished PhD dissertation, Stanford University, Stanford.

Demoule, J.-P., and C. Perlès 1993 The Greek Neolithic: A New Review. *Journal of World Prehistory* 7(4):355–416.

Digerfeldt, G., S. Olsson, and P. Sandgren 2000 Reconstruction of Lake-level Changes in Lake Xinias, Central Greece, during the Last 40,000 Years. *Palaeogeography, Palaeoclimatology, Palaeoecology* 158:65–92.

Galanidou, N., and C. Perlès (editors) 2003 *The Greek Mesolithic. Problems and Perspectives*. BSA Studies 10. British School at Athens, London.

Gallis, K. 1992 *Atlas proïstorikon oikismon tis anatolikis Thessalikis pediadas*. Ephoria of Antiquities, Larisa.

Gimbutas, M., S. Winn, and D. Shimabuku (editors) 1989 *Achilleion, a Neolithic Settlement in Thessaly, Greece, 6400–5600 B.C.* Monumenta Archaeologica 14. UCLA Institute of Archaeology, Los Angeles.

Giosan, L., F. Filip, and S. Constatinescu 2009 Was the Black Sea Catastrophically Flooded in the Early Holocene? *Quaternary Science Reviews* 28:1–6.

Greig, J. R. A., and J. Turner 1974 Some Pollen Diagrams from Greece and their Archaeological Significance. *Journal of Archaeological Science* 1:177–194.

Griffiths, S. J., F. A. Street-Perrott, J. A. Holmes, M. J. Leng. and P. C. Tzedakis 2002 Chemical and Isotopic Composition of Modern Water Bodies in the Lake Kopais Basin, Central Greece: Analogues for the Interpretation of the Lacustrine Sedimentary Sequence. *Sedimentary Geology* 148:79–103.

Halstead, P. 1995 From Sharing to Hoarding: The Neolithic Foundations of Aegean Bronze Age Society? In *Politeia. Society and State in the Aegean Bronze Age. Proceedings of the 5th Aegean Conference,* edited by R. Laffineur and W.-D. Niemeier, pp. 11–21. Aegaeum 12, Université de Liège, Liège.

Halstead, P. 2000 Land Use in Postglacial Greece: Cultural Causes and Environmental Effects. In *Landscape and Use in Postglacial Greece,* edited by P. Halstead and C. Frederick, pp. 77–95. Sheffield University Press, Sheffield.

Halstead, P. 2005 Resettling the Neolithic: Faunal Evidence for Seasons of Consumption and Residence at Neolithic Sites in Greece. In *(Un)settling the Neolithic,* edited by D. Bailey, A. Whittle, and V. Cummings, pp. 38–50. Oxbow Books, Oxford.

Halstead, P. 2006 Sheep in the Garden: The Integration of Crop and Livestock Husbandry in Early Farming Regimes of Greece and Southern Europe. In *Animals in the Neolithic of Britain and Europe,* edited by D. Serjeantson and D. Field, pp. 42–55. Oxbow Books, Oxford.

Halstead, P. 2008 Between Rock and a Hard Place: Coping with Marginal Colonisation in the Later Neolithic and Early Bronze Age of Crete and the Aegean. In *Escaping the Labyrinth: The Cretan Neolithic in Context,* edited by V. Isaakidou and P. Tomkins, pp. 229–257. Oxbow Books, Oxford and Oakville.

Halstead, P. 2011 Farming, Material Culture, and Ideology: Repackaging the Neolithic of Greece (and Europe). In *The Dynamics of Neolithisation in Europe. Studies in Honour of Andrew Sherratt,* edited by A. Hadjikoumis, E. Robinson, and S. Viner, pp. 131–151. Oxbow Books, Oxford and Oakville.

Hansen, J. M. 1988 Agriculture in the Prehistoric Aegean: Data versus Speculation. *American Journal of Archaeology* 92:39–52.

Hansen, J. M. 1991 *The Palaeoethnobotany of Franchthi Cave*. Excavations at Franchthi Cave, Greece, Fascicle 7. Indiana University Press, Bloomington and Indianapolis.

Hayden, B. 2001 Fabulous Feasts: A Prolegomenon to the Importance of Feasting. In *Feasts: Archaeological and Ethnographic Perspectives on Food, Politics, and Power*, edited by M. Dietler and B. Hayden, pp. 23–64. Smithsonian Institution Press, Washington, D.C.

Hondroyanni-Metoki, A. 2009 Aliakmon 1985–2005. I archaiologiki erevna stin periohi tis technitis limnis Polyphytou (koilada mesou rou tou Aliakmona), apotelesmata kai prooptikes. *To Archaiologiko Ergo sti Makedonia kai Thraki*, 20 Chronia, pp. 449–462.

Kotthoff, U., J. Pross, U. C. Muller, O. Peyron, G. Schmiedl, H. Schulz, and A. Bordon 2008 Climate Dynamics in the Borderlands of the Aegean Sea during Formation of Sapropel S1 Deduced from a Marine Pollen Record. *Quaternary Science Reviews* 27:823–845.

Kyparissi-Apostolika, N. 2000 *Theopetra Cave. Twelve Years of Excavation and Research 1097–1998. Proceedings of the International Conference, Trikala, 6–7 November 1998*. Athens.

Kyparissi-Apostolika, N. 2003 The Mesolithic in Theopetra Cave: New Data on a Debated Period of Greek Prehistory. In *The Greek Mesolithic, Problems and Perspectives*, edited by N. Galanidou and C. Perlès, pp. 189–198. BSA Studies 10. British School at Athens, London.

Laskaris, N., A. Sampson, F. Mavridis, and I. Liritzis 2011 Late Pleistocene/Early Holocene Seafaring in the Aegean: New Hydration Dates with the SIMS-SS Method. *Journal of Archaeological Science* 38:2475–2479.

Lawson, I., M. Frogley, C. Bryant, R. Preece, and P. Tzedakis 2004 The Lateglacial and Holocene Environment History of the Ioannina Basin, North-West Greece. *Quaternary Science Reviews* 23:1599–1625.

Lespez, L., Z. Tsirtsoni, P. Darcque, H. Koukouli-Chryssanthaki, D. Malamidou, R. Treuil, R. Davidson, G. Kourtessi-Philippakis, and C. Oberlin 2012 Filling the Gap in Northern Greece: New Evidence on Early and Middle Neolithic Occupation at Dikili Tash. *Antiquity* 87:30–45.

Magri, D. and P. C. Tzedakis 2000 Orbital Signatures and Long-Term Vegetation Patterns in the Mediterranean. *Quaternary International* 73/74:69–78.

Marino, G., E. J. Rohling, F. Sangiorgi, A. Hayes, J. L. Casford, A. F. Lotter, M. Kucera, and H. Brinkhuis 2009 Early and Middle Holocene in the Aegean Sea: Interplay between High and Low Latitude Climate Variability. *Quaternary Science Reviews* 28:3246–3262.

Miracle, P. 2001 Feast or Famine? Epipaleolithic Subsistence Strategies in the Northern Adriatic Basin. *Documenta Praehistorica* XXVIII:177–197.

Mylona, D. 2003 The Exploitation of Fish Resources in Mesolithic Sporades: Fish Remains from the Cave of Kyplops, Yioura. In *The Greek Mesolithic, Problems and Perspectives*, edited by N. Galanidou and C. Perlès, pp. 181–188. BSA Studies 10. British School at Athens, London.

Okuda, M., Y. Yasuda, and T. Setoguchi 1999 Latest Pleistocene and Holocene Pollen Records from Lake Kopaïs, Southeast Greece. *Geological Society of Japan* 105:450–455.

Papathanasiou, A. 2003 Stable Isotope Analysis in Neolithic Greece and Possible Implications on Human Health. *International Journal of Osteoarchaeology* 13:314–324.

Perlès, C. 1990 *Les industries lithiques taillées de Franchthi (Argolide, Grèce). Tome II: Les Industries du Mésolithique et du Néolithique initial.* Excavations at Franchthi Cave, Fascicle 5. Indiana University Press, Bloomington and Indianapolis.

Perlès, C. 1992 Systems of Exchange and Organization of Production in Neolithic Greece. *Journal of Mediterranean Archaeology* 5(2):115–164.

Perlès, C. 2001 *The Early Neolithic in Greece. The First Farming Communities in Europe*. Cambridge University Press, Cambridge.

Perlès, C. 2010a Is the Dryas the Culprit? Socio-Economic Changes during the Final Pleistocene and Early Holocene at Franchthi Cave (Greece). *Journal of the Israel Prehistoric Society* 40:113–129.

Perlès, C. 2010b Grèce et Balkans : deux voies de pénétration distinctes du Néolithique en Europe? In *La révolution néolithique dans le monde*, edited by J.-P. Demoule, pp. 263–281. CNRS Éditions, Paris.

Perlès, C. 2016 Food and Ornaments: Diachronic Changes in the Exploitation of Littoral Resources at Franchthi Cave (Greece) during the Upper Palaeolithic and the Mesolithic (39,000–7000 cal BC). *Quaternary International* 407:45–58.

Perlès, C. In prep. The Turn of the Millennium in Greece: a Quiet Transition. In *Times of Change. The Turn from the 7th to the 6th Millennium BC in the Near East and Southeast Europe*, edited by P. Biehl and E. Rosenstock. Institute for European and Mediterranean Archaeology, State University of New York Press, Albany.

Peyron, O., S. Goring, I. Dormoy, U. Kotthoff, J. Pross, J.-L. de Beaulieu, R. Drescher-Schneider, B. Vannière, and M. Magny 2011 Holocene Seasonality Changes in the Central Mediterranean Region Reconstructed from the Pollen Sequences of Lake Accesa (Italy) and Tenaghi Philippon (Greece). *The Holocene* 21(1):131–146.

Polunin, O. 1987 *Flowers of Greece and the Balkans. A Field Guide*. Oxford University Press, Oxford.

Powell, J. 2003 The Fish Bone Assemblage from the Cave of Kyklops, Yioura: Evidence for Continuity and Change. In *The Greek Mesolithic, Problems and Perspectives*, edited by N. Galanidou and C. Perlès, pp. 173–179. BSA Studies 10. British School at Athens, London.

Pross, J., U. Kotthoff, U. C. Müller, O. Peyron, I. Dormoy, G. Schmiedl, S. Kalaitzidis, and A. M. Smith 2009 Massive Perturbation in Terrestrial Ecosystems of the Eastern Mediterranean Region Associated with the 8.2 kyr B.P. Climatic Event. *Geology* 37(10):887–890.

Renssen, H., H. Goosse, and T. Fichefet 2002 Modeling the Effect of Freshwater Pulses on Early Holocene Climate: The Influence of High-Frequency Variability. *Paleoceanography* 17:1029–1035.

Roberts, N., M. D. Jones, A. Benkaddour, W. J. Eastwood, M. L. Filippi, M. R. Frogley, H. F. Lamb, M. J. Leng, J. M. Reed, M. Stein, L. Stevens, B. Valero-Garcés, and G. Zanchetta 2008 Stable Isotope Records of Late Quaternary Climate and Hydrology from Mediterranean Lakes: The ISOMED Synthesis. *Quaternary Science Reviews* 27:2426–2441.

Roberts, N., D. Brayshaw, C. Kuzucuoğlu, R. Perez, and L. Sadori 2011 The Mid-Holocene Climatic Transition in the Mediterranean: Causes and Consequences. *The Holocene* 21(1):3–13.

Rose, M. 1995 Fishing at Franchthi Cave, Greece: Changing Environments and Patterns of Exploitation. *Old World Archaeology Newsletter* 18(3):21–26.

Rose, M. In prep. *Franchthi fish remains. The Aegean and Mediterranean context*. Excavations at Franchthi Cave, Indiana University Press, Bloomington and Indianapolis.

Rossignol-Strick, M. 1997 Paléoclimats de la Méditerranée orientale et de l'Asie du Sud-Ouest de 15.000 à 6.000 B.P. *Paléorient* 23(2):175–186.

Rossignol-Strick, M., V. Nesteroff, P. Olive, and C. Vergnaud-Grazzini 1982 After the Deluge: Mediterranean Stagnation and Sapropel Formation. *Nature* 295:105–110.

Runnels, C. 1995 Review of Aegean Prehistory IV: The Stone Age of Greece from the Palaeolithic to the Advent of the Neolithic. *American Journal of Archaeology* 99:699–728.

Runnels, C. 2004 A Mesolithic Landscape in Southern Greece. *Context* 18(1):1–5.

Runnels, C. 2009 Mesolithic Sites and Surveys in Greece: A Case Study from the Southern Argolid. *Journal of Mediterranean Archaeology* 22(1):57–73.

Sampson, A. (editor) 2011 *The Cave of the Cyclops. Mesolithic and Neolithic Networks in the Northern Aegean, Greece*. Instap Academic Press, Philadelphia, 2 vols.

Schmiedl, G., T. Kuhnt, W. Ehrmann, K.-C. Ameis, Y. Hamann, U. Kotthof, P. Dulski, and J. Pross 2010 Climatic Forcing of Eastern Mediterranean Deep-Water Formation and Benthic Ecosystems during the Past 22,000 Years. *Quaternary Science Reviews* 29:3006–3020.

Shackleton, N. J. 1969 Appendix I: Preliminary Observations on the Marine Shells. In T. W. Jacobsen, Excavations at Porto Cheli and Vicinity, Preliminary Report, II: The Franchthi Cave, 1967–1968. *Hesperia* XXXVIII(3):379–380.

Shackleton, N. J. 1988 Oxygen Isotope Analysis of Marine Molluscs from Franchthi Cave. In *Marine Molluscan Remains from Franchthi Cave*, edited by J. C. Shackleton, pp. 133–156. Excavations at Franchthi Cave, Greece, Fascicle 4. Indiana University Press, Bloomington and Indianapolis.

Stiner, M., and N. Munro 2011 On the Evolution of Diet and Landscape During the Upper Palaeolithic through Mesolithic at Franchthi Cave (Peloponnese, Greece). *Journal of Human Evolution* 60:618–636.

Turner, J., and J. R. A. Greig 1975 Some Holocene Pollen Diagrams from Greece. *Review of Palaeobotany and Palynology* 20:171–204.

Tzedakis, P. C. 2007 Seven Ambiguities in the Mediterranean Palaeoenvironmental Narrative. *Quaternary Science Reviews* 26:2042–2066.

van Andel, T., K. Gallis, and G. Toufexis 1995 Early Neolithic Farming in a Thessalian River Landscape, Greece. In *Mediterranean Quaternary River Environments*, edited by J. Lewin, M. G. Macklin, and J. C. Woodward, pp. 131–143. A. A. Balkema, Rotterdam.

Van Zeist, W., and S. Bottema 1982 Vegetational History of the Eastern Mediterranean and the Near East During the Last 20,000 years. In *Palaeoclimates, Palaeoenvironments, and Human Communities in the Eastern Mediterranean Region in Later Prehistory*, edited by J. L. Bintliff and W. van Zeist, pp. 277–321. BAR International Series 133. Archaeopress, Oxford.

Voltas, J., J. P. Ferrio, and J. L. Araus 2008 Palaeoenvironmental Reconstruction: The Results of Carbon Isotope Composition (Delta ^{13}C) in Charred Wood. In *MENMED. From the Adoption of Agriculture to the Current Landscape: Long Term Interaction Between Men and Environment in the East Mediterranean Basin*, edited by R. Buxo, and M. Molist, pp. 35–37. Museu d'Arqueologia de Catalunya, Monografies 9.

von den Driesch, A. 1987 Haus- und Jagdtiere im vorgeschichtlichen Thessalien. *Praehistorische Zeitschrift* 62:1–21.

Weninger, B., E. Alram-Stern, E. Bauer, L. Clare, U. Danzeglocke, O. Jöris, C. Kubatski, G. Rollefson, H. Todorova, and T. van Andel 2006 Climate Forcing due to the 8200 Cal Yr BP Event Observed at Early Neolithic Sites in the Eastern Mediterranean. *Quaternary Research* 66:401–420.

Wilkie, N. C., and M. E. Savina 1997 The Earliest Farmers in Macedonia. *Antiquity* 71:201–207.

Willcox, G. 2007 The Adoption of Farming and the Beginnings of the Neolithic in the Euphrates Valley: Cereal Exploitation Between the 12th and 8th Millennia Cal B.C. In *The Origins and Spread of Domestic Plants in Southwest Asia and Europe*, edited by S. Colledge and J. Conolly, pp. 21–36. Left Coast Press, Walnut Creek, CA.

Willis, K. J. 1992a The Late Quaternary Vegetational History of Northwest Greece. I. Lake Gramousti. *New Phytologist* 121:101–117.

Willis, K. J. 1992b The Late Quaternary Vegetational History of Northwest Greece. II. Rezina Marsh. *New Phytologist* 121:119–138.

Willis, K. J. 1994 The Vegetational History of the Balkans. *Quaternary Science Reviews* 13:769–788.

Wilson, G. P., J. M. Reed, I. T. Lawson, M. R. Frogley, R. C. Preece, and P. C. Tzedakis 2008 Diatom Response to the Last Glacial-Interglacial Transition in the Ionannina Basin, Northwest Greece: Implications for Mediterranean Palaeoclimate Reconstruction. *Quaternary Science Reviews* 27:428–440.

Yonko-Hombach, V., A. S. Gilbert, and P. Dolukhanov 2007 Controversy over the Great Flood Hypotheses in the Black Sea in Light of Geological, Paleontological, and Archaeological Evidence. *Quaternary International* 167–168:91–113.

CHAPTER NINE

Rapid Climate Change and Radiocarbon Discontinuities in the Mesolithic–Early Neolithic Settlement Record of the Iron Gates

Cause or Coincidence?

*Clive Bonsall, Mark Macklin,
Adina Boroneanț, Catriona Pickard,
László Bartosiewicz, Gordon Cook, Thomas Higham*

Abstract *The Mesolithic–Early Neolithic radiocarbon record for the Iron Gates is compared against the regional paleoclimatic record. Well-marked minima in the frequency of radiocarbon dates at ca. 9.5–9.0 ka, 8.65–8.0 ka and after 7.8 ka cal B.P. coincide with "rapid climate change events" recorded in Greenland ice cores and paleoclimate archives from the Danube catchment. Four possible explanations of the observed radiocarbon discontinuities are considered: dwindling fish resources, changes in the social environment linked to the spread of farming, flood-induced settlement relocations, and taphonomic effects.*

INTRODUCTION

Summed calibrated probability distributions (SCPDs) of radiocarbon dates have become a popular tool in archaeology for investigating long-term demographic patterns in relation to paleoenvironmental trends. Peaks and troughs in SCPDs are frequently interpreted as evidence of population fluctuations.

Particularly researchers interested in the demographic effects of extreme climatic events have used this approach. Various studies have argued for a relationship between Holocene rapid climate change (RCC) events and prehistoric settlement in Europe and the Near East. Many of these studies have focused on the consequences of rapid climate change for early farming communities (e.g., Bonsall et al. 2002b; Gronenborn 2007,

2009; Macklin et al. 2011; Weninger et al. 2006, 2009). Relatively few studies have considered the effects of climate shifts on Mesolithic hunter-gatherer populations.

In this paper, we compare the largest Mesolithic–Early Neolithic radiocarbon dataset from Southeast Europe—that from the Iron Gates section of the lower Danube Valley (Figure 9.1)—against the early-mid Holocene regional climatic record.

THE IRON GATES RADIOCARBON DATASET

There are now more than 320 radiocarbon dates from Mesolithic and Neolithic sites in the Iron Gates region spanning the time range from the Late Glacial to mid-Holocene (ca. 15–5 ka cal B.P.). Excluding from consideration cave sites in the mountainous hinterland, there are 319 ^{14}C dates from sites along the main trunk of the Danube, extending over a distance of more than 200 km. Of these, 20 dates are unusable in the present context because of very large errors, concerns over sample integrity, or because of suspected reservoir offsets that cannot be corrected for (e.g., human bone collagen dates with no corresponding δ^{15}N values). The remaining 299 dates comprise:

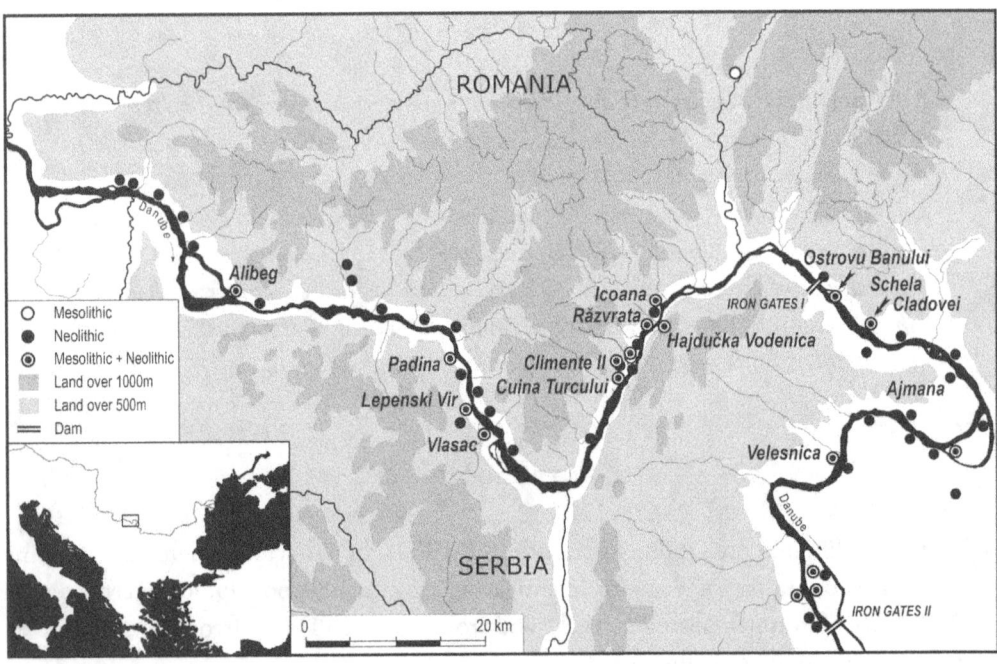

FIGURE 9.1. Mesolithic and Early Neolithic sites in the Iron Gates. Virtually all sites along the main trunk of the Danube were located immediately adjacent to the river and would have been vulnerable (in whole or part) to inundation at times of unusually high river flows (4 m+ floods). Named sites have ^{14}C dates that were used to generate the summed probability distributions in Figures 9.2–9.4.

1. 59 radiometric ¹⁴C dates produced in the 1960s and 1970s (57 on bulk charcoal samples, and one each on human bone and terrestrial animal bone samples); and

2. 240 single-entity AMS ¹⁴C dates on human bone (86 dates), terrestrial animal bone (153 dates) and plant macro-remains (1 date)—all obtained since 1996

Graphs of the summed calibrated probability distributions were generated in *CalPal* (Weninger et al. 2007) and are presented in Figure 9.2 for the full dataset (dataset A), the AMS dates on terrestrial animal/plant material and human bone combined (dataset B), and the AMS dates for terrestrial samples only (dataset C).

Charcoal and human bone samples can give anomalously old radiocarbon ages, and thus introduce "noise" into the analysis. Charcoal samples potentially carry an "old wood effect" resulting in ¹⁴C ages that overestimate the date of a cultural event, sometimes by up to hundreds of years. Similarly, human bone ¹⁴C ages from the Iron Gates sites often include a "reservoir effect" resulting from consumption of freshwater and anadromous fish from the Danube (Cook et al. 2001). A reservoir correction has been applied to the human bone dates prior to calibration based on the bone collagen δ¹⁵N value (Cook et al. 2002, 2009). The resulting calibrated age is a more accurate estimate of the true age, but is less precise (i.e., has a larger error).

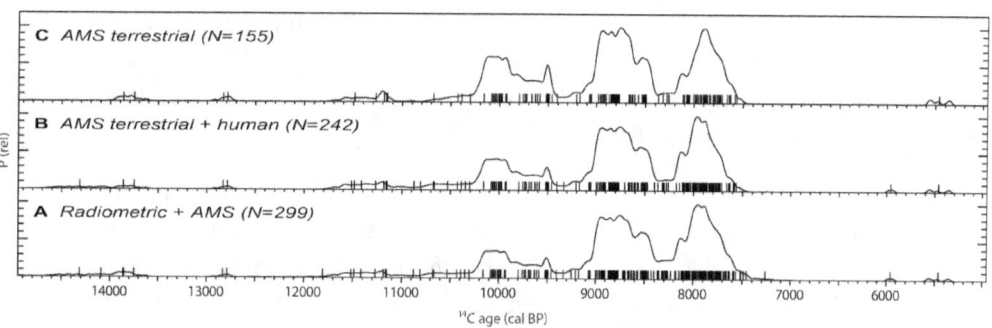

FIGURE 9.2. Cumulative calibrated dating probability of radiocarbon data from Mesolithic and Early Neolithic sites along the Danube main channel in the Iron Gates. Datasets: A—all usable radiometric and AMS dates; B—AMS dates on terrestrial and human bone samples; C—AMS dates on terrestrial animal and plant samples. Calibrations performed with *CalPal* (29 May 2007) [http://www.calpal.de] and the IntCal04 dataset. ¹⁴C data from Bonsall 2008; Bonsall et al. 1997, 2008, 2012, 2015, unpublished; Borić and Miracle 2004; Borić and Dimitrijević 2009; Borić 2011; Borić and Price 2013; Dinu et al. 2007. Prior to calibration human bone ¹⁴C ages were corrected for the "freshwater reservoir effect" using Method 1 of Cook et al. (2002), assuming δ¹⁵N endpoint values for purely terrestrial and purely aquatic diets of +8.3‰ and +17.0‰, respectively (cf. Cook et al. 2009).

Radiocarbon age measurements carried out on the bones of terrestrial herbivores and short-lived plant remains do not suffer from the potential biases that can affect human bones and wood charcoal as described above. Many dates included in our analysis are single-entity measurements on humanly modified bones (artifacts, débitage, or cut-marked bones from the processing of animal carcasses), which can be considered to date human activity directly (cf. Ashmore 1999; Tolan-Smith and Bonsall 1999). Where unmodified animal bones were dated, they come from well-documented archaeological contexts and have a high probability of association with human activity. Arguably, therefore, the most reliable picture is provided by the "terrestrial" subset (Figure 9.2, dataset C), although all the curves show broadly the same pattern.

Discussion

There are three well-marked troughs in the summed calibrated probability distribution for the 15–5 ka cal B.P. time range in Iron Gates (Figure 9.2), representing phases with relatively few ^{14}C dates. There is a rough coincidence between these reductions in ^{14}C date frequency and climate shifts recorded in proxies from Greenland (Mayewski et al. 2004; Rasmussen et al. 2007), the North Atlantic (Bond et al. 1997), and Europe (Figure 9.3).

- The trough in the SCPD at 9.5–9.0 ka cal B.P. correlates fairly closely with the climate anomaly known as the "9.3 ka cold event."

- The major ^{14}C discontinuity at 8.65–8.0 ka cal B.P. corresponds roughly with the general climatic deterioration recognized by Rohling and Pälike (2005) between 8.6 and 7.95 ka cal B.P., which culminated in a sudden cold event ("8.2 ka event") between c. 8175–8025 cal B.P. (Kobashi et al. 2007). Within this phase the lowest frequency of ^{14}C dates occurs between c. 8475 and 8100 cal B.P.

- The dip in the SCPD immediately after 7.8 ka cal B.P. is the prelude to a long lacuna in the ^{14}C record of the Iron Gates. The initial reduction in ^{14}C dates coincides with a significant decline in sea surface temperatures (SSTs) in the southern Adriatic (Figure 9.5) between 7.8 and 7.5 ka cal B.P. This was followed by a more pronounced SST cooling between 7.3 and 6.3 ka cal B.P., which at its nadir was characterized by SSTs lower than those registered during the 8.2 ka event (Siani et al. 2013).

The statistical reliability of SCPDs depends largely on sample size relative to the time interval examined. The number of ^{14}C dates now available for the Mesolithic-Neolithic time range in the Iron Gates satisfies the minimum reliability criteria proposed by Michczyńska and Pazdur (2004) and Williams (2012).

An analysis of ^{14}C dates from Mesolithic and Neolithic sites in Belgium and northeast France led Crombé and Robinson (2014) to conclude that SCPDs may be biased in other ways. Three factors in particular were seen as having skewed their data:

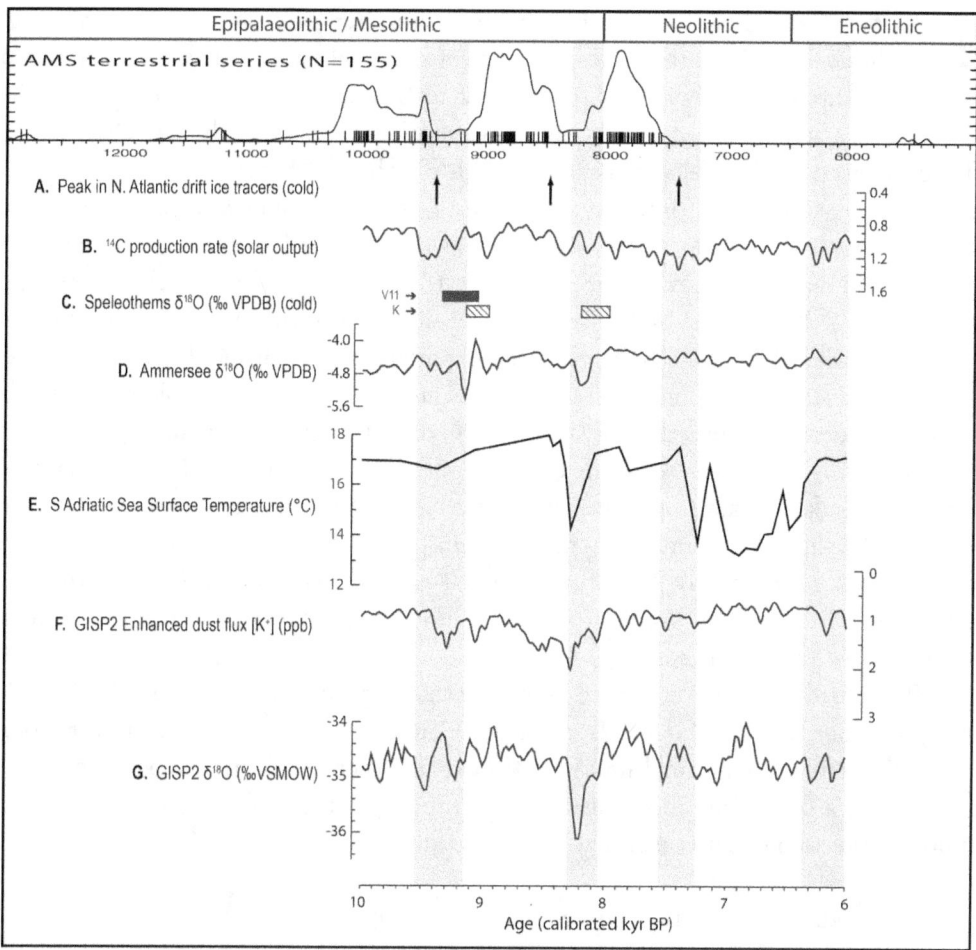

FIGURE 9.3. Cumulative calibrated dating probability of radiocarbon data (terrestrial series) from Mesolithic and Early Neolithic sites along the Danube main channel in the Iron Gates, compared to climate proxy records from the North Atlantic and Europe 10–6 ka cal B.P. A—after Bond et al. (1997); B, D, F, G—smoothed records redrawn from Rohling and Pälike (2005); E—after Siani et al. (2013); C—horizontal bars represent cold phases recorded in $\delta^{18}O$ records from V11 Cave, NW Romania (Tămaş et al. 2005) and Katerloch Cave, Austria (Boch et al. 2009); vertical grey bars represent higher lake-level events in the Alps–Jura region (Magny 2004).

1. Differences in research intensity (e.g., frequency and scale of excavations) between archaeological periods and geographical areas;

2. Differences between sites in the density of archaeological features; and

3. Taphonomic processes that reduce the number of secure contexts with reliable samples for dating, e.g., mixing of datable material between contexts through anthropogenic and/or pedogenic bioturbation.

All these factors also operate in the Iron Gates, but arguably have less impact on the Mesolithic–Early Neolithic SCPD because of the emphasis on single-entity dating of human remains and bone artifacts, in contrast to Belgium, where ^{14}C dating has relied more on "ecofacts" (charred wood, seeds, and nutshells), which without a clear archaeological context are less reliably associated with past human behavior.

Variations in research intensity in the Iron Gates have resulted in some sites having many more ^{14}C dates than others, although no one site predominates overwhelmingly. Lepenski Vir has the largest number of dates (82), nearly half of which fall between 8.6–8.0 ka cal B.P. Yet the SCPD still shows a distinct trough during that period.

Bamforth and Grund (2014) made the important point that the shape of a summed probability distribution can be affected by irregularities ("cliffs" and "plateaus") in the calibration curve. In Figure 9.4 the SCPD of ^{14}C dates from the Iron Gates for the period 12–7 ka B.P. is superimposed on that of radiocarbon dates simulated for calendar dates spaced at regular intervals over the same time range. While there is a close correspondence in the shapes of the two graphs at certain periods, they differ significantly during other periods, most notably 9.5–9.0 ka, 8.6–8.0 ka and after 7.8 ka B.P., suggesting that the trends evident in the Iron Gates SCPD during those periods are not primarily an artifact of the calibration curve.

The broad similarity in timing between centennial scale climate anomalies and the troughs in our SCPDC for the Iron Gates (Figure 9.3) is suggestive of a correlation. However, establishing a causal link between climatic events and archaeological phenomena requires a convincing mechanism for transmitting cause to effect (cf. Behrensmeyer 2006). Various potential causes may be proposed (Bonsall et al. 2014).

1. Climate-related Fluctuations in Fish Resources

Climate-related changes in river conditions can have a negative impact on fish stocks. Spawning and feeding behavior can be significantly affected by sudden and/or prolonged changes in water temperature. Spawning in salmonids and wels catfish (*Silurus glanis*) may be inhibited by sudden drops in water temperature; while catfish cease to feed at temperatures below ca. 10°C and growth is impeded (Copp et al. 2009; David 2006; Omarov and Popova 1985). Similarly, river icing and sudden or prolonged changes in flow conditions can have an impact on fish populations. Ice jams and high river flows cause "scouring" of the riverbed destroying aquatic insects and underwater plants on which fish feed and even killing young fish, resulting in the phenomenon of "winterkill." The severity of winterkill varies year by year. One "bad" year is unlikely to have a significant impact on the fish population of a river, but a succession of bad years could interfere with the reproduction of fish and lead to a sustained reduction in fish catches. But periods of high flow are also an important trigger for sturgeon migrations, and higher water levels allow fish to pass through river stretches containing rapids or shallows. Conversely, any reduction in river discharge during the period of migration diminishes the attractiveness of the river and the number of migrants.

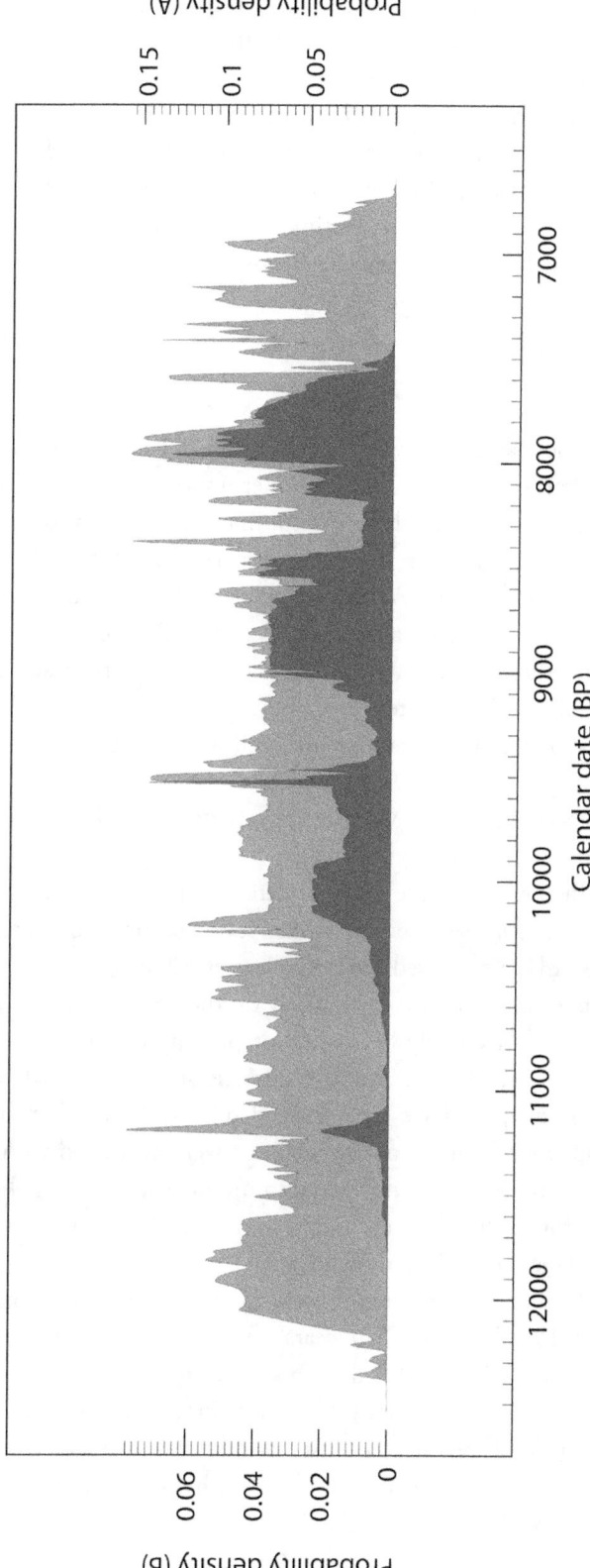

FIGURE 9.4. Summed calibrated probability distribution of radiocarbon dates from the Iron Gates for the period 12–7 ka B.P. (A, black, n = 143), superimposed on a summed probability distribution of radiocarbon dates simulated for calendar dates spaced every 25 years (B, grey, n = 213). Summed probability distributions produced with OxCal 4.2.4 (Bronk Ramsey 2009), using IntCal13 (Reimer et al. 2013) and ±25yr standard errors.

However, two things argue against changes in river conditions as the cause of the discontinuities observed in the radiocarbon record of the Iron Gates Mesolithic–Early Neolithic. On the one hand, no significant temporal fluctuations in the exploitation of fish resources are reflected in either the archaeofaunal or human bone stable isotope records prior to the appearance of agriculture ca. 8.1/8.0 ka cal B.P. On the other hand, there is no firm evidence for prolonged major reductions in fish stocks or catches along the Danube or other major European rivers during the most recent RCC event of the Holocene (the *Little Ice Age*) that can be unambiguously linked to climate change (Bonsall et al. 2014).

2. Change in the Social Environment

Farming spread through the Balkans between ca. 8.6 and 8.0 ka cal B.P., and several authors have speculated that as the agricultural frontier approached the Danube, interaction with farmers could have disrupted the Mesolithic settlement-subsistence system in the Iron Gates (Borić 2011; Tringham 2000; see also Bonsall 2008:277). While this might explain the ^{14}C discontinuity centered on 8.3 ka cal B.P., it could not account for the reduction in the number of ^{14}C ages around 9.3 ka cal B.P.; while the ^{14}C discontinuity commencing after 7.8 ka cal B.P., which appears to have marked a sharp decline in Neolithic occupation of the Iron Gates sites (Bonsall et al. 2002a; Bonsall 2008), occurred several centuries after farming had been established in the region.

3. Changes in Flood Frequency and Magnitude along the Danube

Climate-related changes in the magnitude and frequency of floods on the Danube, and the variability of flow regime, still seem the most likely explanation of the discontinuities in the Iron Gates ^{14}C record, as proposed by Bonsall et al. (2002a).

In historical times floods have occurred regularly on the Danube in spring and early summer. The main causes are excessively heavy and/or prolonged rainfall and snowmelt from upland areas within the catchment. Another historically documented cause is the buildup and then sudden collapse of ice jams formed during the breakup of river ice. Some of the biggest floods ever recorded on the Danube were associated with the ponding and/or release of water by ice jams (which were more prevalent during the Little Ice Age), including the catastrophic flood of 1838 that devastated large areas of what today is the eastern part of Budapest (Smith and Ward 1998).

According to river flow data from gauging stations along the Danube, major floods have occurred in at least 15 of the last 150 years. Two recent examples illustrate the severity and duration of some of these flood events. The flood that also affected large areas of Central Europe in August 2002 was the result of intense and prolonged summer rainfall over the eastern Alps and the upper Danube catchment. Two periods of sustained rainfall several days apart generated successive flood waves that traveled down the Danube at rates of up to 74 km/day (Mihailova et al. 2012). During the combined snowmelt

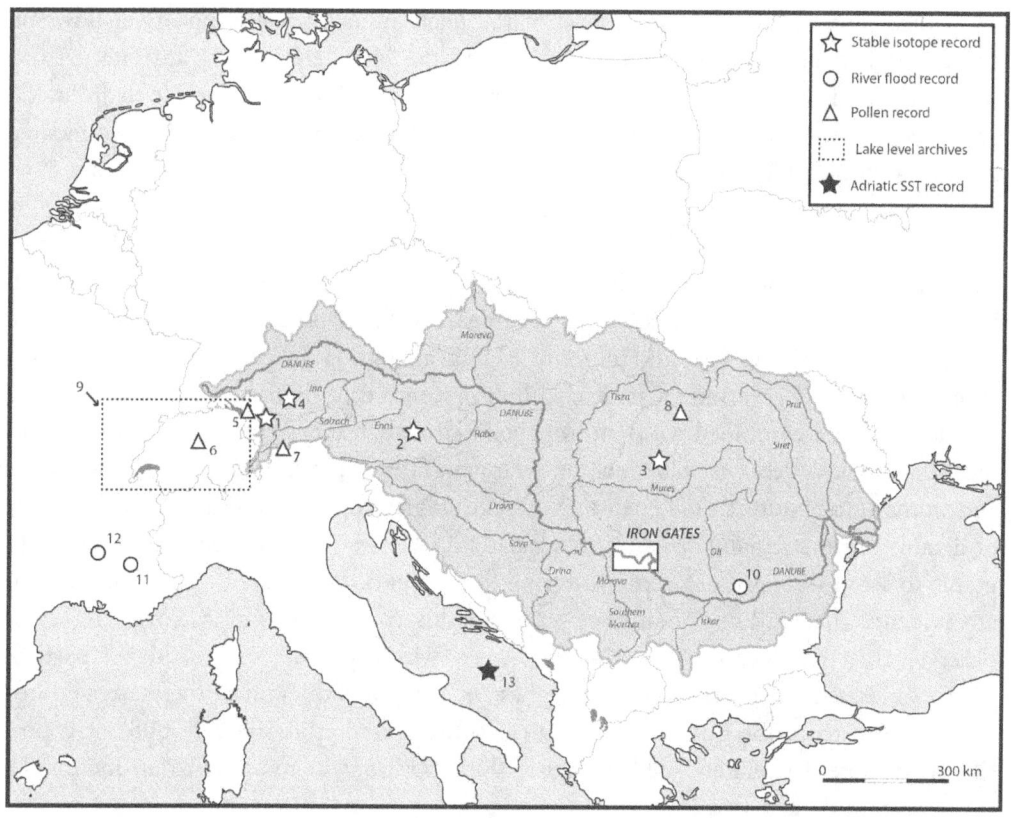

FIGURE 9.5. The Danube catchment showing the Iron Gates and key localities with climate proxy records for the Early to Middle Holocene: 1—Hölloch Cave (Wurth et al. 2004); 2—Katerloch Cave (Boch et al. 2009); 3—V11 Cave (Tămaş et al. 2007); 4—Lake Ammersee (von Grafenstein et al. 1998, 1999); 5—Lake Schleinsee (Tinner and Lotter 2001); 6—Lake Soppensee (Tinner and Lotter 2001); 7—Brunnboden and Krummgampen peat bogs (Kofler et al. 2005); 8—Preluca Tiganului and Steregoiu peat bogs (Feurdean 2005); 9—Alps-Jura lakes study region (Magny 2004); 10—Teleorman Valley (Macklin et al. 2011); 11—Durance Valley (Miramont et al. 2001); 12—Middle Rhône Valley (Berger et al. 2002); 13—South Adriatic Sea, core MD90-917 (Siani et al. 2013).

and rainfall spring-summer flood of 2006 the Danube at the Orşova gauging station (inside the Iron Gates gorge) underwent stage changes of 6–7 m on several occasions between late January and mid-July. Stage changes were rapid (occurring within a few days) and high river flows were maintained for periods lasting up to six weeks (Mihailova et al. 2012:Fig. 5c). More extreme variations in river discharge can perhaps be expected during Holocene RCC events (especially the 8.2 ka event), given the evidence of major snowmelt and ice jam floods during the Little Ice Age.

These observations give some idea of the likely impact of very high river flows on Stone Age settlements in the Iron Gates. Within the Iron Gates gorge, any site or zone within 10 m of the mean river level could have been at risk during extreme hydrological events, while a rise in the level of the Danube of 4 m or more would have inundated the site at Schela Cladovei and some other sites in the downstream area (Figure 9.1). Moreover, flood conditions may have persisted for weeks.

Borić and Miracle (2004:362) questioned whether such extreme hydrological events were characteristic of the Iron Gates reach of the Danube before the historic period. Various lines of evidence contradict this argument. At Schela Cladovei, repeated overbank flooding during the early-mid Holocene is indicated by 1.5–2 m of alluvial deposits (Boroneanț et al. 1999) and an associated luminescence date (Fuller et al. 1994); however soil development combined with anthropogenic disturbance has erased any original stratification in these deposits, blurring the stratigraphic relationships between flood events and human occupations. At Lepenski Vir, poor archaeological survival (including a lack of datable organic remains) in lower-lying parts of the site nearer the river may be attributable to scour (removal of material) during flood events. Remains of stone-lined hearths are evidence that buildings had once stood in this part of the site. Srejović (1972:53) observed that these hearths differ from those set within plaster floors in higher parts of the site and assigned them to an earlier phase of the Mesolithic. Since the plaster-floored structures started to be built after 8.3 ka cal B.P. (Bonsall et al. 2002a, 2008), it is possible they were deliberately sited in higher slope positions in response to an increase in flood magnitude and frequency.

The historically documented floods in the Iron Gates and lower Danube were caused by climatic events in the middle and upper reaches of the Danube and its major tributaries (Figure 9.5). Knowledge of Holocene climatic variations in the upper-middle Danube catchment has developed significantly over the past 15 years. Various proxy indicators reflect climatic trends during the early and middle Holocene that mirror those seen in climatic archives from the North Atlantic region:

- Episodes of cooler climate at 9.3–9.1 and ca. 8.2 ka cal B.P. are indicated by O-isotope studies of speleothems from caves in Austria (Wurth et al. 2004; Boch et al. 2009) and northwest Romania (Tămaș et al. 2005);

- Negative O-isotope excursions indicative of cooling and increased precipitation are registered in sediment cores from Lake Ammersee in southern Germany at ca. 9.2 ka and 8.2 ka cal B.P. (von Grafenstein et al. 1998, 1999);

- Vegetation responses to climatic cooling have been recorded in high altitude peat bogs and lake sediments in the Swiss Alps (Kofler et al. 2005; Tinner and Lotter 2001) at 8.2 ka cal B.P. and in northwest Romania at 9.3 and 8.2 ka cal B.P. (Feurdean 2005); and

- Research by Magny (2004) has documented 15 episodes of higher lake level in the Alps and Jura mountains reflecting increases in annual precipitation that were broadly synchronous with RCC events recorded in Greenland ice cores, including 9.55–9.15, 8.3–8.05, and 7.55–7.25 ka cal B.P. (Figure 9.3).

From a pan-European study of climate proxy records, Magny et al. (2003) concluded that mid-latitude Europe between 43° and 50° N experienced significantly wetter (and cooler) conditions during the RCC events of the Holocene.

Rivers tend to be very sensitive indicators of short-term and rapid climate change (Macklin et al. 2006, 2012) and a likely consequence of cooler conditions and increased precipitation over the upper Danube catchment such as characterized the 9.3 and 8.2 ka events and the period after 7.8 ka cal B.P. would have been an increase in seasonal and/or annual discharges along the Danube and its major tributaries in Central Europe, upstream of the Iron Gates. Mathematical modeling has suggested that water discharge along the Danube and some of its major tributaries in Central Europe during the most recent RCC event of the Holocene, the Little Ice Age, was on average 10–15 percent higher than at present (McCarney-Castle et al. 2012).

Considering its size and importance in Europe, remarkably little work has been done on Holocene fluvial development and flood histories of the Danube and its tributaries. Research by Howard et al. (2004) and Macklin et al. (2011) along the Teleorman River in southern Romania, which flows into the Danube east of the Iron Gates, found no evidence of enhanced river activity corresponding to RCC events of the earlier Holocene between 11.6 and 5.8 ka cal B.P. However, the Teleorman is a relatively small low-gradient catchment tributary of the lower Danube with very different hydro-climatic characteristics to the Danube catchment upstream of the Iron Gates.

A better proxy for hydrological patterns in the Danube catchment upstream of the Iron Gates is the behavior of rivers draining the western Alps, where several studies have shown substantially increased discharges around the time of the 9.3 and 8.2 ka events (Berger et al. 2002; Miramont et al. 2000).

4. Taphonomic Processes

Taphonomic processes (e.g., destruction, movement, and burial of artifacts by river processes, and archaeologists' choices of sites for investigation and samples for dating) may also have contributed to the fluctuations in the radiocarbon time series for the Iron Gates. Overbank flows could have resulted in the removal of artifacts and ecofacts from occupation surfaces—charcoal, since it floats, would be particularly vulnerable—which then would not be available for dating. However, many of the ^{14}C-dated animal and human bones from the Iron Gates sites (cf. Figure 9.2) were recovered from pit features (including graves) where they would have been relatively protected from river scour. Therefore, it seems unlikely that the ca. 300-year gap in the radiocarbon record at Schela Cladovei (Bonsall et al. 2002a:Fig. 2), for example, was a consequence of removal of material from the site during flood events.

Conclusions

A large number of radiocarbon dates (>300) are available for Mesolithic and Neolithic sites in the Iron Gates reach of the Danube. They include radiometric ^{14}C dates on bulk charcoal, soil organic matter, a human bone and a terrestrial animal bone, as well as AMS ^{14}C dates on human and terrestrial animal bones and a plant macroremain. Dates obtained from terrestrial mammals (ungulates) and short-lived plant remains can be considered the most accurate, as they are not affected by old carbon or aquatic reservoir effects.

The summed calibrated probability distribution of ^{14}C dates of terrestrial samples (Figure 9.2, dataset C) shows three distinct probability troughs (periods with relatively few or no ^{14}C dates) between 9.5–9.0 and 8.65–8.0 ka cal B.P. and after 7.8 ka cal B.P. These "discontinuities" in the radiocarbon record are broadly coincident with periods of rapid climate change (specifically, cold events) recorded in Greenland ice cores and paleoclimate archives from the Danube catchment.

Changing frequencies of ^{14}C dates over time have tended to be used by archaeologists as a proxy for fluctuations in population size. In the case of the Iron Gates, however, we suggest that other factors have strongly influenced the temporal pattern. Our preferred explanation is that the ^{14}C discontinuities reflect periods of higher annual river discharge and an increase in flood magnitude during Holocene "neoglacial" events, associated with generally cooler, wetter conditions in the Danube catchment upstream of the Iron Gates, and that the increased flood risk led to a reduction in the intensity with which people used certain sites or the lower parts of sites bordering the river. River scour may also have reduced the archaeological visibility of "neoglacial" periods by removing material from occupation surfaces.

Acknowledgments

We thank Christopher Bronk Ramsey and Derek Hamilton for advice on the use of OxCal 4.2.4.

References Cited

Ashmore, P. 1999 Single Entity Dating. *Mémoires de la Société préhistorique française* 26:65–71.

Bamforth, D. B., and B. Grund 2014 Radiocarbon Calibration Curves, Summed Probability Distributions, and Early Paleoindian Population Trends in North America. *Journal of Archaeological Science* 39:1768–1774.

Behrensmeyer A. K. 2006 Climate Change and Human Evolution. *Science* 311:476–478.

Berger, J. F., C. Delhon, S. Bonté, S. Thiébault, D. Peyric, A. Beeching, and J. Vital 2002 Paléodynamique fluviale, climat, action humaine et évolution des paysages du bassin versant de la Citelle (moyenne vallée du Rhône, Drôme) à partir de l'étude de la séquence alluviale d'Espeluche-Lalo. In *Les fleuves ont une histoire. Paléoenvironnement des rivières et des lacs français depuis 15 000 ans*, edited by J. P. Bravard and M. Magny, pp. 223–237. Errance, Paris.

Boch, R., C. Spötl, and J. Kramers 2009 High-Resolution Isotope Records of Early Holocene Rapid Climate Change from Two Coeval Stalagmites of Katerloch Cave, Austria. *Quaternary Science Reviews* 28:2527–2538.

Bond, G., W. Showers, M. Cheseby, R. Lotti, P. Almasi, P. deMenocal, P. Priore, H. Cullen, I. Hajdas, and G. Bonani 1997 A Pervasive Millennial-Scale Cycle in North Atlantic Holocene and Glacial Climates. *Science* 278:1257–1266.

Bonsall, C. 2008 The Mesolithic of the Iron Gates. In *Mesolithic Europe*, edited by G. Bailey and P. Spikins, pp. 238–279. Cambridge University Press, Cambridge.

Bonsall, C., R. J. Lennon, K. McSweeney, C. Stewart, D. D. Harkness, V. Boroneanţ, R. W. Payton, L. Bartosiewicz, and J. C. Chapman 1997 Mesolithic and Early Neolithic in the Iron Gates: A Palaeodietary Perspective. *Journal of European Archaeology* 5(1):50–92.

Bonsall, C., M. G. Macklin, R. W. Payton, and A. Boroneanţ 2002a Climate, Floods and River Gods: Environmental Change and the Meso-Neolithic Transition in South-East Europe. *Before Farming: The Archaeology of Old World Hunter-Gatherers* 3–4(2):1–15.

Bonsall, C., M. G. Macklin, D. E. Anderson, and R. W. Payton 2002b Climate Change and the Adoption of Agriculture in North-West Europe. *European Journal of Archaeology* 5(1):7–21.

Bonsall, C., I. Radovanović, M. Roksandic, G. T. Cook, T. Higham, and C. Pickard 2008 Dating Burial Practices and Architecture at Lepenski Vir. In *The Iron Gates in Prehistory*, edited by C. Bonsall, V. Boroneanţ, and I. Radovanović, pp. 175–204. Archaeopress, Oxford.

Bonsall, C., A. Boroneanţ, A. Soficaru, K. McSweeney, T. Higham, N. Mirițoiu, C. Pickard, and G. T. Cook 2012 Interrelationship of Age and Diet in Romania's Oldest Human Burial. *Naturwissenschaften* 99:321–325.

Bonsall, C., M. G. Macklin, A. Boroneant, C. Pickard, L. Bartosiewicz, G. T. Cook, and T. F. G. Higham 2014 Holocene Climate Change and Prehistoric Settlement in the Lower Danube Valley. *Quaternary International* (2014) http://dx.doi.org/10.1016/j.quaint.2014.09.031.

Bonsall, C., R. Vasić, A. Boroneanţ, M. Roksandic, A. Soficaru, K. McSweeney, A. Evatt, Ü. Aguraiuja, C. Pickard, V. Dimitrijević, T. Higham, D. Hamilton, and G. Cook 2015 New AMS ^{14}C Dates for Human Remains from Stone Age Sites in the Iron Gates Reach of the Danube, Southeast Europe. *Radiocarbon* 57(1):33–46.

Borić, D. 2011 Adaptations and Transformations of the Danube Gorges Foragers (c. 13.000–5500 BC): An Overview. In *Beginnings—New Research in the Appearance of the Neolithic between Northwest Anatolia and the Carpathian Basin; Papers of the International Workshop 8th–9th April 2009, Istanbul*, edited by R. Krauß, pp. 157–203. Leidorf, Rahden/Westf.

Borić, D., and V. Dimitrijević 2009 Apsolutna hronologija i stratigrafija Lepenskog Vira. *Starinar* 62(2007):9–55.

Borić, D. and P. Miracle 2004 Mesolithic and Neolithic (Dis)continuities in the Danube Gorges: New AMS Dates from Padina and Hajdučka Vodenica (Serbia). *Oxford Journal of Archaeology* 23(4):341–371.

Borić, D. and T. D. Price 2013 Strontium Isotopes Document Greater Human Mobility at the Start of the Balkan Neolithic. *Proceedings of the National Academy of Sciences of the USA* 110(9):3298–3303.

Boroneanţ, V., C. Bonsall, K. McSweeney, R. Payton, and M. Macklin 1999. A Mesolithic Burial Area at Schela Cladovei, Romania. In *L'Europe des Derniers Chasseurs: Épipaléolithique et Mésolithique, Actes du 5e colloque international UISPP, commission XII, Grenoble, 18–23 septembre 1995*, edited by A. Thévenin, pp. 385–390. Comité des Travaux Historiques et Scientifiques, Paris.

Bronk Ramsey, C. 2009 Bayesian Analysis of Radiocarbon Dates. *Radiocarbon* 51(1):337–360.

Cook, G. T., C. Bonsall, R. E. M. Hedges, K. McSweeney, V. Boroneanţ, and P. B. Pettitt 2001 A Freshwater Diet–Derived ^{14}C Reservoir Effect at the Stone Age Sites in the Iron Gates Gorge. *Radiocarbon* 43:453–460.

Cook, G. T., C. Bonsall, R. E. M. Hedges, K. McSweeney, V. Boroneanţ, L. Bartosiewicz, and P. Pettitt 2002 Problems of Dating Human Bones from the Iron Gates. *Antiquity* 76:77–85.

Cook, G. T., C. Bonsall, C. Pickard, K. McSweeney, L. Bartosiewicz, and A. Boroneanţ 2009 The Mesolithic-Neolithic Transition in the Iron Gates, Southeast Europe: Calibration and Dietary Issues. In *Chronology and Evolution within the Mesolithic of North-West Europe*, edited by P. Crombé, M. Van Strydonck, J. Sergant, M. Bats, and M. Boudin, pp. 497–515. Cambridge Scholars Publishing, Newcastle upon Tyne.

Copp, G. H., J. R. Britton, J. Cucherousset, E. García-Berthou, R. Kirk, E. Peeler, and S. Stakenas 2009 Voracious Invader or Benign Feline? A Review of the Environmental Biology of European Catfish *Silurus glanis* in Its Native and Introduced Ranges. *Fish and Fisheries* 10:252–282.

Crombé, P., and E. Robinson 2014 ^{14}C Dates as Demographic Proxies in Neolithisation Models of Northwestern Europe: A Critical Assessment Using Belgium and Northeast France as a Case-study. *Journal of Archaeological Science* 52:558–566.

David, J. A. 2006 Water Quality and Accelerated Winter Growth of European Catfish Using an Enclosed Recirculating System. *Water and Environment Journal* 20(4):233–239.

Dinu, A., A. Soficaru, and N. Miriţoiu 2007 The Mesolithic at the Danube's Iron Gates: New Radiocarbon Dates and Old Stratigraphies. *Documenta Praehistorica* 34:31–52.

Feurdean A. 2005 Holocene Forest Dynamics in Northwestern Romania. *The Holocene* 15:435–446.

Fuller, I. C., A. G. Wintle, and G. A. T. Duller 1994 Test of the Partial Bleach Methodology as Applied to the Infra-Red Stimulated Luminescence of an Alluvial Sediment from the Danube. *Quaternary Science Reviews* 13:539–543.

Gronenborn, D. 2007 Climate Change and Socio-Political Crises: Some Cases from Neolithic Central Europe. In *War and Sacrifice. Studies in the Archaeology of Conflict*, edited by T. Pollard and I. Banks, pp. 13–32. Brill, Leiden/Boston.

Gronenborn, D. 2009 Climate Fluctuations and Trajectories to Complexity in the Neolithic: Towards a Theory. *Documenta Praehistorica* 36:97–110.

Howard, A. J., M. G. Macklin, D. W. Bailey, S. Mills, and R. Andreescu 2004 Late Glacial and Holocene River Development in the Teleorman Valley on the Southern Romanian Plain. *Journal of Quaternary Science* 19:217–280.

Kobashi, T., J. P. Severinghaus, E. J. Brook, J-M. Barnola, and A. M. Grachev 2007 Precise Timing and Characterization of Abrupt Climate Change 8200 Years Ago from Air Trapped in Polar Ice. *Quaternary Science Reviews* 26:1212–1222.

Kofler, W., V. Krapf, W. Oberhuber, and S. Bortenschlager 2005 Vegetation Responses to the 8200 cal. BP Cold Event and to Long-Term Climatic Changes in the Eastern Alps: Possible Influence of Solar Activity and North Atlantic Freshwater Pulses. *The Holocene* 15:779–788.

McCarney-Castle, K., G. Voulgaris, A. J. Kettner, and L. Giosan 2012 Simulating Fluvial Fluxes in the Danube Watershed: The "Little Ice Age" Versus Modern Day. *The Holocene* 22(1):91–105.

Macklin, M. G., G. Benito, K. J. Gregory, E. Johnstone, J. Lewin, D. J. Michczyńska, R. Soja, L. Starkel, and V. R. Thorndycraft 2006 Past Hydrological Events Reflected in the Holocene Fluvial Record of Europe. *Catena* 66(1–2):145–154.

Macklin, M. G., D. W. Bailey, A. J. Howard, S. Mills, R. A. J. Robinson, P. Mirea, and L. Thissen 2011 River Dynamics and the Neolithic of the Lower Danube Catchment. In *The Lower*

Danube in Prehistory: Landscape Changes and Human-Environment Interactions, edited by S. Mills and P. Mirea, pp. 9–14. Editura Renaissance, București.

Macklin, M. G., J. Lewin, and J. C. Woodward 2012 The Fluvial Record of Climate Change. *Philosophical Transactions of the Royal Society A* 370:2143–2172.

Magny, M. 2004 Holocene Climatic Variability as Reflected by Mid-European Lake-Level Fluctuations, and Its Probable Impact on Prehistoric Human Settlements. *Quaternary International* 113:65–79.

Magny, M., C. Bégeot, J. Guiot, and O. Peyron 2003 Contrasting Patterns of Hydrological Changes in Europe in Response to Holocene Climate Cooling Phases. *Quaternary Science Reviews* 22:1589–1596.

Mayewski, P. A., E. J. Rohling, J. C. Stager, W. Karlén, K. A. Maasch, L. D. Meeker, E. A. Meyerson, F. Gasse, S. Van Kreveld, K. Holmgrend, J. Lee-Thorp, G. Rosqvist, F. Racki, M. Staubwasser, R, R. Schneider, and E. J. Steig 2004 Holocene Climate Variability. *Quaternary Research* 62:243–255.

Michczyńska, D. J., and A. Pazdur 2004 Shape Analysis of Cumulative Probability Density Function of Radiocarbon Dates Set in the Study of Climate Change in the Late Glacial and Holocene. *Radiocarbon* 46(2):733–744.

Mihailova, M. V., V. N. Mikhailov, and V. N. Morozov 2012 Extreme Hydrological Events in the Danube River Basin over the Last Decades. *Water Resources* 39(2):161–179.

Miramont, C., O. Sivan, T. Rosique, J. L. Edouard, and M. Jorda 2000 Subfossil Tree Deposits in the Middle Durance (Southern Alps, France): Environmental Changes from Allerød to Atlantic. *Radiocarbon* 42:423–435.

Omarov, O. P., and O. A. Popova 1985 Feeding Behavior of Pike, *Esox lucius,* and Catfish, *Silurus glanis,* in the Arakum Reservoirs of Dagestan. *Journal of Ichthyology* 25(1):25–36.

Rasmussen, S. O., B. M. Vinther, H. B. Clausen, and K. K. Andersen 2007 Early Holocene Climate Oscillations Recorded in Three Greenland Ice Cores. *Quaternary Science Reviews* 26:1907–1914.

Reimer, P. J., E. Bard, A. Bayliss, J. W. Beck, P. G. Blackwell, C. Bronk Ramsey, C. E. Buck, H. Cheng, R. L. Edwards, M. Friedrich, P. M. Grootes, T. P. Guilderson, H. Haflidason, I. Hajdas, C. Hatté, T. J. Heaton, D. L. Hoffman, A. G. Hogg, K. A. Hughen, K. F. Kaiser, B. Kromer, S. W. Manning, M. Niu, R. W. Reimer, D. A. Richards, E. M. Scott, J. R. Southon, R. A. Staff, C. S. M. Turney, and J. van der Plicht 2013 IntCal13 and Marine13 Radiocarbon Age Calibration Curves 0–50,000 Years Cal BP. *Radiocarbon* 55(4):1869–1887.

Rohling, E. J., and H. Pälike 2005 Centennial-Scale Climate Cooling with a Sudden Cold Event Around 8,200 Years Ago. *Nature* 434:975–979.

Siani, G., M. Magny, M. Paterne, M. Debret, and M. Fontugne 2013 Paleohydrology Reconstruction and Holocene Climate Variability in the South Adriatic Sea. *Climate of the Past* 9:499–515.

Smith, K., and R. Ward 1998 *Floods: Physical Processes and Human Impacts.* Wiley, New York.

Srejović, D. 1972 *Europe's First Monumental Sculpture. New Discoveries at Lepenski Vir.* Thames and Hudson, London.

Tămaş, T., B. P. Onac, and A-V. Bojar 2005 Lateglacial–Middle Holocene Stable Isotope Records in Two Coeval Stalagmites from the Bihor Mountains, NW Romania. *Geological Quarterly* 49(2):185–194.

Tinner, W., and A. F. Lotter 2001 Central European Vegetation Response to Abrupt Climate Change at 8.2 ka. *Geology* 29:551–554.

Tolan-Smith, C., and C. Bonsall 1999 Stone Age Studies in the British Isles: The Impact of Accelerator Dating. In *¹⁴C et Archéologie. Actes du 3ème congrès international, Lyon, 6–10 avril 1998*, edited by J. Evin, C. Oberlin, J. P. Daugas, and J. F. Salles, pp. 249–257. Société Préhistorique Française, Paris.

Tringham, R. 2000 Southeastern Europe in the Transition to Agriculture in Europe: Bridge, Buffer, or Mosaic. In *Europe's First Farmers*, edited by T. D. Price, pp. 19–56. Cambridge University Press, Cambridge.

von Grafenstein, U., H. Erlenkeuser, J. Müller, J. Jouzel, and S. J. Johnsen 1998 The Cold Event 8200 Years Ago Documented in Oxygen Isotope Records of Precipitation in Europe and Greenland. *Climate Dynamics* 14:73–81.

von Grafenstein, U., H. Erlenkeuser, A. Brauer, J. Jouzel, and S. J. Johnsen 1999 A Mid-European Decadal Isotope-Climate Record from 15,500 to 5000 Years B.P. *Science* 284:1654–1657.

Weninger, B., E. Alram-Stern, E. Bauer, L. Clare, U. Danzeglocke, O. Jöris, C. Kubatzki, G. Rollefson, H. Todorova, and T. Van Andel 2006 Climate Forcing Due to the 8200 Cal BP Event Observed at Early Neolithic Sites in the Eastern Mediterranean. *Quaternary Research* 66:401–420.

Weninger, B., O. Jöris, and U. Danzeglocke 2007 CalPal-2007 *Cologne Radiocarbon Calibration & Palaeoclimate Research Package*. http://www.calpal.de/.

Weninger, B., L. Clare, E. J. Rohling, O. Bar-Yosef, U. Boehner, M. Budja, M. Bundschuh, A. Feurdean, H-G. Gebe, O. Jöris, J. Lindstaedter, P. Mayewski, T. Muehlenbruch, A. Reingruber, G. Rollefson, D. Schyle, L. Thissen, H. Todorova, and C. Zielhofer 2009 The Impact of Rapid Climate Change on Prehistoric Societies during the Holocene in the Eastern Mediterranean. *Documenta Praehistorica* 36:7–59.

Williams, A. N. 2012 The Use of Summed Radiocarbon Probability Distributions in Archaeology: A Review of Methods. *Journal of Archaeological Science* 39:578–589.

Wurth, G., S. Niggemann, D. K. Richter, and A. Mangini 2004 The Younger Dryas and Holocene Climate Record of a Stalagmite from Hölloch Cave (Bavarian Alps, Germany). *Journal of Quaternary Science* 19:291–298.

Chapter Ten

Climate Fluctuations, Human Migrations, and the Spread of Farming in Western Eurasia

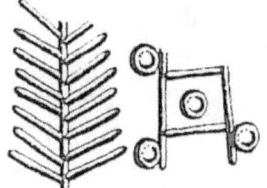

Refining the Argument

Detlef Gronenborn

Abstract *It has been suggested that the expansion of farming in western Eurasia was paced by Holocene cooling phases. However, the immediate effects and the mechanisms could not be explained. Based on a fine-graded data set of the West-Central European early Neolithic Linear pottery culture, a refined model of the interaction between climate variability and both population growth as well as decline is presented. The changing interrelation between these fluctuations may more thoroughly explain the step-wise advance of farming in western Eurasia.*

Introduction

The spread of farming has been and continues to be one of the major research topics in prehistoric archaeology worldwide. The roots of academic inquisitiveness lie in the nineteenth century (Roth 1887), but with Gordon Childe's (1936) demonstration of the "revolutionary" character of this socioeconomic shift the fundamental significance of sedentism and food production for any subsequent period of human history have become fully realized. Indeed, more recently the onset of farming is being discussed as the onset of global anthropogenic environmental and climate change (Lemmen 2009; Kaplan et al. 2010; Ruddiman 2003; but for an even broader approach see Foley et al. 2014).

Countless studies are devoted to the phenomena in every location on the globe where plants or animals had been domesticated (e.g., Barker 2006; Bellwood 2005). Earlier studies were based on archaeological, paleobotanical, or archaeozoological methodology, more recently archaeogenetics and mathematical modeling have been applied (e.g., Bocquet-Appel et al. 2009; Bramanti et al. 2009; Edwards et al. 2007; Haak et al. 2010; Lemmen et al. 2011).

While the emergence of farming has been studied in every possible environment—from the tropics to the Arctic—one of the major study regions remains western Eurasia, with the Fertile Crescent as its core and Europe, North Africa, and Central Asia as the fringes (e.g., Diamond 2002). Within this broad region, the most detailed body of data is available from temperate Europe where the Linear Pottery Culture (LBK, Germ. *Linienbandkeramik*) constitutes the classic textbook example for early farming societies in temperate woodland environments (e.g., Bánffy 2009; Bánffy and Oross 2010; Gronenborn 2007a; Gronenborn and Dolukhanov 2013; Pavúk 2004). This chapter will focus on the emergence and spread of farming in western Eurasia with a detailed focus on western Central Europe. Particular attention is given to early and mid-Holocene migrations and to the question as to what degree might climate fluctuations have played a role in this great transitional period.

Scales of Resolution

Any comprehensive analysis of the spread of farming requires a geographically and temporally multiscalar approach. This is simply dictated by the fact that this process covered both a broad supraregional territory but also many millennia (Figure 10.1). Still, the necessary detail for a full comprehension has to be extracted from regional and local high-resolution sample data sets. Hence, upscaling and downscaling between coarse supraregional and fine-grained local and regional data sets is a methodological prerequisite. This approach taken in geography and mathematical modeling (Bierkens et al. 2000; Burt 2003) is equally applicable in archaeology and may also be taken in archaeological narratives.

Such a multiscalar approach is also indispensable, if climate fluctuations are to be included in the discussion. Climate operates at different spatial levels and timescales down to weather phenomena (Bradley 1999:32–34). Thus, climate-proxies of supraregional significance may sometimes be regarded when local phenomena are to be understood.

The frame of the approach taken here thus reaches from single regions in the range of a few hundred square kilometers up to the entire landmass of western Eurasia (Figure 10.1). The temporal resolution is limited by the dating uncertainties of the archaeological chronologies and the paleoclimatic age-models for proxies. The finest resolution is provided by the ^{14}C production curve (Kromer and Friedrich 2007) or other dendrochronological data sets with a resolution of usually 10 (± 5) years, and the archaeological resolution for the LBK of about 25 years (one house generation).

Brief Excursus into Early to Mid-Holocene Paleoclimatology

The Younger Dryas ended around 10,000 ^{14}C years B.P., equivalent to about 9500 cal B.C. (Muscheler et al. 2008). A rapid global climate change is visible in records worldwide; rather abrupt warming is accompanied by the northward shift of the Intertropical

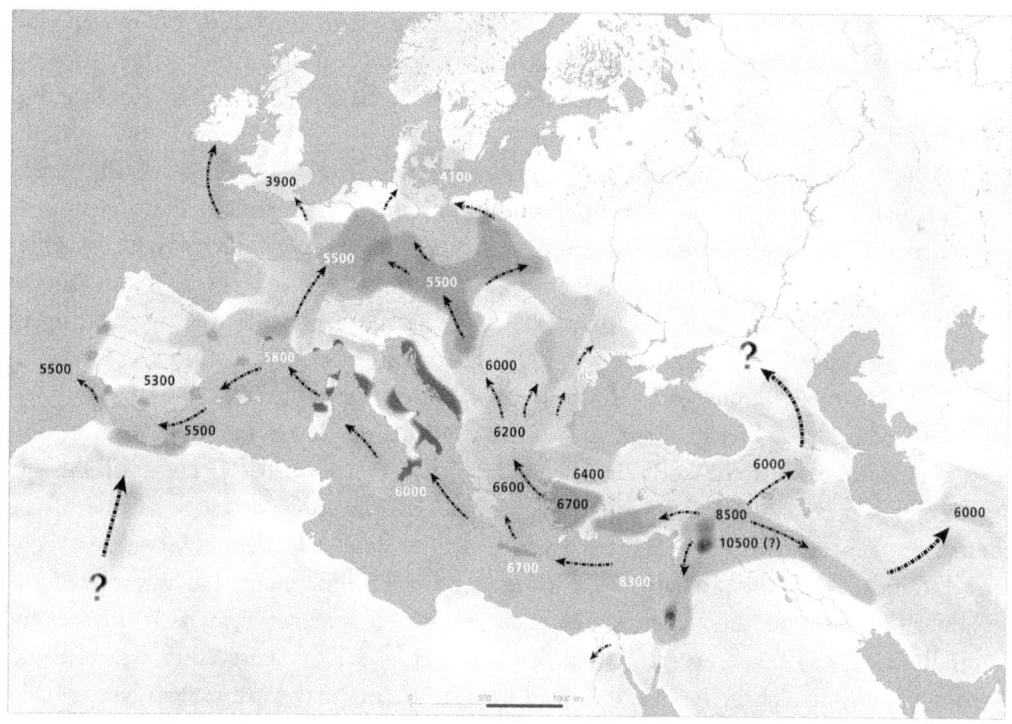

FIGURE 10.1. Comprehensive map of the spread of farming in western Eurasia; immediate study area is outlined by box (modified after Gronenborn 2010).

Convergence Zone leading to considerable changes in rainfall patterns, particularly across the Arabian Peninsula and the Levant (Migowski et al. 2006). The monsoon intensified (e.g., Garcin et al. 2007), and global temperatures rose within decades (Dansgaard et al. 1989). With temperatures and rainfall increasing, farming commenced in the "Fertile Crescent" (Bar Yosef 2006; Belfer-Cohen and Goring-Morris 2011; Weninger et al. 2009; Willcox 2005), while in the more northerly regions precipitation and warmer temperatures caused an increase in vegetation (Lang 1994), leading to the temperate woodland adaptations of European hunter-gatherer societies.

At the same time, the melting polar icecaps resulted in a rise of the global water table, and for the North Atlantic in a series of meltwater outbursts and also massive iceberg discharges (IRD-Ice rafting detritus events), which had effects on the North Atlantic thermohaline circulation and on climate fluctuations in Western Eurasia (Figure 10.2) (Bond et al. 2001; Jennings et al. 2015; Rasmussen et al. 2007; Teller and Leverington 2004; Wanner et al. 2008): Early Holocene climate is punctuated by a series of anomaly periods such as the Preboreal Oscillation (PBO), the 7.3 ka cal B.C. event (9.3 ka B.P.), the 6.2 ka cal B.C. event (8.2 ka B.P.), the 5.1 ka cal B.C. event, and the 4.2 cal B.C. (e.g., Barber et al. 2004; Fisher et al. 2002; Fleitmann et al. 2008; Luterbacher et al. 2001; Rasmussen et al. 2007; Schulz and Paul 2002; Wanner et al. 2008). The most prominent effects were reached during the 8.2 ka event, a roughly 200-year-long period of considerable climate degradations triggered by a combination of meltwater outbursts and IRD event 5a (e.g., Alley and Ágústsdóttir 2005; Barber et al. 1999; Rohling and Pälike 2005; Roy et al. 2011).

The underlying mechanism behind the centennial-long IRD-events is still unclear but fluctuations in solar activity may be one important component: In the ^{14}C production curve an increase in ^{14}C production indicates a decrease in solar activity (Kromer and Friedrich 2007; Stuiver and Quay 1980) and periods of a less active sun can be paralleled with the ice rafting detritus events suggesting a possible forcing mechanism (Bond et al. 2001; Davies and Brewer 2009; Fleitmann et al. 2007; Gupta et al. 2003; Wang et al. 2005; Wanner et al. 2008; but see Renssen et al. 2007).

If "solar forcing" (e.g., Bard and Frank 2006; Rind 2002; van Geel et al. 1999) is accepted, one gains the additional advantage that the residual ^{14}C curve or the ^{14}C production curve is applicable for any region and can therefore be taken as one of the best-dated general climate proxies currently available. Based on these observations, Central European Neolithic archaeology, informed by paleoclimatology, has interpreted the residual ^{14}C curve or the ^{14}C production curve as a climate proxy in itself, mainly for rainfall and less so for temperature (e.g., Arbogast et al. 2006; Maise 2005; Schibler and Jacomet 2010; Schlichtherle 2011). Linked to solar forcing may for instances be the "cold events" in the central Alps (Haas et al. 1998), the glacial advances in the western Alps (Burga and Perret 1998), lake level increases proposed by Magny (2004; but see Bleicher 2013 and refs. therein), wet-shifts in bogs (Blaauw et al. 2004), and fluctuations in the Main river oak curve (Spurk et al. 2002). In that sense the ^{14}C production curve is also used here, with the addition of other proxy records.

For a number of the above listed fluctuations and anomalies (Figure 10.2), impacts on human societies have been discussed (e.g., van Geel et al. 1998; Hassan 1997; Pfister and Brázdil 2006; Weninger et al. 2009); those that might have had an impact on the spread of farming will be treated next in more detail.

Climate Fluctuations and the Spread of Farming

The Broad Scale: Out of Asia

In recent years, it has become obvious that the expansion of farming was not a steady and continuous process but happened in stages, with periods of increased dynamics being intercalated by periods of stasis (e.g., Colledge et al. 2004; Gronenborn 2009a; Guilaine 2001; Schier 2009). A step-wise advance is also reproduced by recent mathematical models of the Neolithic expansion (Bocquet-Appel et al. 2009; Lemmen et al. 2011). If these phases of expansion are plotted against paleoclimatological data, it becomes obvious that there is a coarse temporal overlap between the steps of advance and paleoclimatic fluctuations, particularly phases of decreased solar activity and associated IRD events (Figure 10.3).

How may this contemporaneity be interpreted? A first and simple hypothesis was that IRD events had somehow paced the expansion of farming societies (Gronenborn

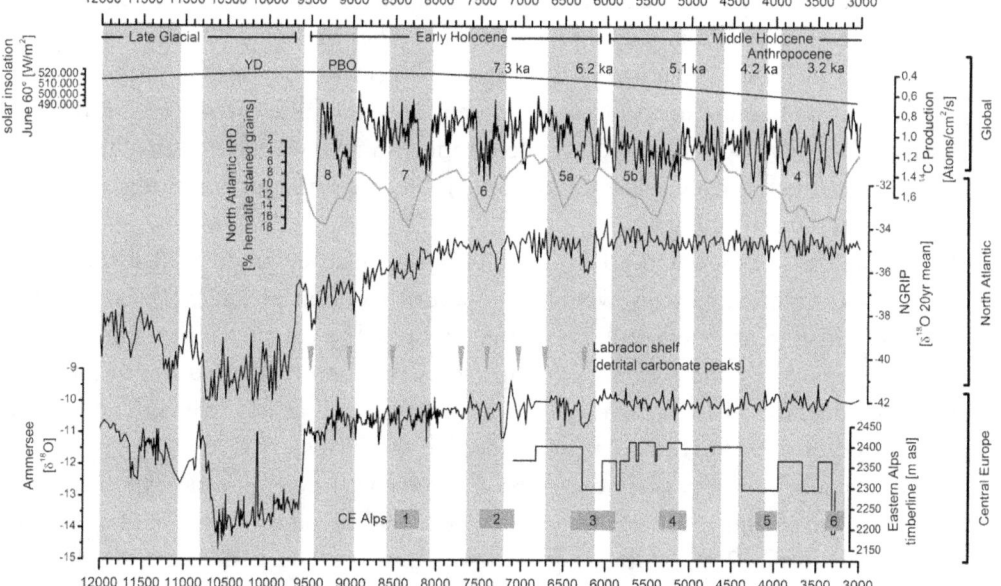

FIGURE 10.2. Paleoclimatic proxy data with global, North Atlantic, and Central European significance. Solar insolation: Berger and Loutre 1991; ¹⁴C production rate: Kromer and Friedrich 2007; IRD events: Bond et al. 2001; NGRIP: Vinther et al. 2006; Labrador shelf freshwater forcing: Jennings et al. 2015; Ammersee: von Grafenstein et al. 1999; timberline eastern Alps: Nicolussi et al. 2005; Cold events Alps: Haas et al. 1998).

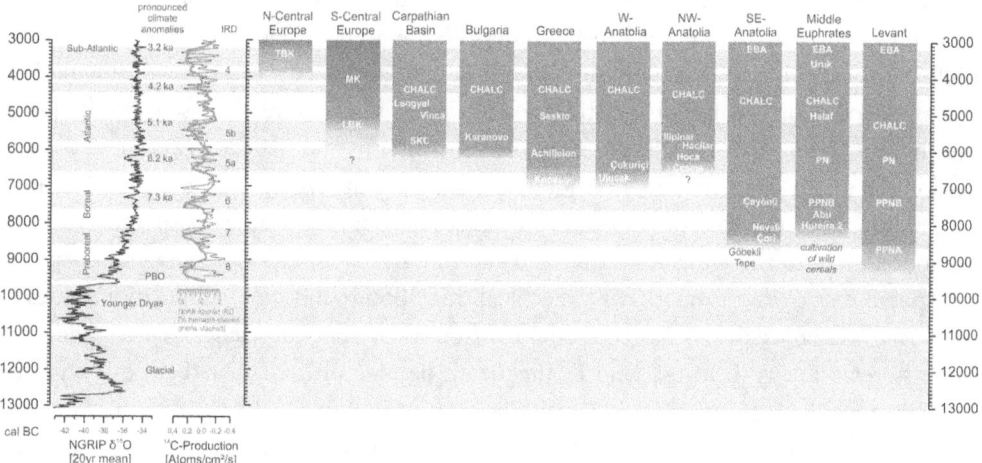

FIGURE 10.3. Simplified chronological table of the earliest appearance of domesticated cereals in selected regions across western Eurasia. PBO—Preboreal Oscillation; IRD—Ice Rafting detritus events, TBK—Funnel Beaker Culture; MK—Michelsberg Culture; LBK—Linear Pottery Culture; CHALC—Chalcolithic; PN—Pottery Neolithic; PPN—Pre-Pottery Neolithic (sources: NGRIP—Vinther et al. 2006; Rasmussen et al. 2007; ¹⁴C production—Kromer and Friedrich 2007; Lang 1994; Stahm 2010; Böhner and Schyle 2006; Colledge et al. 2004; Willcox 2005; Willcox et al. 2009; Jacomet online; Reingruber 2011).

2010, 2012), an idea that was also promoted by Weninger et al. (2009) by using a slightly different data set and terminology. While these two lines of thought consider a series of droughts and increased precipitation as triggering forces for increases in socioeconomic dynamics and lastly movements of people, others looked at different aspects. Turney and Brown (2007:2040), for instance, considered the rising water table during the earlier Holocene and the subsequent loss of coastal lands as the cause for "mass migration and cultural change on a regional scale."

On the basis of these and other presumptions it might be proposed that the expansion phases of farming societies would have occurred during periods of socioeconomic uncertainty, mainly as a response to external threats. The search for new lands would have been dictated by the attempt to escape from environmental problems at home. A major component of these and other scenarios was to look at how climate fluctuations had triggered or even forced societies into decline or collapse.

Increasingly, however, these one-sided and somewhat deterministic hypotheses had become challenged and also the concept of collapse had become debated in general (e.g., McAnany and Yoffee 2009). Other aspects were brought in, such as population fluctuations, associated sociopolitical complexity and internal conflicts, networks, identity, and systems theory (e.g., Bocquet-Appel 2011; Düring 2013; this volume; Reingruber 2011; Scheffran et al. 2012). These approaches are in accordance with nonlinear dynamics and chaos theory, following which endogenous processes force systems toward increasing complexity until a phase of instability is reached and complexity cascades into a lower-ranking or alternate state (e.g., Brunk 2002; Coombes and Barber 2005). Hence, the causes for when and why a tipping point leading to a subsequent downturn or change occurs in social cycles may be difficult to single out by just focusing on external parameters; internal dynamics and their nature need to be understood as well. In fact, Coombes and Barber (2005) basically question any alleged simplistic causalities between climate fluctuations and the decline or collapse of societies, as any society might have collapsed simply due to internal processes. This view is also promoted by Turchin and Nefedov (2009) who disregard climate as a triggering or even forcing parameter altogether. In that sense, Shennan (2013) discusses the role of population fluctuations for the unfolding of the Neolithic without touching the climate debate, and in Shennan et al. (2013) climate is dismissed as a parameter in the formation of the Neolithic "boom-and-bust" population cycles.

Also, resilience theory has been adopted from the ecological sciences. By applying this more systems theory–oriented field of theories it became obvious that societies only react gravely to external threats such as climate fluctuations when no other solution is possible and internal social and economic mechanisms fail (e.g., Gronenborn et al. 2013). Following Holling and Gunderson (2002), any system—humans included—undergoes a cyclical series of changes along a theoretical loop spanning between the parameters resilience, vulnerability, potential, and complexity. This trajectory was termed the "adaptive cycle," indicating that systems attempt to stay in a given stage as long as they are able to maintain it. If internal or external impulses make it necessary, a system will move to the next stage. Gunderson and Holling (2002) gave these stages symbols indicating their characteristics: the loop sets out with the α-stage and terminates with the Ω-stage. The space between the beginning and end of each loop is filled by the r- and the K-stage, termed after the r- and K-strategies

CLIMATE FLUCTUATIONS, HUMAN MIGRATIONS   217

of biogeography. The r-stage is characterized by an extensive expansion, and the K-stage by the conservation and intensification of the previously achieved.

Gunderson and Holling (2002) depicted these cycles in the form of a "∞," thereby symbolizing the eternal nature of the process. However, a mathematical model developed by Bub (2011) has produced a visualization in the form of a triangle, which is adopted here (Figure 10.4a). The visualization in the shape of a triangle is more helpful for

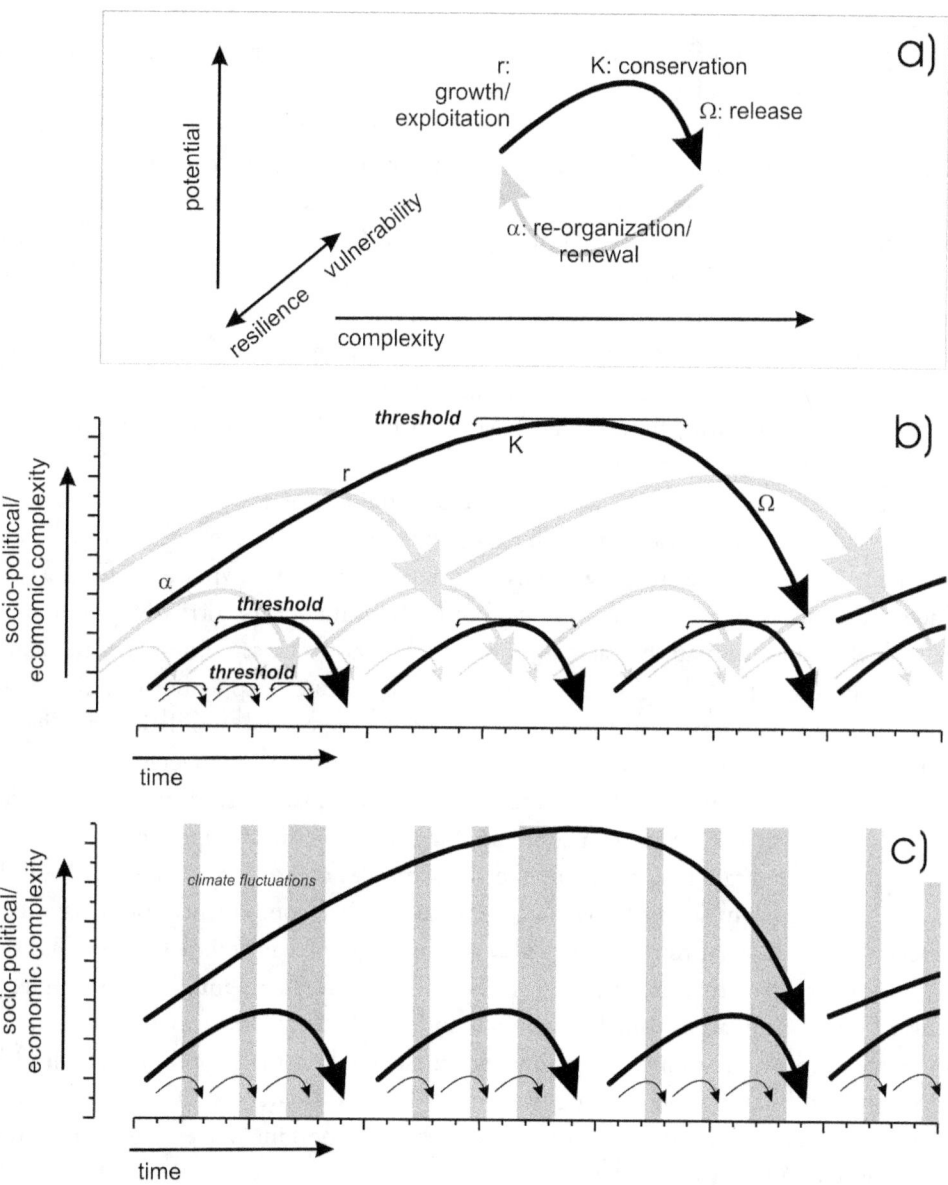

FIGURE 10.4. Adaptive cycles from Resilience Theory (modified after Holling and Gunderson [2002] and Bub [2011]): (a) basic build-up of adaptive cycles; (b) nested cycles in time with threshold values; (c) cycles and climate fluctuations.

historians and archaeologists because it reflects the shape of the distribution curve of objects in time and space, a pattern underlying many archaeological chronologies and methodologies, for example: correspondence analysis or seriation (e.g., Greenacre 2007). Cycles are nested, with long-term sequences being composed of short-term ones (Figure 10.4b). Information is transferred between these different levels as well as between earlier and subsequent cycles, as a memory effect regulates information flow between entities (Gunderson and Holling 2002).

During the earlier phases within a cycle, external threats would not cause any major changes in societies. Slight alterations in agrarian, social, and/or political strategies may be sufficient to absorb harvest losses. Only when societies have reached certain threshold levels of complexity and vulnerability may external punctuations lead to fundamental consequences with relocation, socioeconomic reorganization, and possibly ultimate collapse (Figure 10.4c). It is difficult to define and quantify factors determining adaptive cycles, particularly resilience, for societies without textual documentation, but one proxy might be population estimations or rather proxies for human activity as these are usually measured and then transferred into population estimates (e.g., Hinz et al. 2012). Thus, tipping points in population curves may not indicate worsening external conditions only but also—and maybe foremost—changes in the degree of internal resilience (Zimmermann 2012). Thus, decline phases are linked to impermeable internal and external threshold levels; obviously, these would have been reached more often during periods of increasing climatic deterioration.

However, initial phases of adaptive cycles should have formed during periods of beneficial climate, "good years" that allowed populations to increase, societies to become more complex, economies to grow. Thus, contrary to previous hypotheses formulation, which had focused on crises alone (Gronenborn 2010), it appears currently more rewarding to look at how climate has actually brought forward socioeconomic growth than at how climate fluctuations might have contributed to decline (Gronenborn et al. 2014).

This reverse in arguments is supported by mathematical simulation applied by Lemmen et al. (2011). Successful runs suggest that in a fixed homogeneous mid-Holocene environment farming would emerge under any circumstances, regardless of whether people migrated or not, at least in southern Europe. Once the components "technology exchange" and "migration" are added, the simulation is able to reproduce the Neolithization process equally successfully, suggesting that either component would have been sufficient to produce farming societies also in temperate and northern Europe. The simulations also suggest that an indigenous population component must not be neglected (ibid.). Further simulation runs by Lemmen and Wirtz (2014) then focused on the role of climate degradation events, suggesting that these would actually prolong the Neolithization process, bringing the model closer to the empirical data set. In general, however, disturbing climate events appear not to "have been as important" as "endogenous factors" (Lemmen and Wirtz 2013).

Beyond such principal and basic contemplations, the simulation result that farming hypothetically may have spread without any human migrations (Lemmen et al. 2011) adds a further component to the mid-Holocene changes: Mid-Holocene migrations, as an empirically observed historic phenomenon, have to be logically decoupled from the process of the spread of farming. Or, to say it the other way around, while active migrations of people as part of the Neolithic expansion can no longer be denied, they were nevertheless not a mandatory component of the Neolithization process and would have taken place regardless of whether farming had spread or not.

A logical decoupling of migrations from the process of the spread of farming then raises the question as to whether the Mid-Holocene migrations really are fully contemporaneous to the spread of farming, or whether they occurred at other times as well. One possible indication for pre-Neolithic migrations might be the appearance of the Late Mesolithic blade-and-trapeze industries (Gehlen 2010; Gronenborn 1997). The origin of the specific technology is sought in Central Eurasia (see, e.g., Biagi and Kiosak 2010 with further references). Groups of western Central Asian origin may have arrived in Temperate Europe as early as 7000–6700 cal B.C., at least in certain regions. These advances may have been patchy, undertaken by small bands and might have taken several centuries as trapezes in Europe do not appear within a confined period (e.g., Bokelmann 1999; Eichmann 2010; Gehlen 2010; Gronenborn 1997). Interestingly, the technological change from the Early to the Late Mesolithic is chronologically associated to a prominent climate fluctuation, namely the so-called Early Holocene event (EHE), or 9.3 ka event. This anomaly is increasingly better understood (Fleitmann et al. 2008; Hou et al. 2012; Lang et al. 2010; von Grafenstein et al. 1999; Yu et al. 2010) and currently linked to a freshwater pulse into the North Atlantic. It dates to about 7300/7200 cal B.C., with an overall duration of forty to one hundred years (Rasmussen et al. 2007).

This significant fluctuation predates the advent of the new lithic technology in Central Europe by possibly 100 years but is contemporaneous to its appearance in the eastern Ukraine (Biagi and Kiosak 2010). Yet, how these fluctuations before and after 7000 cal B.C. may be linked to possible migrations of hunter-gatherer populations during the latter half of the Boreal and the incipient Atlantic is presently entirely unclear. It may, however be noted that in the Levant the settlement size of 'Ain Ghazal suddenly decreases around 7000 cal B.C., and the Dead Sea levels also drop; this is coincident with the transition from Middle PPNB to Late PPNB (Weninger et al. 2009:27–32). At Çayönü the cobble building phase changes into the cell building phase, indicating architectural change (Özdogan 1999). New excavations show that pre-pottery farming communities had established themselves along the western Anatolian coast and on Crete around 6700 cal B.C earliest (Düring 2013; Efstratiou et al. 2004; Reingruber 2011). Apparently, the 7.2 event does mark an archaeologically visible change and/or disruption in sequences throughout western Eurasia.

By applying the above outlined model we may then understand the latter eight millennium or late Boreal as a period of comparable prosperity both for farming societies in eastern Anatolia but also for hunter-gatherers in what today is the Russian steppe zones and eastern Ukraine. The 7.2 event then would have forced groups of both farming as well as hunting and gathering societies into moving westward, obviously not in great numbers, as the Neolithic along the Turkish coast is not uniform and also the appearance of the blade-and-trapeze industries in Temperate Europe appears to be spotty at first. In the sense of the Lemmen et al. (2011) simulation, indigenous components in the process have to be acknowledged, which is also supported by archaeology (Düring 2013).

By 6500 cal B.C. Neolithic settlements had firmly established themselves along the western Anatolian coastline, shortly thereafter they appear in Greece and the Balkans (Reingruber 2011). Interestingly, contact between these early farming communities and Late Mesolithic hunter-gatherers rather quickly reaches as far west as the Swiss Jura: a clay *pintadera,* likely of southeastern European origin, found in the Late Mesolithic layers of the rockshelter of Arconciel-la-Souche and dating to 6200 cal B.C., is hard proof for intensive contacts between the northern Alpine region and the Balkans (Mauvilly et al. 2008).[1]

Furthermore, judging from the Late Mesolithic Ofnet burials (Orschiedt 2005), mollusks of both southern French or Mediterranean and Mid-Danube origin were brought to southern Central Europe, this in greater numbers than during the Early Mesolithic indicating an increase in supraregional communication and exchange (Gronenborn 1999). Ofnet dates to the beginning of the 6.2 event and is one of a number of Late Mesolithic burials that culminate exactly during this particular 200-year period (Gehlen and Schön 2005). This may be an indication of increased warfare and violence during that time (Gronenborn 2007b). Given these indications of violence, may we then have to account for early migrations from western Anatolia and from Eurasia centuries before the advent of the full-scale farming communities? This may be one of the most interesting questions for future research, particularly archaeogenetics, but what is already quite clear from archaeological evidence today: the pathways that were taken around the mid-sixth millennium by the early farmers of the LBK and the early agro-pastoralists of La Hoguette appear to have been known and traversed many centuries before their arrival.

The Regional Scale: IRD 5b and the LBK

A bit clearer, as far as migrations are concerned, is the picture for the next advance of people, associated with the LBK. The comparatively good archaeological visibility of the LBK and more than a century of research have resulted in a fine-grained chronology on the level of decades (e.g., Bánffy and Oross 2010; Gronenborn et al. 2014; Strien 2000). LBK thus serves as a perfect example for the detailed study of the interrelationship between climate fluctuations and preindustrial simple farming societies in a temperate woodland environment.

LBK emergence, expansion, and decline occurred during IRD event 5b (Figures 10.2 and 10.5). This period of a generally calmer sun and presumably generally cooler and wetter conditions (Heiri et al. 2004)—within the Altithermal (Figure 10.2)—is visible in a number of proxy data sets from southern Central Europe. IRD 5b begins around 5800 cal B.C. after a warming period following the 8.2 ka event and terminates around 5100 cal B.C. The terminal decades during the fifty-second century cal B.C. are characterized by strong precipitation fluctuations (see below). IRD 5b is contemporaneous to cold event 4 of the Central Alps (Haas et al. 1998) and the Cerin lake-level rise in the western Alps and the Jura (Magny 2004). IRD 5b is quite visible in the ^{14}C production curve (Kromer and Friedrich 2007; Sirocko et al. 2009) and may thus be dated with a precision of ±5 years. A rise of the ^{14}C production signifies a decrease of solar activity (Stuiver and Quay 1980) and several solar minima are visible: 5550, 5410, 5240, 5150, 5085, and 4985 den B.C. (Figure 10.5). Each minimum is followed by a maximum of solar activity starting with 5520 and ending with 5105 den B.C. (Figure 10.5).

The most notable amplitudes of proxies occur toward the end of the sixth millennium and start with a maximum of solar activity around 5100 den B.C., an intermediate phase with a weaker maximum at 5080 den B.C., followed by longer period of lesser ^{14}C production during the fifty-first century. These fluctuations reflect water influx of the Ulmen Maar in the Eifel hill region (Sirocko 2009). This data set provides a high-resolution and well-dated curve for the Eifel, the Hunsrück, the Rhenish Bay, and

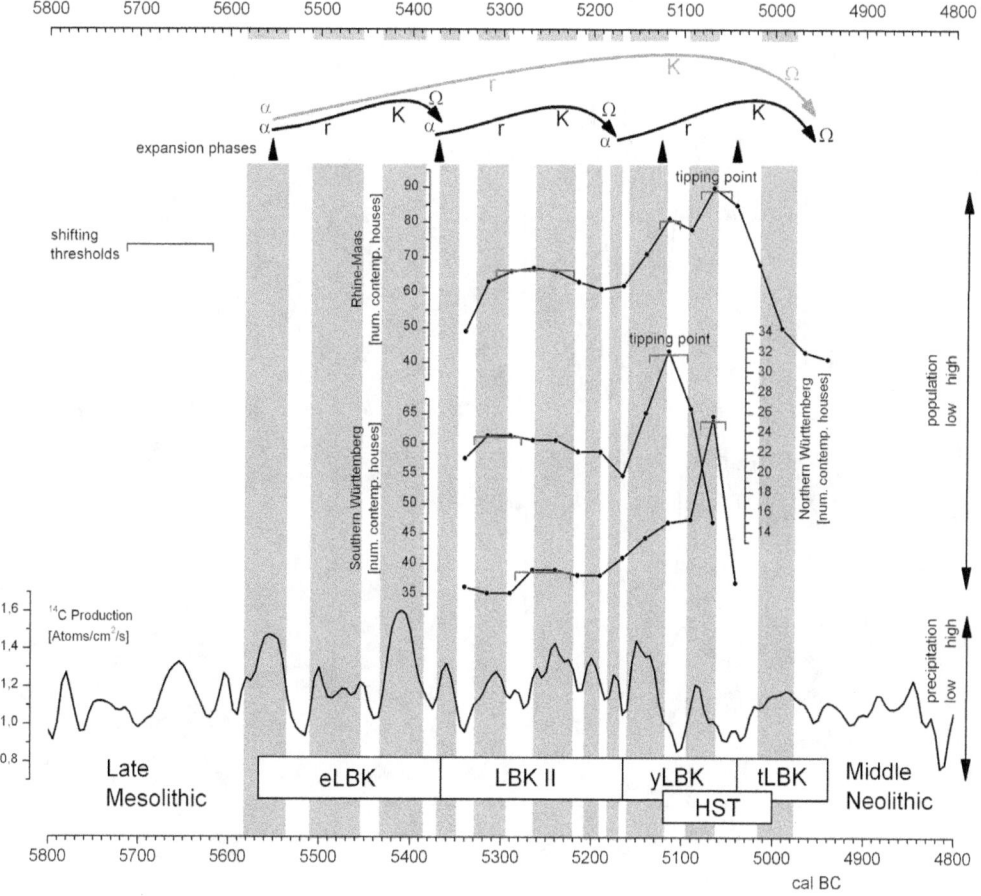

FIGURE 10.5. Paleoclimatic proxy-data for period of IRD 5b and archaeological chronology of LBK. IRD events: Bond et al. 2001; Cold events Alps: Haas et al. 1998; Main oak deposition rate: Spurk et al. 2002; Main oak ring-width index: Spurk personal communication 2004; archaeological chronology: Gronenborn et al. 2013.

the Hessian hill regions up to the Odenwald (Wernli and Pfahl 2009). However, the fine-tuning of the amplitudes is still ongoing and the curve depicted does have a dating uncertainty on the level of decades. Still, the highly fluctuating sequence of phases with strong water influx and phases with little to no water influx is obvious (Figure 10.6). Lastly, this water influx reflects mostly winter precipitation intensity.

Independently, the growth pattern of oaks from the LBK well of Kückhofen have been interpreted by Schmidt (et al. 1998:287f.) as an indication of a continuous decrease of growth as from 5210 den B.C. onward; after a period of intensive fluctuations, this trend then changed to the opposite from 5090 den B.C. onward. An increase of pre-

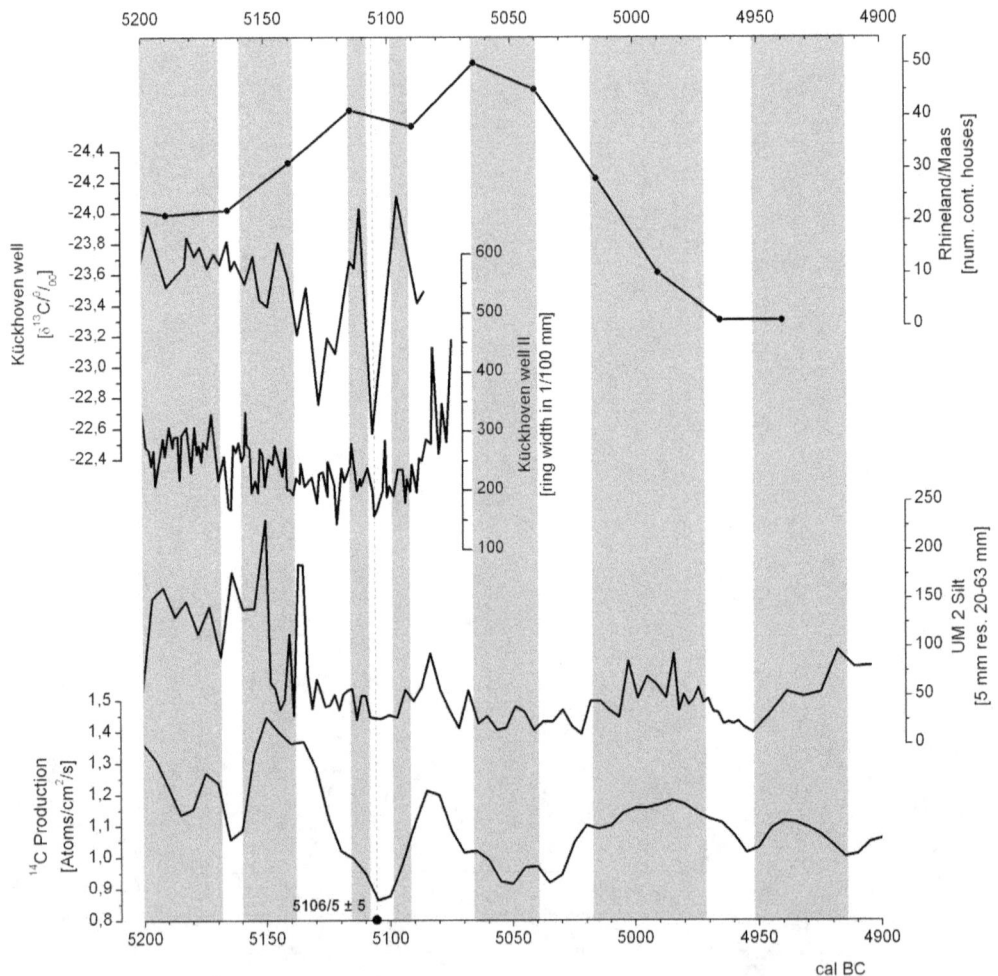

FIGURE 10.6. Paleoclimatic proxy-data for the end of IRD 5b and archaeologically dated events toward the termination of LBK in the Rhineland and Dutch Limburg. $\delta^{13}C$ Kückhoven: Helle and Schleser 1989; tree ring-width Kückhoven: Schmidt et al. 1998; ^{14}C production rate: Kromer and Friedrich 2007; house numbers Rheinland/Maas region (dASIS data bank RGZM).

cipitation from 5089 den B.C. onward may equally be read from the Ulmener Maar data (Figure 10.6). Helle and Heinrich (2012) analyzed the carbon isotopes from the Kückhofen well and noted strong fluctuations between 5150 and 5089 den B. C. While the years 5126 den B.C. and 5106/5 den B.C. appear to have been unusually dry and warm. The following periods (5126–5113 den B.C. and 5106–5098 den B.C.) are characterized by sharp trends to rather wet conditions (Helle and Heinrich 2012:60). These are followed by a trend to dryer conditions, particularly the year 5106/5 den B.C. stands out; Helle and Heinrich (2012:60) interpret these signals as an "extreme drought."

For this period the Eifel Maar curve shows the transition from a phase of intensive water influx to weak water influx. A slight increase is visible as from 5100 den B.C. onward (Figure 10.6). As in the Kückhoven well data, strong fluctuations are visible in the silt curve from 5150 den B.C. until the end of the fifty-first century. The second half of the fiftieth century is calmer but with less water influx into the Ulmen Maar, hence drier conditions predominate the following centuries. With the onset of this dry period IRD 5b has ended.

In a number of attempts climate fluctuations as inferred from various proxy data have been linked to archaeological chronologies of the western LBK, sometimes to shifts in LBK expansion (e.g., Gronenborn 2012; Schmidt et al. 2004; Strien and Gronenborn 2005). However, a recently applied and refined archaeological chronology (see online material in Gronenborn et al. 2014) has shown that only for the youngest phase of the LBK may climate fluctuations have had critical consequences on otherwise internally determined population dynamics. Most transition periods may foremost be explained by internal cyclical dynamics, with the addition of climatically determined destabilization factors (Figure 10.5). However, climate does not only operate adversely but beneficially: there does appear to exist a correlation between periods of increased humidity and population increases. This becomes apparent for the decades after 5340 and 5160 cal B.C., which witness population increases, particularly the latter phase as it is also correlated with warm temperatures. These decades may be understood as climate-supported boom phases. Periods with decreasing rainfall, like the decades around 5200 cal B.C., are parallel to a population decrease. This interrelation changes, though, after 5140 cal B.C. when precipitation seems to have decreased. This decrease with an increase in temperatures has apparently caused the opening of new farming lands: LBK moves to upland locations exactly during this period (Lefranc 2007:159ff.; Gronenborn et al. 2014). With a general decrease in rainfall these elevated plots could be farmed allowing a further population increase. The general trend to less rainfall is disturbed by a period of extreme fluctuations between 5113 and 5089 den B.C. (Figure 10.6). These fluctuations and maybe the draught of 5106/5 den B. C. may have caused the population curve to stabilize, even decrease. The Rhine-Maas population curve rises again with a return to an apparent wetter period between 5090 and 5060 den B.C. Thereafter, population in the Rhineland declines during a possibly less humid period, which is however not documented in the Kückhofen well any more. LBK in Württemberg decreases slightly earlier as here the culture change to the Early Middle Neolithic Hinkelstein Group (HST) has already begun (Figure 10.5).

The correlation between decreasing rainfall may be explained by Bleicher's (2011) suggestion that a precipitation decrease in the early and mid-Holocene Central Europe, particularly during spring and summer, would have resulted in negative harvest results.

Cultural and economic changes are most fundamental in the western regions of LBK, along the Rhine and in the Belgian and French provinces (e.g., Hauzeur 2006; Jeunesse 2010). In the northern Upper Rhine Valley, the earliest Middle Neolithic societies emerge with Hinkelstein (HST) around 5125 den B.C. (Jeunesse and Strien 2009: 243). Burial rites, pottery decoration modes, and the settlement system change; Spatz (2003) had interpreted these changes—particularly the ones in the iconography—as resulting from a new belief system. Also the Paris Basin experienced major transformations, as the appearance of the Groupe de Cerny may be connected to population shifts from the south (Jeunesse and Van Willigen 2010). With Cerny in the Paris Basin, Großgartach and Rössen as the successors of HST in Central Europe, the rearrangement of the early Neolithic societies was completed and a new cycle was in its early phase. However, changes are much less marked in the east (Kaufmann 2009), yet may have been inspired by transitions in the west (Jeunesse and Strien 2009), a typical spillover scenario where changes in one region affect the trajectories in the surrounding ones.

Conclusions

Farming in Western Eurasia, particularly in Europe seems to have spread in the course of an early to mid-Holocene extended period of increased sociopolitical dynamics that might be termed the "Neolithic Expansion Period." It is characterized by migratory dynamics along three major contact network streams, which have their origins in the Paleolithic routes of the repopulation of Europe (Gronenborn 2011; Lazaridis et al. 2014). Migrations might have set in already during the eighth millennium cal B.C., and would initially have been undertaken by small bands of hunter-gatherers and possible horticulturalists. Only with the second half of the seventh millennium cal B.C. did repeated and more massive population shifts bring the change to fully evolved and sedentary agricultural village societies. The migrants first came to southeastern Europe and later, during the sixth millennium, to temperate central and western Europe. These populations then became the foundation of any later developments to complex agricultural manifestations with hierarchical sociopolitical organizations and multitiered settlement hierarchies as observable in the Middle Neolithic.

Empirical archaeological and paleoclimatological studies as well as mathematical simulations indicate that migration and exchange as well as population increase (Bocquet-Appel 2011; Shennan et al. 2013) may have to be considered as the prime movers of the process, although an indigenous trajectory to farming is theoretically possible for parts of southern Europe (Lemmen et al. 2011). Mathematical simulation runs are still based on simple frameworks and do not include small to medium-scale population fluctuations. Thus, how these may have influenced the mechanisms of the Neolithic expansion has not yet been tested. However, with the data presented above, I hereby would like to suggest a basic module for coupled climate-population interaction that could well have been at the basis for the population dynamics behind the Neolithic expansion, in temperate Europe but also beyond (Figure 10.7). It is based on the fine-graded data set of the west-central European LBK as described above.

The most important component of that module is not collapse but rather the period during which societies emerge and flourish, lastly the period of the positive feedback between climate proxy and population curve (see also Lemmen 2014:88). Speaking in Shennan's (2013) terms, this would be the onset of the boom phase, or r phase within adaptive cycle terminology, which necessarily precedes the conservation phase (K) and ultimately the Ω or bust phase. Such a population module is driven by the Malthusian assumption that all societies grow until they exceed their economic basis. In systems theory terminology, any system may cross a threshold line above which it is no longer able to maintain the achieved level of complexity and is forced back into simpler states (see above). In simple farming societies within a set environment such as the LBK (Bogaard 2004; Kreuz 2010) the vertical position of the shifting threshold line is very much dictated by climate fluctuations, as these societies do not have the technical means to rapidly adapt or to import foods in times of want. They immediately react positively or negatively to any climate change that affects harvest results. A positive effect—in case of LBK—seems to have been a rather rapid population increase from 5170 cal B.C. onward, a negative effect population decline from 5120 or 5070 cal B.C. onward. Migration would have been an even more immediate reaction as people might have expanded during population increase when fissioning was one solution to increasing social problems. Such fissioning might have been regional or supraregional. An interesting example for regional fissioning occurs during the dry phase toward the end of LBK when uplands were settled. The case of the LBK may thus be taken as a detailed template of how humans and climate may have interacted and how climate fluctuations may have influenced the way people migrated and farming spread during the earlier Holocene in western Eurasia.

Lastly, the entire Neolithic expansion might have been composed of these intercalated basic modules with periods of rather rapid population growth being followed by a possibly rather brief consolidation period and a decline. As Lemmen (2014) has suggested, driving forces would have been complex, encompassing interrelations between population, technological innovation, the environment (climate), and cultural peculiarities. In time, these components would also change in intensity as does their interrelation. Simplistic cause-and-effect scenarios do not operate, at least not on levels of finer chronological resolution (decades). Nevertheless, the role of climate is crucial in this coupled model as it allows for social and economic growth—"good years"—particularly in the temperate zones of western Eurasia during the earlier Holocene (Figure 10.7). During periods of climate extremes or increased anomalies, climate may either foster rapid growth or rapid decline, the latter particularly after a period of rapid growth. In order to fully understand the dynamics of the Neolithic expansion we have to analyze the entire process, as each phase is linked to the previous and succeeding ones just as in adaptive cycle theory each cycle is linked to the previous and succeeding one (Gunderson and Holling 2002).

In a recent, much debated contribution (Gronenborn 2010:72), I had ended with this statement: "The Neolithic Expansion (and, associated with it, also culture change

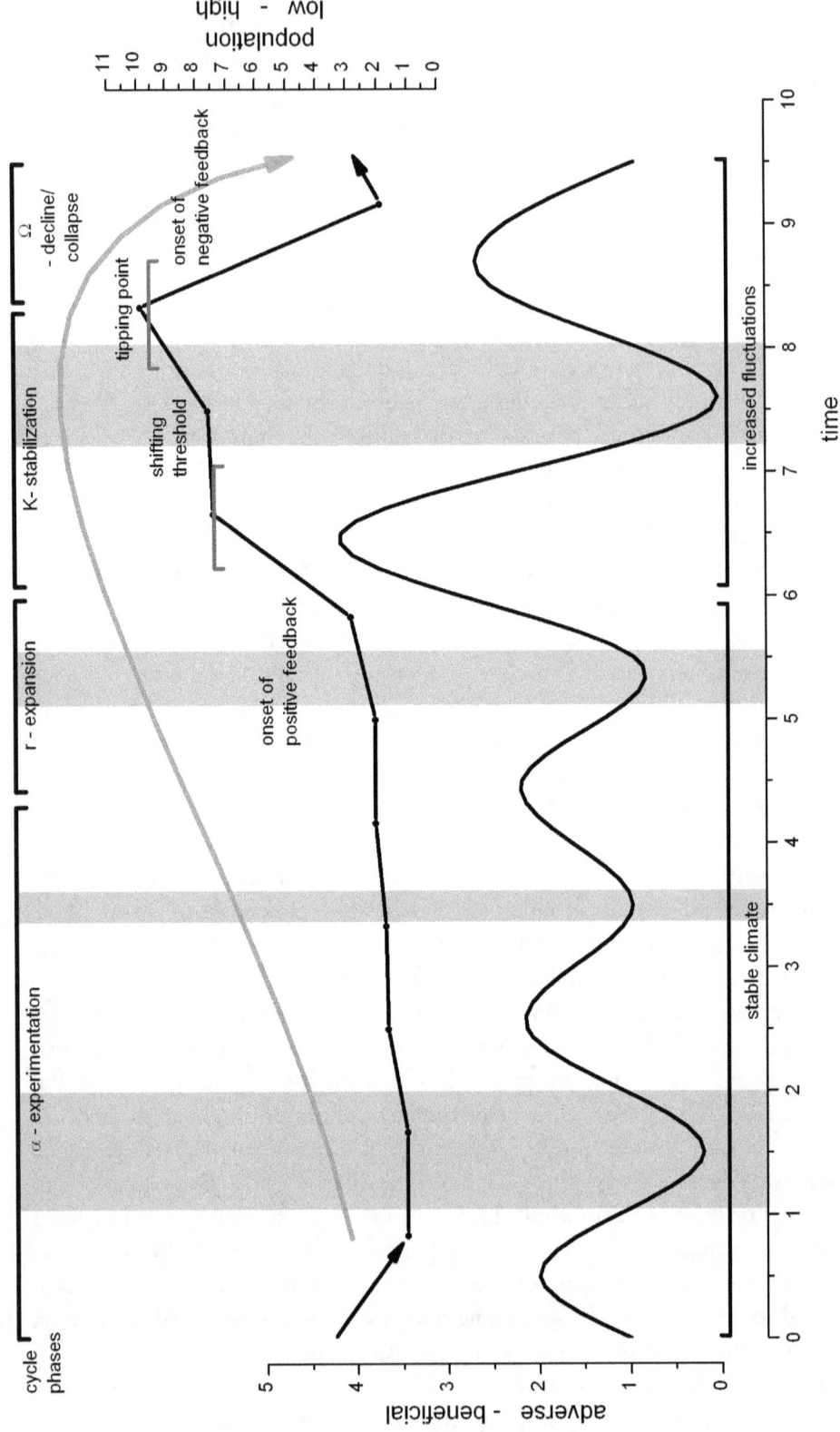

FIGURE 10.7. Coupled climate and population dynamics model for early farming societies in western Eurasia. Shaded areas denote periods of adverse climate.

within Neolithic societies) was paced by climatic fluctuations; in this long-term process IRD-events 6 and 4 mark the outer chronological margins of the " 'neolithisation' in Central Europe."

Mathematical simulation and empirical archaeology, as presented above, have so far shown that the basic message of this statement must not be changed; climate fluctuations did have pacing effects on the Neolithic expansion, and Lemmen and Wirtz (2013) have successfully simulated the interrelation. But, additionally to previous scenarios, not only did adverse periods pace the Neolithic expansion, but maybe even moreso the beneficial ones. Indeed, this might be the most important climate effect during the Holocene (Richerson et al. 2001).

Acknowledgments

I am indebted to Bernd Kromer, Mannheim, for allowing me to use the ^{14}C production curve, Frank Sirocko, Mainz, for the access to the silt curve from the Ulmener Maar, and Stephan Dietrich, Koblenz, for his expertise in fine-tuning the silt curve. Christoph Strien, Ahrweiler, kindly provided the data on the LBK.

Note

1. The debate around possible early indications for horticulture is ongoing but evidence is increasingly viewed with much criticism (e.g., Tinner et al. 2007, but see Behre 2007).

References Cited

Alley, R. B., and A. M. Ágústsdóttir 2005 The 8k Event: Cause and Consequences of a Major Holocene Abrupt Climate Change. *Quaternary Science Reviews* 24:1123–1149.

Arbogast, R.-M., S. Jacomet, M. Magny, and J. Schibler 2006 The Significance of Climate Fluctuations for Lake Level Changes and Shifts in Subsistence Economy during the Late Neolithic (4300–2400 B.C.) in Central Europe. *Vegetation History and Archaeobotany* 15:403–418.

Bánffy, E. 2009 Variations on the Neolithic Transition in Eastern and Western Hungary. In *Creating Communities: New Advances in Central European Neolithic Research*, edited by Daniela Hofmann and Penny Bickle, pp. 44–62. Oxbow, Oxford.

Bánffy, E., and K. Oross 2010 The Earliest and Earlier Phase of the LBK in Transdanubia. In *The Spread of the Neolithic to Central Europe. International Symposium, Mainz 24 June–26 June 2005*, edited by D. Gronenborn and J. Petrasch, pp. 255–272. RGZM-Tagungen 4,1. Römisch-Germanisches Zentralmuseum, Mainz.

Barber, D. C., A. Dyke, C. Hillaire-Marcel, A. E. Jennings, J. T. Andrews, M. W. Kerwin, G. Bilodeau, R. McNeely, J. Southon, M. D. Morehead, and J.-M. Gagnon 1999 Forcing of the Cold Event of 8,200 Years Ago by Catastrophic Drainage of Laurentide Lakes. *Nature* 400: 344–348.

Barber, K., B. Zolitschka, P. Tarasov, and A. F. Lotter 2004 Atlantic to Urals—The Holocene Climatic Record of Mid-Latitude Europe. In *Past Climate Variability through Europe and Africa*, edited by R. W. Battarbee, pp. 417–442. Kluwer, Dordrecht.

Bard, E., and M. Frank 2006 Climate Change and Solar Variability: What's New under the Sun? *Earth and Planetary Science Letters* 248:1–14.

Barker, G. 2006 *The Agricultural Revolution in Prehistory. Why did Foragers become Farmers?* Oxford University Press, Oxford.

Bar Yosef, O. 2006 L'impact des changements climatiques du Dryas récent et de l'Holocène inférieur sur les sociétés de chasseurs-cueilleurs et d'agriculteurs au Proche-Orient. In *L'Homme face au climat*, edited by E. Bard, pp. 283–301. Jacob, Paris.

Behre, K. E. 2007 Evidence for Mesolithic Agriculture in and around Central Europe? *Vegetation History and Archaeobotany* 16:203–219.

Belfer-Cohen, A., and A. Goring-Morris 2011 Becoming Farmers: The Inside Story. *Current Anthropology* 52:209–220.

Bellwood, P. 2005 *First Farmers: The Origins of Agricultural Societies*. Blackwell, Oxford.

Biagi, P., and D. Kiosak 2010 The Mesolithic of the Northwestern Pontic Region. New AMS Dates for the Origin and Spread of the Blade and Trapeze Industries in Southeastern Europe. *Eurasia Antiqua* 16:21–41.

Bierkens, M. F. P., P. A. Finke, and P. De Willigen 2000 *Upscaling and Downscaling: Methods for Environmental Research*. Kluwer Academic Publishers, Dordrecht, Boston, and London.

Blaauw, M., B. van Geel, and. J. van der Plicht 2004 Solar Forcing of Climatic Change during the Mid-Holocene: Indications from Raised Bogs in The Netherlands. *The Holocene* 14:35–44.

Bleicher, N. 2011 Einige kritische Gedanken zur Erforschung des Zusammenhangs von Klima und Kultur in der Vorgeschichte. In *Strategien zum Überleben. Umweltkrisen und ihre Bewältigung. Tagung des Römisch-Germanischen Zentralmuseums, 19./20. September 2008*, edited by Falko Daim, Detlef Gronenborn, and Rainer Schreg, pp. 67–80. RGZM-Tagungen 11, Römisch-Germanisches Zentralmuseum, Mainz.

Bleicher, N. K. 2013 Orbital, Ice-Sheet, and Possible Solar Forcing of Holocene Lake-Level Fluctuations in West-Central Europe: A Reply to Magny. *The Holocene* 23(8):1213–1215.

Bocquet-Appel, J.-P. 2011 When the World's Population Took Off: The Springboard of the Neolithic Demographic Transition. *Science* 333:560–561.

Bocquet-Appel, J.-P., S. Naji, M. Vander Linden, and J. K. Kozlowki 2009 Detection of Diffusion and Contact Zones of Early Farming in Europe from the Space-Time Distribution of ^{14}C Dates. *Journal of Archaeological Science* 36:807–820.

Bogaard, A. 2004 *Neolithic Farming in Central Europe. An Archaeobotanical Study of Crop Husbandry Practices*. Routledge, London.

Bokelmann, K. 1999 Zum Beginn des Spätmesolithikums in Südskandinavien. Geweihaxt, Dreieck und Trapez, 6100 cal BC. *Offa* 56:183–197.

Bond, G., B. Kromer, J. Beer, R. Muscheler, M. N. Evans, W. Showers, S. Hoffmann, R. Lotti-Bond, I. Hajdas, and G. Bonani 2001 Persistent solar influence on North Atlantic climate during the Holocene. *Science* 294:2130–2136.

Bosquet, D., and M. Golitko In press Highlighting and Characterizing the Pioneer Phase of the Hesbayan LBK (Liège Province, Belgium) In *Siedlungsstrukturen und Kulturwandel in der Bandkeramik. Beiträge der Internationalen Tagung "Neue Fragen zur Bandkeramik oder Alles beim Alten?," Leipzig 23.–24. September 2010*. Arbeits- und Forschungsberichte zur sächsischen Bodendenkmalpflege, Beiheft 24, Dresden.

Boulestin, B., A. Zeeb-Lanz, C. Jeunesse, F. Haack, R.-M. Arbogast, and A. Denaire 2009 Mass Cannibalism in the Linear Pottery Culture at Herxheim (Palatinate, Germany). *Antiquity* 83:968–982.

Bradley, R. S. 1999 *Paleoclimatology: Reconstructing Climates of the Quaternary.* Second Edition. Harcourt Academic Press, Amsterdam.

Bramanti, B., M. G. Thomas, W. Haak, M. Unterlaender, P. Jores, K. Tambets, I. Antanaitis-Jacobs, M. N. Haidle, R. Jankauskas, C.-J. Kind, F. Lueth, T. Terberger, J. Hiller, S. Matsumura, P. Forster, and J. Burger 2009 Genetic Discontinuity between Local Hunter-Gatherers and Central Europe's First Farmers. *Science* 326:137–140.

Brunk, G. G. 2002 Why Do Societies Collapse?: A Theory Based on Self-Organized Criticality. *Journal of Theoretical Politics* 14: 195–230.

Bonsall, C. 2008 The Mesolithic of the Iron Gates. In *Mesolithic Europe*, edited by Geoff Bailey and Penny Spikins, pp. 238–279. Cambridge University Press, Cambridge.

Bub, S. 2011 Long-Term Transformation Processes in Complex Systems: Transforming the Adaptive Cycle into a Minimal Model Using a Genetic Modelling Approach. Unpublished Diploma-Thesis, Fachbereich 7 Natur- und Umweltwissenschaften, University of Koblenz-Landau, Landau.

Burga, C. A., and R. Perret 1998 *Vegetation und Klima der Schweiz seit dem jüngeren Eiszeitalter.* Ott Verlag, Thun.

Burt, T. 2003 Scale: Upscaling and Downscaling in Physical Geography. In *Key Concepts in Geography*, edited by Sarah L. Holloway, Stephen P. Rice, and Gill Valentine, pp. 209–228. Sage, London.

Childe, G. 1936 *Man Makes Himself.* Watts, London.

Colledge, S., J. Conolly, and S. Shennan 2004 Archaeobotanical Evidence for the Spread of Farming in the Eastern Mediterranean. *Current Anthropology* 45(Suppl.):35–58.

Coombes, P., and K. Barber 2005 Environmental Determinism in Holocene Research: Causality or Coincidence? *Area* 37:303–311.

Dansgaard, W., J. W. C. White, and S. J. Johnsen 1989 The Abrupt Termination of the Younger Dryas Climate Event. *Nature* 339:532–534.

Davies, B. A. S., and S. Brewer 2009 Orbital Forcing and Role of the Latitudinal Insolation/Temperature Gradient. *Climate Dynamics* 32:143–165.

Diamond, J. 2002 Evolution, Consequences, and Future of Plant and Animal Domestication. *Nature* 418:700–707.

Düring, B. S. 2013 Breaking the Bond: Investigating The Neolithic Expansion in Asia Minor in the Seventh Millennium BC. *Journal of World Prehistory* 26(2):75–100.

Ebersbach, R., and C. Schade 2005 Modelle zur Intensität der bandkeramischen Landnutzung am Beispiel der Altsiedellandschaft Mörlener Bucht / Wetterau. In *Die Bandkeramik im 21. Jahrhundert. Symposium in der Abtei Brauweiler bei Köln vom 16.9-19.9.2002*, edited by Jens Lüning, Christiane Frirdich, and Andreas Zimmermann, pp. 259–274. Internationale Archäologie Arbeitsgemeinschaft, Symposium, Tagung, Kongress Band 7. Marie Leidorf, Rahden/Westf.

Edwards, C. J., R. Bollongino, A. Scheu, A. Chamberlain, A. Tresset, J.-D. Vigne, J. F. Baird, G. Larson, S. Y. W. Ho, T. H. Heupink, B. Shapiro, A. R. Freeann, M. G. Thomas, R. Arbogast, B. Arndt, and L. Bartosiewicz 2007 Mitochondrial DNA Analysis Shows a Near Eastern Neolithic Origin for Domestic Cattle and No Indication of Domestication of European Aurochs. *Proceedings of the Royal Society* B 274:1377–1385.

Efstratiou, N., A. Karetsou, E. S. Banou and D. Margomenou 2004 The Neolithic Settlement of Knossos: New Light on an Old Picture. In *Knossos: Palace, City State. Proceedings of the Conference in Herakleion Organised by the British School at Athens and the 23rd Ephoreia of*

Prehistoric and Classical Antiquities of Herakleion, in November 2000, for the Centenary of Sir Arthur Evan's Excavations at Knossos, edited by Gerald Cadogan, E. H. and A. Vasilakis, pp. 37–49. British School at Athens Studies 12. The British School at Athens, London.

Eichmann, W. J., R. Kertész, and T. Marton 2010 Mesolithic in the LBK Heartland of Transdanubia, Western Hungary. In Gronenborn, Detlef / Petrasch, Jörg (Hrsg.), *The Spread of the Neolithic to Central Europe*, pp. 211–234. International Symposium, Mainz 24 June–26 June 2005. RGZM-Tagungen 4,1. Römisch-Germanisches Zentralmuseum, Mainz.

Fisher, T. G., Derald G. Smith, and John T. Andrews 2002 Preboreal Oscillation Caused by a Glacial Lake Agassiz Flood. *Quaternary Science Reviews* 21:873–878.

Fleitmann, D., S. J. Burns, A. Mangini, M. Mudelsee, J. Kramers, I. Villa, U. Neff, A. A. Al-Subary, A. Buettner, D. Hippler, and A. Matter 2007 Holocene ITCZ and Indian Monsoon Dynamics Recorded in Stalagmites from Oman and Yemen (Socotra). *Quaternary Science Reviews* 26:170–188.

Fleitmann, D., M. Mudelsee, S. J. Burns, R. S. Bradley, J. Kramers, and A. Matter 2008 Evidence for a Widespread Climatic Anomaly at Around 9.2 ka before Present. *Paleoceanography* 23, PA1102, doi:10.1029/2007PA001519.

Foley, S. F., D. Gronenborn, M. O. Andreae, J. W. Kadereit, J. Esper, D. Scholz et al. 2013 The Palaeoanthropocene—The Beginnings of Anthropogenic Environmental Change. *Anthropocene* 3:83–88.

Garcin, Y., A. Vincens, D. Williamson, G. Buchet, and J. Guiot 2007 Abrupt Resumption of the African Monsoon at the Younger Dryas—Holocene Climatic Transition. *Quaternary Science Reviews* 26:690–704.

Gehlen, B. 2010 *Innovationen und Netzwerke. Das Spätmesolithikum vom Forggensee (Südbayern) im Kontext des ausgehenden Mesolithikums und des Altneolithikums in der Südhälfte Europas.* Welt und Erde, Loogh.

Gehlen, B., and W. Schön 2005 Klima und Kulturwandel—Mögliche Folgen des "6200-Events" in Europa. In *Klimaveränderung und Kulturwandel in neolithischen Gesellschaften Mitteleuropas, 6700–2200 v. Chr.*, edited by Detlef Gronenborn, pp. 53–74. RGZM-Tagungen 1. Römisch-Germanisches Zentralmuseum, Mainz.

Greenacre, M. 2007 *Correspondence Analysis in Practice*. Second Edition. Chapman and Hall, London.

Gronenborn, D. 1997 Sarching 4 und der Übergang vom Früh- zum Spätmesolithikum im südlichen Mitteleuropa. *Archäologisches Korrespondenzblatt* 27:387–402.

Gronenborn, D. 2007a Beyond the Models: "Neolithisation" in Central Europe. In *Going Over: The Mesolithic-Neolithic Transition in North-West Europe*, edited by Alasdair Whittle and Vicky Cummings, pp. 73–98. Proceedings of the British Academy 144, London.

Gronenborn, D. 2007b Climate Change and Socio-Political Crises: Some Cases from Neolithic Central Europe. In *War and Sacrifice: Studies in the Archaeology of Conflict*, edited by Tony Pollard and Ian Banks, pp. 13–32. Brill, Leiden.

Gronenborn, D. 2009a Climate Fluctuations and Trajectories to Complexity in the Neolithic: Towards a Theory. In *Neolithic Studies 16*, edited by Michael Budja, pp. 97–110. Documenta Praehistorica 36, University of Lubljana, Ljubljana.

Gronenborn, D. 2009b Transregional Culture Contacts and the Neolithization Process in Northern Central Europe. In *Ceramics before Farming: The Dispersal of Pottery among Prehistoric Eurasian Hunter-Gatherers*, edited by Peter Jordan and Marek Zvelebil, pp. 527–550. Left Coast Press, Walnut Creek.

Gronenborn, D. 2010 Climate, Crises, and the "Neolithisation" of Central Europe between IRD-Events 6 and 4. In *Die Neolithisierung Mitteleuropas—The Spread of the Neolithic to*

Central Europe. *Internationale Tagung Mainz, 24.–26. Juni 2005*, edited by Detlef Gronenborn and Jörg Petrasch, pp. 61–80. RGZM-Tagungen 4,1/2, Römisch-Germanisches Zentralmuseum, Mainz.

Gronenborn, D. 2011 Early Pottery in Afroeurasia: Origins and Possible Routes of Dispersal. In *Early Pottery in the Baltic—Dating, Origin, and Social Context. International Workshop at Schleswig from 20th to 21st October* 2006, edited by Sönke Hartz, Friedrich Lüth and Thomas Terberger, pp. 59–88. Bericht der Römisch-Germanischen Kommission 89, Römisch-Germanische Kommission, Frankfurt.

Gronenborn, D. 2012 Das Ende von IRD 5b: Abrupte Klimafluktuationen um 5100 denBC und der Übergang vom Alt- zum Mittelneolithikum im westlichen Mitteleuropa. In *Siedlungsstrukturen und Kulturwandel in der Bandkeramik. Beiträge der Internationalen Tagung "Neue Fragen zur Bandkeramik oder Alles beim Alten?" Leipzig 23.—24. September 2010*, pp. 241–250. Arbeits- und Forschungsberichte zur sächsischen Bodendenkmalpflege, Beiheft 24, Dresden.

Gronenborn, D., and P. Dolukhanov n.d. Early Neolithic Manifestations in Central and Eastern Europe. D. Hoffmann u. a. (Hrsg), *Oxford Handbook of Neolithic Archaeology*. Oxford University Press, Oxford. (DOI:10.1093/oxfordhb/9780199545841.013.005).

Gronenborn D., H.-C. Strien, S. Dietrich, and F. Sirocko 2014 "Adaptive Cycles" and Climate Fluctuations—A Case Study from Linear Pottery Culture in Western Central Europe. *Journal of Archaeological Science* 51:73–83.

Guilaine, J. 2001 La diffusion de l'agriculture en Europe: une hypothèse arythmique. *Zephyrus* 53:267–272.

Gupta, A. K., D. M. Anderson, and J. T. Overpeck 2003 Abrupt Changes in the Asian Southwest Monsoon during the Holocene and Their Links to the North Atlantic Ocean. *Nature* 421:354–357.

Haak, W., O. Balanovsky, J. J. Sanchez, S. Koshel, V. Zaporozhchenko, C. J. Adler, C. S. I. Der Sarkissian, G. Brandt, C. Schwarz, N. Nicklisch, V. Dreseley, B. Fritsch, E. Balanovska, R. Villems, H. Meller, K. W. Alt, A. Cooper, and the Genographic Consortium 2010 Ancient DNA from European Early Neolithic Farmers Reveals Their Near Eastern Affinities. *PLoS Biol* 8(11):e1000536. doi:10.1371/journal.pbio.1000536.

Haas, J. N., I. Richoz, W. Tinner, and L. Wick 1998 Synchronous Holocene Climatic Oscillations Recorded on the Swiss Plateau and at Timberline in the Alps. *The Holocene* 8:301–309.

Hassan, F. A. 1997 Nile Floods and Political Disorder in Early Egypt. In *Third Millennium BC Climate Change and Old World Collapse*, edited by H. N. Dalfes, G. Kukla, and H. Weiss, pp. 1–23. NATO AS1 Series 149. Springer, Berlin.

Hauzeur, A. 2006 *Le Rubané au Luxembourg: Contribution à l'étude du Rubané du Nord-Ouest européen*. Dossiers d'Archéologie du Musée National d'Histoire et d'Art X / Études et Recherches Archéologiques de l'Universitè de Liège 114. Musée national d'histoire et d'art, Luxembourg.

Heiri, O., W. Tinner, Lotter, and F. André 2004 Evidence for Cooler European Summers during Periods of Changing Meltwater Flux to the North Atlantic. *PNAS* 101:15285–15288.

Helle, G., and I. Heinrich 2012 Baumjahresringe als chemisch-physikalischer Datenträger für Umwelt- und Klimainformationen der Vergangenheit. *System Erde* 2,1, 2012, DOI: 10.2312/GFZ.syserde.02.01.11.

Hinz, M., I. Feeser, K.-G. Sjögren, and J. Müller 2012 Demography and the Intensity of Cultural Activities: An Evaluation of Funnel Beaker Societies (4200–2800 cal BC). *Journal of Archaeological Science* 39 (10): 3331–3340.

Holling, C. S., and L. H. Gunderson 2002 Resilience and Adaptive Cycles. In *Panarchy: Understanding Transformations in Human and Natural Systems*, edited by Lance H.Gunderson and C. S. Holling, pp. 25–62. Island Press, Washington.

Hoppe, W. In press Flomborn in der hessischen Chronologie—Übergang oder Neuanfang? In *Siedlungsstrukturen und Kulturwandel in der Bandkeramik. Beiträge der Internationalen Tagung "Neue Fragen zur Bandkeramik oder Alles beim Alten?"* Leipzig 23.–24. September 2010. Arbeits- und Forschungsberichte zur sächsischen Bodendenkmalpflege, Beiheft 24, Dresden.

Hou, J., Y. Huang, B. N. Shuman, W. W. Oswald, and D. R. Foster 2012 Abrupt Cooling Repeatedly Punctuated Early-Holocene Climate in Eastern North America (5), zuletzt geprüft am 13.11.2014.

Jäger, K.-D., and D. Kaufmann 1989 Zur frühneolithischen Besiedlung der naturräumlichen Einheit um Eilsleben, Kreis Wanzleben. In *Bylany Seminar* 1987: *Collected Papers*, edited by J. Rulf, pp. 305–313. Archaeological Institute of the Czechoslovak Academy of Sciences, Praha.

Jennings, A., J. Andrews, C. Pearce, L. Wilson, S. Ólfasdótttir 2015 Detrital Carbonate Peaks on the Labrador Shelf, a 13–7ka Template for Freshwater Forcing from the Hudson Strait Outlet of the Laurentide Ice Sheet into the Subpolar Gyre. *Quaternary Science Reviews* 107:62–80.

Jeunesse, C. 2010 Die Michelsberger Kultur. In *Jungsteinzeit im Umbruch. Die "Michelsberger Kultur" und Mitteleuropa im Umbruch vor 6000 Jahren*, edited by C. Lichter, pp. 46–55. Badisches Landesmuseum, Karlsruhe.

Jeunesse, C., and H.-C. Strien 2009 Bemerkungen zu den stichbandkeramischen Elementen in Hinkelstein. In *Krisen-Kulturwandel-Kontinuitäten. Zum Ende der Bandkeramik in Mitteleuropa. Beiträge der internationalen Tagung in Herxheim bei Landau (Pfalz) vom 14.–17. 06. 2007*, edited by Andrea Zeeb-Lanz, pp. 241–247. Internationale Archäologie. Arbeitsgemeinschaft, Symposium, Tagung, Kongress 10, Rahden/Westf.

Jeunesse, C., and S. van Willigen 2010 Westmediterranes Frühneolithikum und westliche Linearbandkeramik: Impulse, Interaktionen, Mischkulturen. In *Die Neolithisierung Mitteleuropas—The Spread of the Neolithic to Central Europe*, edited by Detlef Gronenborn and Jörg Petrasch, pp. 569–606. RGZM-Tagungen 4,2. Verlag des Römisch-Germanischen Zentralmuseums, Mainz.

Kaplan, M. R. 2010 Glacier Retreat in New Zealand during the Younger Dryas Stadial. *Nature* 467:194–197.

Kaufmann, D. 2009 Anmerkungen zum Übergang von der Linien- zur Stichbandkeramik in Mitteldeutschland. In *Krisen-Kulturwandel-Kontinuitäten: Zum Ende der Bandkeramik in Mitteleuropa*, edited by A. Zeeb-Lanz, pp. 267–282. Internationale Archäologie Arbeitsgemeinschaft, Symposium, Tagung, Kongress 10. Marie Leidorf, Rahden/Westf.

Kerig, T. 2008 *Hanau-Mittelbuchen: Siedlung und Erdwerk der bandkeramischen Kultur*. Universitätsforschungen zur Prähistorischen Archäologie 156. Rudolf Habelt, Bonn.

Kreuz, A. 2010 Die Vertreibung aus dem Paradies? Archäobiologische Ergebnisse zum Frühneolithikum im westlichen Mitteleuropa. Bericht der Römisch-Germanischen Kommission 91:24–196.

Kromer, B., and M. Friedrich 2007 Jahrringchronologien und Radiokohlenstoff: Ein ideales Gespann in der Paläoklimaforschung. *Geographische Rundschau* 59(4):50–55.

Lang, B., A. Bedford, S. J. Brooks, R. T. Jones, N. Richardson, H. John B.Birks, and J. D. Marshall 2010 Early-Holocene Temperature Variability Inferred from Chironomid Assemblages at Hawes Water, Northwest England. *The Holocene* 20(6):943–954.

Lang, G. 1994 *Quartäre Vegetationsgeschichte Europas: Methoden und Ergebnisse.* Gustav Fischer, Jena.

Lazaridis, I., N. Patterson, A. Mittnik, G. Renaud, S. Mallick, K. Kirsanow et al. 2014 Ancient Human Genomes Suggest Three Ancestral Populations for Present-Day Europeans. *Nature* 513 (7518): 409–413.

Lefranc, P. 2007 *La céramique du Rubané en Alsace: Contribution à l'étude des groupes régionaux du Néolithique ancien dans la plaine du Rhin supérieur.* Rhin Meuse Moselle Monographies d'Archéologie du Grand Est. Université Marc-Bloch, Strasbourg.

Lemmen, C. 2009 World Distribution of Land Cover Changes during Pre- and Protohistoric Times and Estimation of Induced Carbon Releases. *Géomorphologie: relief, processus, environnement* 2009/4:303–312.

Lemmen, C., D. Gronenborn, and K. Wirtz 2011 A Simulation of the Neolithic Transition in Western Eurasia. *Journal of Archaeological Science* 38:3459–3470.

Lemmen, C., and K. W. Wirtz 2014 On the Sensitivity of the Simulated European Neolithic Transition to Climate Extremes. *Journal of Archaeological Science* 51:65–72.

Luterbacher, J., R. Rickli, E. Xoplaki, C. Tinguely, C. Beck, C. Pfister, and H. Wanner 2001 The Late Maunder Minimum (1675–1715): A Key Period for Studying Decadal Scale Climatic Change in Europe. *Climatic Change* 49:441–462.

Magny, M. 2004 Holocene Climate Variability as Reflected by Mid-European Lake-Level Fluctuations and Its Probable Impact on Prehistoric Human Settlements. *Quaternary International* 113:65–79.

Maise, C. 2005 Archäoklimatologie neolithischer Seeufersiedlungen. In *Klimaveränderung und Kulturwandel in neolithischen Gesellschaften Mitteleuropas, 6700–2200 v. Chr.*, edited by Detlef Gronenborn, pp. 181–188. Verlag des Römisch-Germanischen Zentralmuseums, Mainz.

Mauvilly, M., C. Jeunesse, and T. Doppler 2008 Ein Tonstempel aus der spätmesolithischen Fundstelle Arconciel/La Souche (Kanton Freiburg/Schweiz). *Quartär* 55:151–157.

McAnany, P. A., and N. Yoffee (editors) 2009 *Questioning Collapse. Human Resilience, Ecological Vulnerabiulity, and the Aftermath of Empire.* Cambridge University Press, Cambridge.

Migowski, C., M. Stein, S. Prasad, J. F. W. Negendank, and A. Agnon 2006 Holocene Climate Variability and Cultural Evolution in the Near East from the Dead Sea Sedimentary Record. *Quaternary Research* 66(3):371–504.

Muscheler, R., B. Kromer, S. Björck, A.Svensson, M.Friedrich, K. F.Kaiser, and J. Southon 2008 Tree Rings and Ice Cores Reveal ^{14}C Calibration Uncertainties during the Younger Dryas. *Nature Geoscience* 1: doi:10.1038/ngeo128.

Nielsen, E. H. 2009 The Mesolithic Background for the Neolithisation Process. In *16th Neolithic Studies*, edited by Michael Budja, pp. 151–158. Documenta Praehistorica 36. University of Lubljana, Ljubljana.

Orschiedt, J. 2005 The Head Burials from Ofnet Cave: An Example of Warlike Conflict in the Mesolithic. In *Warfare, Violence, and Slavery in Prehistory*, edited by Mike Parker Pearson and I. J. N. Thorpe, pp. 67–74. BAR International Series 1374. Archaeopress, Oxford.

Özdogan, A. 1999 Çayönü. In *Neolithic in Turkey: The Cradle of Civilization*, edited by M. Özdogan and N. Basgelen, pp. 35–64. Arkeoloji Sanat Yayinlari, Istanbul.

Pavúk, J. 2004 Early Linear Pottery Culture in Slovakia and the Neolithisation of Central Europe. In *LBK Dialogues. Studies in the Formation of the Linear Pottery Culture*, edited by Alena Lukes and Marek Zvelebil, pp. 71–82. BAR International Series 1304. Archaeopress, Oxford.

Perlès, C. 2001 *The Early Neolithic in Greece: The First Farming Communities in Europe.* Cambridge University Press, Cambridge.

Pfister, C., and R. Brázdil 2006 Social Vulnerability to Climate in the "Little Ice Age": An Example from Central Europe in the Early 1770s. *Climate of the Past* 2:115–129.

Quitta, H. 1969 Zur Deutung bandkeramischer Siedlungsfunde in Auen und grundwassernahen Standorten. In *Siedlung, Burg und Stadt: Studien zu ihren Anfängen. Deutsche Akademie der Wissenschaften zu Berlin*, edited by K.-H. Otto and J. Herrmann, pp. 42–55. Schriften der Sektion für Vor- und Frühgeschichte 25. Akademie-Verlag, Berlin.

Rasmussen, S. O., B. M. Vinther, H. B. Clausen, and K. K. Andersen 2007 Early Holocene Climate Oscillations Recorded in Three Greenland Ice Cores. *Quaternary Science Reviews* 26:1907–1914.

Reingruber, A. 2008 Die deutschen Ausgrabungen auf der Argissa-Magula in Thessalien 2: *Die Argissa-Magula. Das frühe und das beginnende mittlere Neolithikum im Lichte transägäischer Beziehungen*. Beiträge zur ur- und frühgeschichtlichen Archäologie des Mittelmeer-Kulturraumes 35. Habelt, Bonn.

Reingruber, A. 2011 Early Neolithic Settlement Patterns and Exchange Networks in the Aegean. *Documenta Praehistorica* 38:291–305.

Renssen, H., H. Goosse, and T. Fichefet 2007 Simulation of Holocene Cooling Events in a Coupled Climate Model. *Quaternary Science Reviews* 26:2019–2029.

Richerson, P. J., R. Boyd, and R. L. Bettinger 2001 Was Agriculture Impossible during the Pleistocene but Mandatory during the Holocene? A Climate Change Hypothesis. *American Antiquity* 66:1–50.

Rind, D. 2002 The Sun's Role in Climate Variations. *Science* 296:673–677.

Rohling, E. J., and H. Pälike 2005 Centennial-Scale Climate Cooling with a Sudden Cold Event around 8,200 Years Ago. *Nature* 434:975–979.

Roth, H. L. 1887 On the Origin of Agriculture. *The Journal of the Anthropological Institute of Great Britain and Ireland* 16:102–136.

Roy, M., F. Dell'Oste, J. J. Veillette, D. de Vernal, J.-F. Hélie, and M. Parent 2011 Insights on the Events Surrounding the Final Drainage of Lake Ojibway Based on James Bay Stratigraphic Sequences (5–6). *Quaternary Science Reviews* 30:682–692.

Ruddiman, W. F. 2003 The Anthropogenic Greenhouse Era Began Thousands of Years Ago. *Climatic Change* 61:261–293.

Schibler, J., and S. Jacomet 2010 Short Climatic Fluctuations and Their Impact on Human Economies and Societies: The Potential of the Neolithic Lake Shore Settlements in the Alpine Foreland. *Environmental Archaeology* 15(2):173–182.

Schier, W. 2009 Extensiver Brandfeldbau und die Ausbreitung der neolithischen Wirtschaftsweise in Mitteleuropa und Südskandinavien am Ende des 5. Jahrtausends v. Chr. *Praehistorische Zeitschrift* 84(1):15–43.

Schlichtherle, H. 2011 Bemerkungen zum Klima- und Kulturwandel im südwestdeutschen Alpenvorland im 4.-3-Jahrtausend v. Chr. In *Strategien zum Überleben. Umweltkrisen und ihre Bewältigung. Tagung des Römisch-Germanischen Zentralmuseums, 19./20. September 2008*, edited by Falko Daim, Detlef Gronenborn, and Rainer Schreg, pp. 155–168. RGZM-Tagungen 11. Römisch-Germanisches Zentralmuseum, Mainz.

Schmidt, B., E. Höfs, M. Khalessi, and P. Schemainda 1998 Dendrochronologische Befunde zur Datierung des Brunnens von Erkelenz-Kückhoven in das Jahr 5090 vor Christus. In *Brunnen der Jungsteinzeit. Internationales Symposium in Erkelenz 27. bis. 29. Oktober 1997*, edited by Harald Koschik, pp. 279–290. Materialien zur Bodendenkmalpflege im Rheinland 11. Rheinland-Verlag, Köln.

Schmidt, B., W. Gruhle, and O. Rück 2004 Klimaextreme in Bandkeramischer Zeit (5300 bis 5000 v. Chr.). Interpretation dendrochronologischer und archäologischer Befunde. *Archäologisches Korrespondenzblatt* 34:303–307.

Schulz, M., and A. Paul 2002 Holocene Climate Variability on Centennial-to-Millennial Time Scales: 1. Climate Records from the North-Atlantic Realm. In *Climate Development and History of the North Atlantic Realm*, edited by G. Wefer, W. H. Berger, K.-E. Behre, and E. Jansen, pp. 41–54. Springer, Berlin.

Shennan, S., S. S. Downey, A. Timpson, K. Edinborough, S. Colledge, T. Kerig et al. 2013 Regional Population Collapse Followed Initial Agriculture Booms in Mid-Holocene Europe. *Nat Comms* 4 [DOI: 10.1038/ncomms3486].

Sirocko, F. 2009 Bohrungen und Untersuchungsgebiete. In *Wetter, Klima, Menschheitsentwicklung. Von der Eiszeit bis ins 21. Jahrhundert*, edited by Frank Sirocko, pp. 33–36. Wissenschaftliche Buchgesellschaft, Darmstadt.

Sirocko, F., B. Kromer, and H. Wernli 2009 Ursachen von Klimavariabilität in der Vergangenheit. In *Wetter, Klima, Menschheitsentwicklung: Von der Eiszeit bis ins 21. Jahrhundert*, edited by Frank Sirocko, pp. 53–59. Wissenschaftliche Buchgesellschaft, Darmstadt.

Spatz, H. 2003 Hinkelstein: Eine Sekte als Initiator des Mittelneolithikums. In *Archäologische Perspektiven. Analysen und Interpretationen im Wandel. Festschrift für Jens Lüning zum 65. Geburtstag*, edited by Jörg Eckert, Ursula Eisenhauer, and Andreas Zimmermann, pp. 575–587. Marie Leidorf, Rahden/Westfalen.

Spurk, M., H. H. Leuschner, M. G. L. Baillie, K. R. Briffa, and M. Friedrich 2002 Depositional Frequency of German Subfossil Oaks: Climatically and Non-Climatically Induced Fluctuations in the Holocene. *The Holocene* 12(6):707–715.

Strien, H.-C., and D. Gronenborn 2005 Klima- und Kulturwandel während des mitteleuropäischen Altneolithikums (58./57.–51./50. Jahrhundert v. Chr.). In *Klimaveränderung und Kulturwandel in neolithischen Gesellschaften Mitteleuropas, 6700–2200 v. Chr.*, edited by Detlef Gronenborn, pp. 131–150. RGZM-Tagungen 1. Verlag des Römisch-Germanischen Zentralmuseums, Mainz.

Strien, H.-C. 2000 Untersuchungen zur Bandkeramik. In *Württemberg. Universitätsforschungen zur Prähistorischen Archäologie 69*. Habelt (Bonn 2000).

Strien, H.-C. 2007 Archäologische Methoden. In *Tatort Talheim: 7000 Jahre später*, edited by J. Wahl, H.-C. Strien, C. Jacob, A. Golowin, W. Schnaubelt, and N. Kieser, pp. 6–9. Städtische Museen Heilbronn, Heilbronn.

Stuiver, M., and P. Quay 1980 Patterns of Atmospheric ^{14}C changes. *Radiocarbon* 22:166–167.

Tackenberg, K. 1937 Beiträge zur Landschafts- und Siedlungskunde der sächsischen Vorzeit. In *Von Land und Kultur. Beiträge zur Geschichte des mitteldeutschen Ostens*, edited by W. Emmerich, pp. 15–37. Bibliographisches Institut, Leipzig.

Teller, J. T., and D. W. Leverington 2004 Glacial Lake Agassiz: A 5000 yr History of Change and its Relationship to the δ^{18}O Record of Greenland. *GeoScienceWorld Bulletin* 116(5–6):729–742.

Tinner, W., E. H. Nielsen, and A. F.Lotter 2007 Mesolithic Agriculture in Switzerland? A Critical Review of the Evidence. *Quaternary Science Reviews* 26(9–10):1416–1431.

Turchin, P., and S. A. Nefedov 2009 *Secular Cycles*. Princeton University Press, Princeton.

van Geel, B., J. van der Pflicht, M. R. Kilian, E. R. Klaver, J. H. M. Kouwenberg, H. Renssen, I. Reynaud-Farrera, and H. T. Waterbolk 1998 The Sharp Rise of δ^{14}C ca. 800 cal BC: Possible Causes, Related Climatic Teleconnections, and the Impact on Human Environments. *Radiocarbon* 40:535–550.

van Geel B., O. M. Raspopov; H. Renssen; van der J. Plicht, V. A Dergachev, and H. A. J. Meijer 1999 The Role of Solar Forcing upon Climate Change. *Quaternary Science Reviews* 18:331–338.

Vinther, B. M., H. B.Clausen, S. J. Johnsen, S. O. Rasmussen, K. K. Andersen, S. L. Buchardt, D. Dahl-Jensen, I. K. Seierstad, M.-L. Siggaard-Andersen, J. P. Steffensen, A. M. Svensson, J. Olsen, and J. Heinemeier 2006 A Synchronized Dating of Three Greenland Ice Cores throughout the Holocene. *Journal of Geophysical Research* 111 D13102, doi:10.1029/2005JD006921.

von Grafenstein, U., H. Erlenkeuser, A. Brauer, J. Jouzel, and S. J. Johnsen 1999 A Mid-European Decadal Isotope-Climate Record from 15,500 to 5000 years B.P. *Science* 284:1654–1657.

Wang, Y., H. Cheng, L. R. Edwards, Y. He, X. Kong, Z. An, J. Wu, M. J. Kelly, C. A. Dykoski, and X. Li 2005 The Holocene Asian Monsoon: Links to Solar Changes and North Atlantic Climate. *Science* 308(5723):854–857.

Wanner, H., J. Beer, J. Bütikofer, T. J. Crowley, U. Cuibash, J. Fluckiger, H. Goosse, M. Grosjean, F. Joos, J. O. Kaplan, M. Kuttel, S. A. Muller, I. C. Prentice, O. Solomina, T. F. Stocker, P. Tarasov, M. Wagner, and M. Widmann 2008 Mid- to Late Holocene Climate Change: An Overview. *Quaternary Science Reviews* 27(19–20):1791–1828.

WBGU 2007 *World in Transition: Climate Change as a Security Risk*. German Advisory Council on Global Change. Wissenschaftlicher Beirat der Bundesregierung Globale Umweltveränderungen WBGU, Berlin.

Weninger, B., L. Clare, E. J. Rohling, O. Bar-Yosef, U. Böhner, M. Budja, M. Bundschuh, A. Feurdean, H.-G. Gebel, O. Jöris, J. Linstädter, P. Mayewski, T. Mühlenbruch, A. Reingruber, G. Rollefson, D. Schyle, L. Thissen, H. Todorova, and C. Zielhofer 2009 The Impact of Rapid Climate Change on Prehistoric Societies During the Holocene in the Eastern Mediterranean. In *16th Neolithic Studies*, edited by Mihael Budja, pp. 7–59. Documenta Praehistorica 36. University of Lubljana, Ljubljana.

Wernli, H., and S. Pfahl 2009 Grundlagen des Klimas und extremer Wettersituationen. In *Wetter, Klima, Menschheitsentwicklung. Von der Eiszeit bis ins 21. Jahrhundert*, edited by Frank Sirocko, pp. 44–52. Wissenschaftliche Buchgesellschaft, Darmstadt.

Willcox, G. 2005 The Distribution, Natural Habitats, and Availability of Wild Cereals in Relation to their Domestication in the Near East: Multiple Events, Multiple Centres. *Vegetation History and Archaeobotany* 14:534–541.

Willcox, G., R. Buxo, and L. Herveux 2009 Late Pleistocene and Early Holocene Climate and the Beginnings of Cultivation in Northern Syria. *The Holocene* 19(1):151–158.

Yu, S.-Y; S. M. Colman, T. V. Lowell, G. A. Milne, T. G. Fisher, A. Breckenridge et al. 2010 Freshwater Outburst from Lake Superior as a Trigger for the Cold Event 9300 Years Ago. *Science* 328 (5983):1262–1266.

Zimmermann, A. 2012 Cultural Cycles in Central Europe during the Holocene. *Quaternary International* 274:251–258.

CHAPTER ELEVEN

Economic and Social Changes and Climate between 3200 and 2500 B.C.

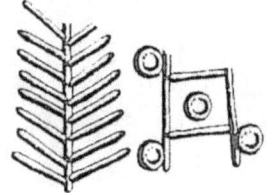

Late Neolithic Transformations in Southeastern Poland

Andrzej Pelisiak

Abstract *One of the important questions of prehistoric research is where, when, and to what extent the climate was an influential factor in changing human ways of life. Results of the archaeological and paleogeographical investigations carried out during recent years in southeastern Poland can be a strong basis for the discussion of these problems. This area consists of several ecological zones: lowland, loess uplands, Carpathian foothills, and mountains. The archaeological database of this area consists of more than 9,000 sites dated from ca. 5300 B.C. to ca. 1600 B.C. The Holocene history of vegetation changes and evidence of human activity comes from pollen analysis of the deposits from several paleobotanical sites located within these zones. Significant changes in human activity (settlement patterns, economy, and social organization) in southeastern Poland took place between ca. 3200 and 2500 B.C. A sedentary way of life was gradually replaced by mobile husbandry and pastoralism. I would like to discuss the questions of how climatic events (fluctuations and anomalies) and to what extent they stimulated cultural and economic changes during the Late Neolithic.*

INTRODUCTION

One of the crucial questions of prehistoric research is where, when, and to what extent the climate was an influential factor in changing the human way of life. Results from recent archaeological and paleogeographical investigations carried out in southeastern Poland can be a basis for discussion on these problems (Figure 11.1). The

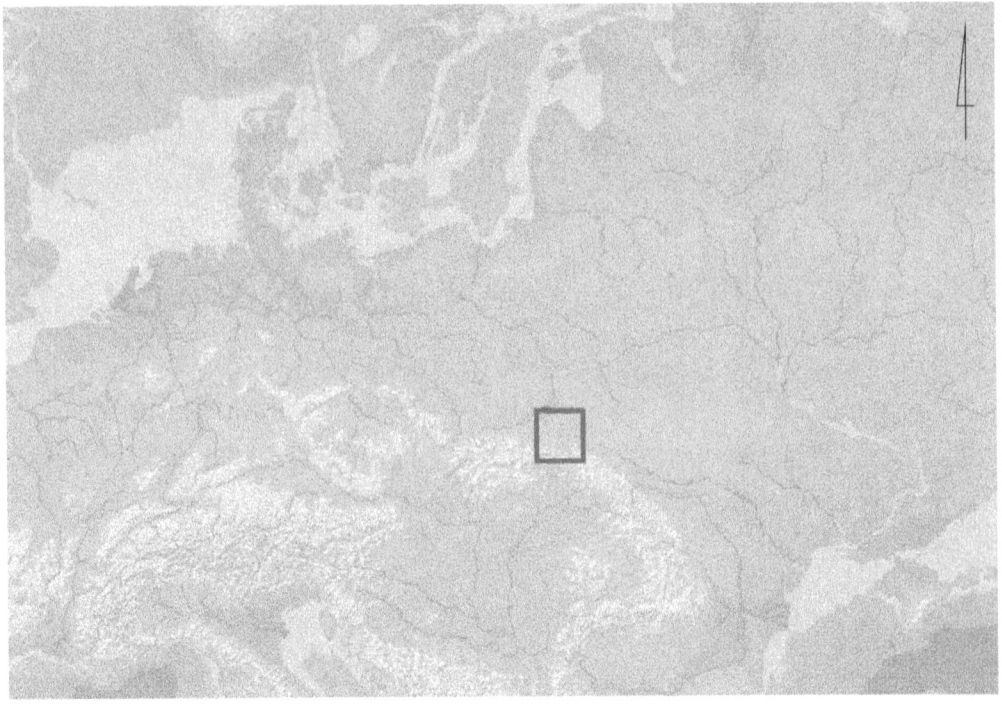

FIGURE 11.1. Map of southeastern Poland showing the area of research.

whole area of southeastern Poland has been covered by systematic surface surveys over the last three decades, as a result of the Archaeological Record of Poland Project (Archeologiczne Zdjęcie Polski; AZP). With regard to this area, the archaeological database already consists of more than 9,500 sites dated from ca. 5300 B.C. (the Linear Pottery culture) to ca. 1600 B.C. (the Early Bronze Age, the Mierzanowice culture), including settlement sites, burial mounds, and single finds (Pelisiak 2005). Numerous sites were excavated, including a large-scale multidisciplinary research project (conducted from 1995 to the present) of rescue excavations on the route of the prospective A-4 motorway (Czopek and Pelisiak 2007, 2008).

This area consists of several ecological zones: the lowland in the north, the loess uplands, the Carpathian foothills, the Jasło-Sanok Depression, and the mountains (the Lower Beskid Mountains and the Bieszczady Mountains). Highly productive soil types in the loess uplands (up to 250 m asl), and fertile soils in the Jasło-Sanok Depression, provided the most favorable conditions for early farming, beginning with plant cultivation. The Carpathian foothills (between 250 and 550 m asl) are partly covered (up to 430 m asl) by loess-like dusty deposits (Gerlach et al. 1991). This region offered much poorer conditions for agriculture but the dusty covers were also suitable for plant cultivation. In higher parts of the foothills (between 430 and 550 m asl) and in the mountains (above 550 m asl), surface deposits consist of thin initial soils with gravels, meaning the environment was unsuitable for early plant cultivation.

The Holocene history of vegetation has changed and evidence of human activity comes from pollen analysis of the deposits from several palaeobotanical sites located within all zones mentioned above: Besko (Koperowa 1970), Cergowa (Szczepanek 2001; Wacnik et al. 2001), Imielty Ług (Mamakowa 1962), Jasiel (Szczepanek 1987; Wacnik et al. 2001), Kępa (Gerlach et al. 1972), Kosobudy (Bałaga 1998), Krasnobród (Bałaga 2002), Kružlova (Wacnik 2001; Wacnik et al. 2001), Obary (Mamakowa 1962), Podbukowina (Mamakowa 1962); Regetovka (Wacnik 1995, 1999; Wacnik et al. 2001), Roztoki and Tarnowiec (Harmata 1987a, 1987b; Wacnik et al. 2001), Rzemień (Mamakowa 1962), Smerek (Ralska-Jasiewiczowa 1969, 1980); Świlcza (Mamakowa 1962); Tarnawa Wyżna (Ralska-Jasiewiczowa 1969, 1980), Tarnawatka (Bałaga 1998), and Wola Ługowa (Madeja 2001). Evidence from pollen diagrams (Figure 11.2) indicates progressive landscape enculturation and landscape transformation from wild to cultural, but this process did not begin in all regions at same time.

This paper discusses the extent and nature of the influence of climate changes and anomalies on the transformations of the human way of life during the Late Neolithic period. There is no doubt that past socioeconomic and cultural changes in southeastern

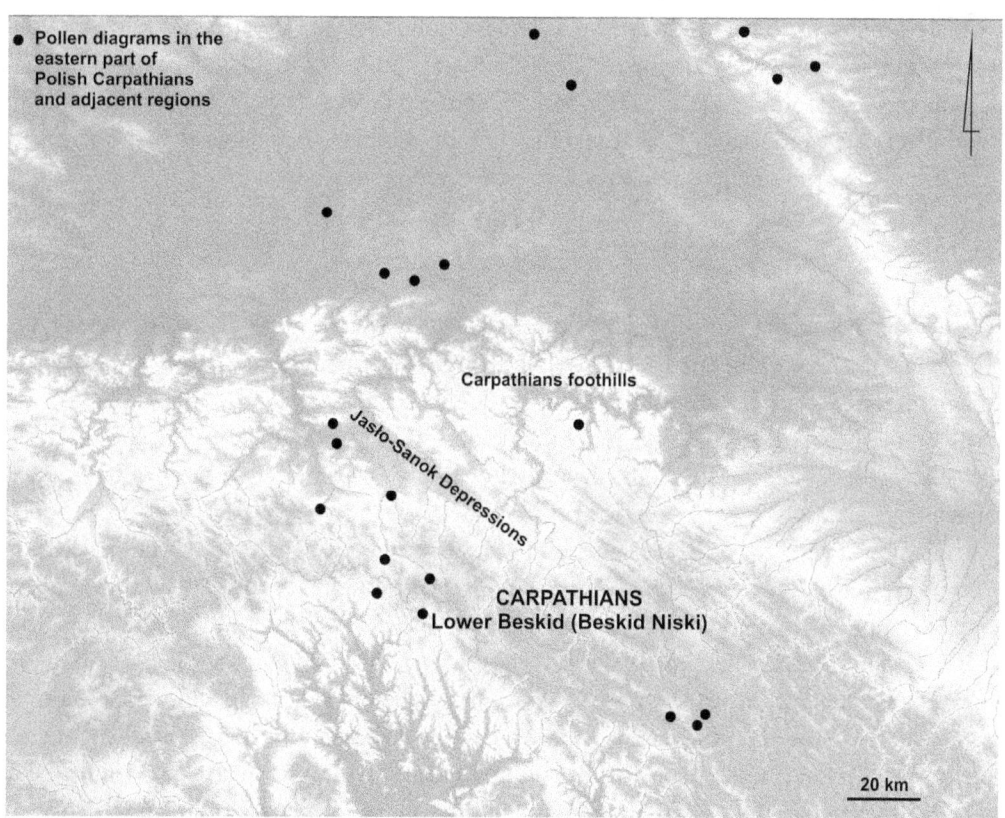

FIGURE 11.2. Map of southeastern Poland showing the locations of pollen diagrams discussed in the text.

Poland were connected with transformations of much wider geographical extent. In this respect, I will use archaeological and palynological data from southeast Poland as well as from other parts of Europe.

Neolithic and Early Bronze Age Settlements in Southeastern Poland

There are more than 9,500 sites found in southeastern Poland dating from the Early Neolithic period (ca. 5300 B.C.) to the Early Bronze Age (ca. 1600 B.C.). These sites are located in different regions and different zones of landscape.

The Danubian Cultures

The first groups of people of the Neolithic Linear Pottery culture arrived in southeast Poland in ca. 5300 B.C. These communities, as well as the younger Danubian cultures (the Malice and Lublin-Volhynian cultures), settled the area of the loess belt north of the Carpathian foothills. More than 400 sites, including 170 settlement sites, were discovered there, and several of them have been excavated (Aksamit 1968; Dębiec 2006; Dzieduszycka-Machnikowa 1960; Gruszczyńska 1991; Kadrow 1990, 1992, 1997; Moskwa 1963; Przybyła 2004). All the settlements, and almost all camps and single finds, are located within this zone. There are no finds from Danube cultures from either the Carpathian foothills or the mountains. The area of distribution of these sites suggests that only more fertile loess zones were settled and utilized by these communities.

The Funnel Beaker Culture

About 900 sites of the Funnel Beaker culture were discovered in the discussed area (Figure 11.3). There are settlements, camp sites, and single finds of flint and stone artifacts. The sites are concentrated within four main ecological zones: lowlands, loess uplands, the Jasło-Sanok Depression and the Carpathian foothills (up to 400 m asl) (Nowak 1998; Pelisiak 2003, 2004; Zych 2008). Several sites located within each of zones mentioned above were excavated, for example: Białobrzegi site 5, located in the lowland (Czopek and Kadrow 1988; Zych 2008); Olchowa site 20, situated in the loess uplands (Mitura 2007; Mitura and Zych 1999; Zych 2008); Przybówka site 1, located in the foothills near the border of the Jasło-Sanok Depression (Gancarski et al. 2008); and Tarnawka site 9 (Zych 2004, 2008), Tarnawka site 13 (Zych 2002, 2008), and Manasterz site 13 (Zych 2003, 2008) within the zone of foothills. The sites located in different ecological zones vary in size and time of occupation. In the lowland, the largest settlements are up to one hectare in size (consisting of four to six households). They could have been inhabited no longer than 40 to 50 years. In addition to permanently occupied settlements, numerous camp sites were recognized in this zone, as well as numerous single finds of flint and stone artifacts. Within the loess upland settlements, camp sites and single finds

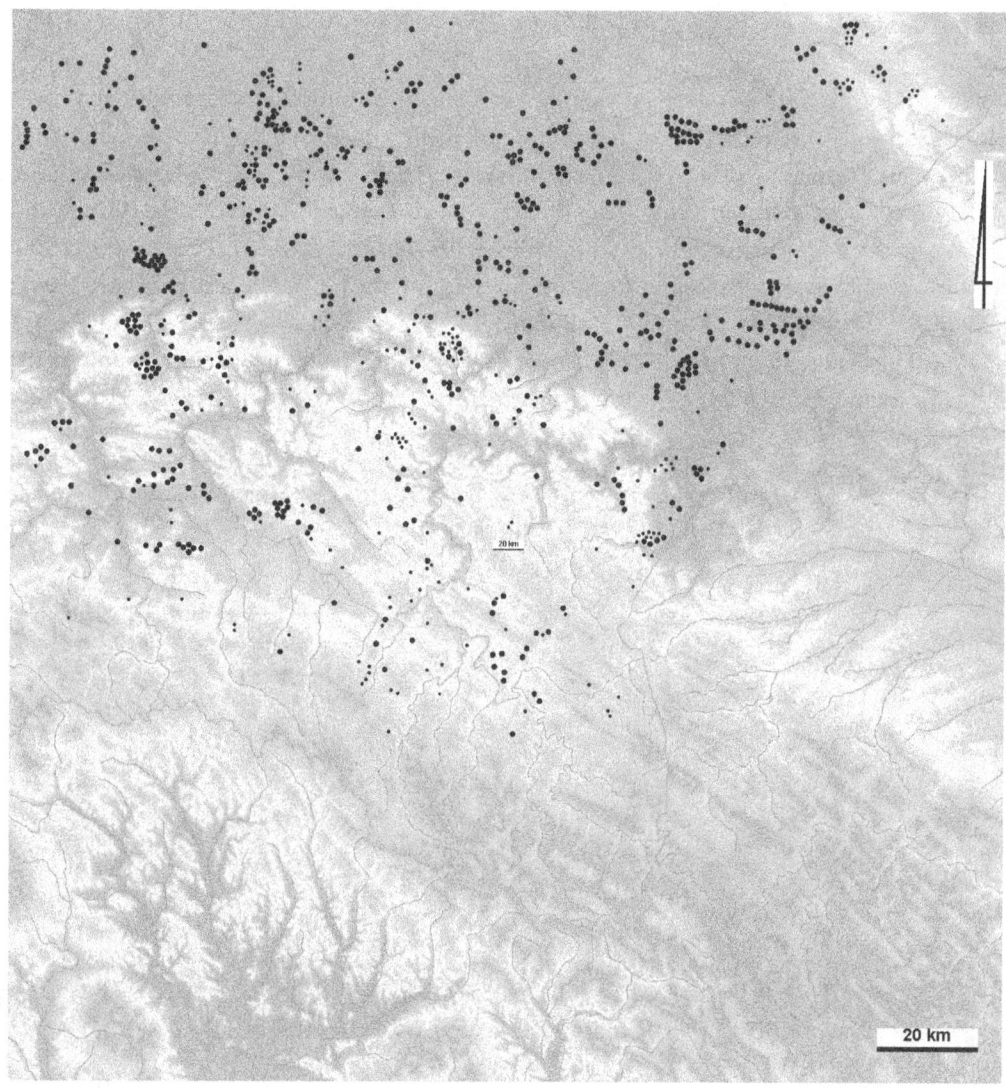

FIGURE 11.3. The distribution of Funnel Beaker culture sites. Large dots: settlement sites; small dots: single finds of stone and chipped artifacts.

were discovered. The largest settlements of the Funnel Beaker culture were more than five hectares in size and were inhabited much longer than those in the lowland. In the Jasło-Sanok Depression and in the lower zone of the foothills (up to 400 m. asl), the settlement sites were probably up to one hectare in size and each of them was inhabited no longer than 20 to 30 years. There are numerous camp sites and single finds of flint and stone artifacts in this zone. Moreover, single finds of flint and stone artifacts appeared in the higher parts of foothills and in the mountains (above 400 m asl).

The Corded Ware Culture

There are about 730 Corded Ware culture sites within the study area, including barrows, single finds of flint and stone artifacts (some of them probably from destroyed barrows), and several camp sites (Figure 11.4). There are no permanently occupied settlements of the Corded Ware culture in this area. The sites are concentrated in the loess uplands and the lower foothills (up to 430 m asl) (Czopek 1997; Gedl 1997; Pelisiak 2003, 2004, 2005; Valde-Nowak 1988), but they are present in the lowland and the Jasło-Sanok Depression as well (Machnik 1989, 1990, 1992, 1998, 2001).

Additionally, single finds were discovered in the higher parts of foothills and in the Beskid Mountains. Barrows are located mainly in the long ridges of uplifts and hills. Many

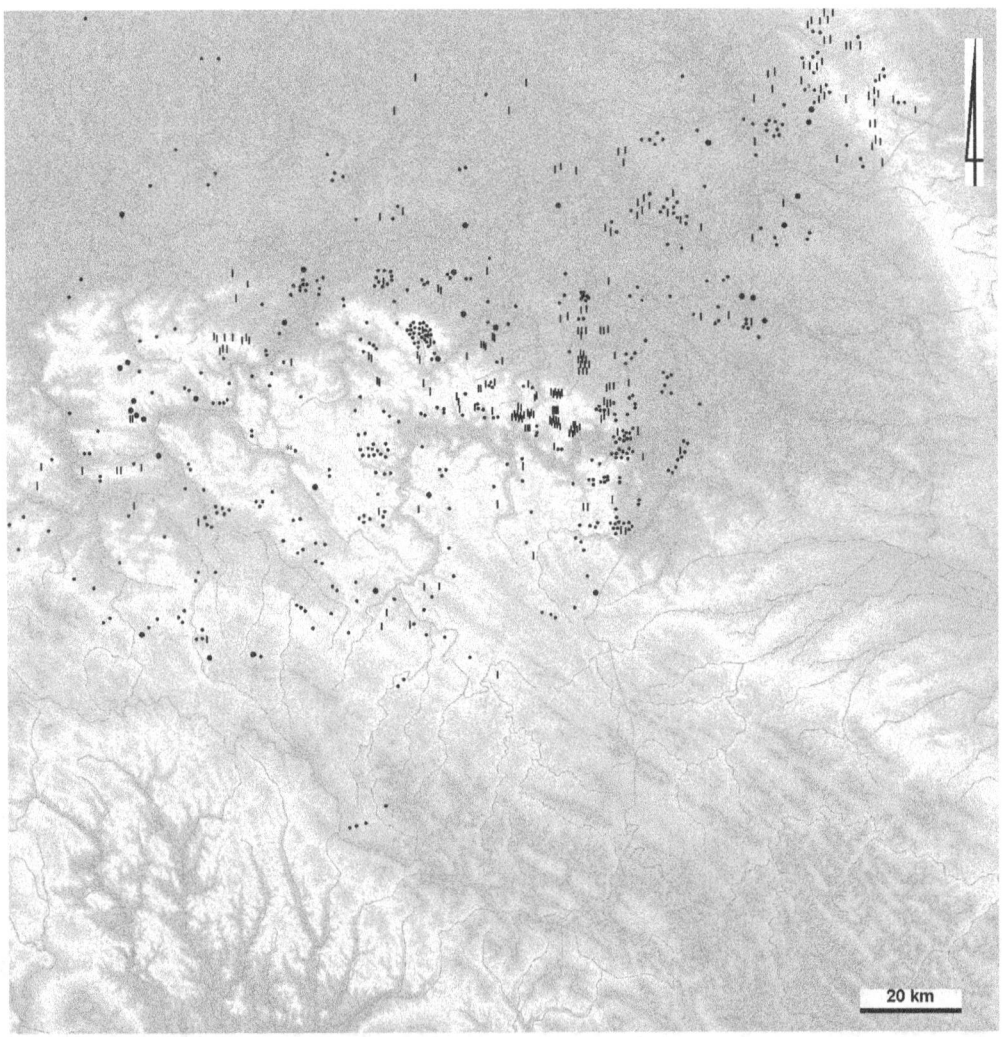

FIGURE 11.4. The distribution of Corded Ware culture sites. Vertical lines: barrows; small dots: single finds; larger dots: camp sites.

barrows have been excavated, such as Cieszacin Wielki site 37 (Łanczont et al.2001, 2003, 2007), Cieszcin Wielki site 38 (Łanczont et al. 2001, 2004), and Morawsko (Machnik 1995) in the loess upland, and Bierówka (Gancarski et al. 1986, 1991), Średnia site 11 (Jarosz 2002), Średnia site 3 (Machnik and Sosnowska 1996, 1999), Wola Węgierska site 3 (Machnik and Sosnowska 1999) in the foothills. The geographic distribution of Corded Ware culture sites is similar to that of the Funnel Beaker culture. Moreover, the Corded Ware barrows were frequently located in the same place where the Funnel Beaker settlements previously existed (Machnik 1990, 1992, 1998; Valde-Nowak 1998, 2001).

The Early Bronze Age Mierzanowice Culture

There are about 1,300 sites of the Mierzanowice culture in southeastern Poland, consisting of settlements, graves, and single finds of flint artifacts (Figure 11.5). The sites are

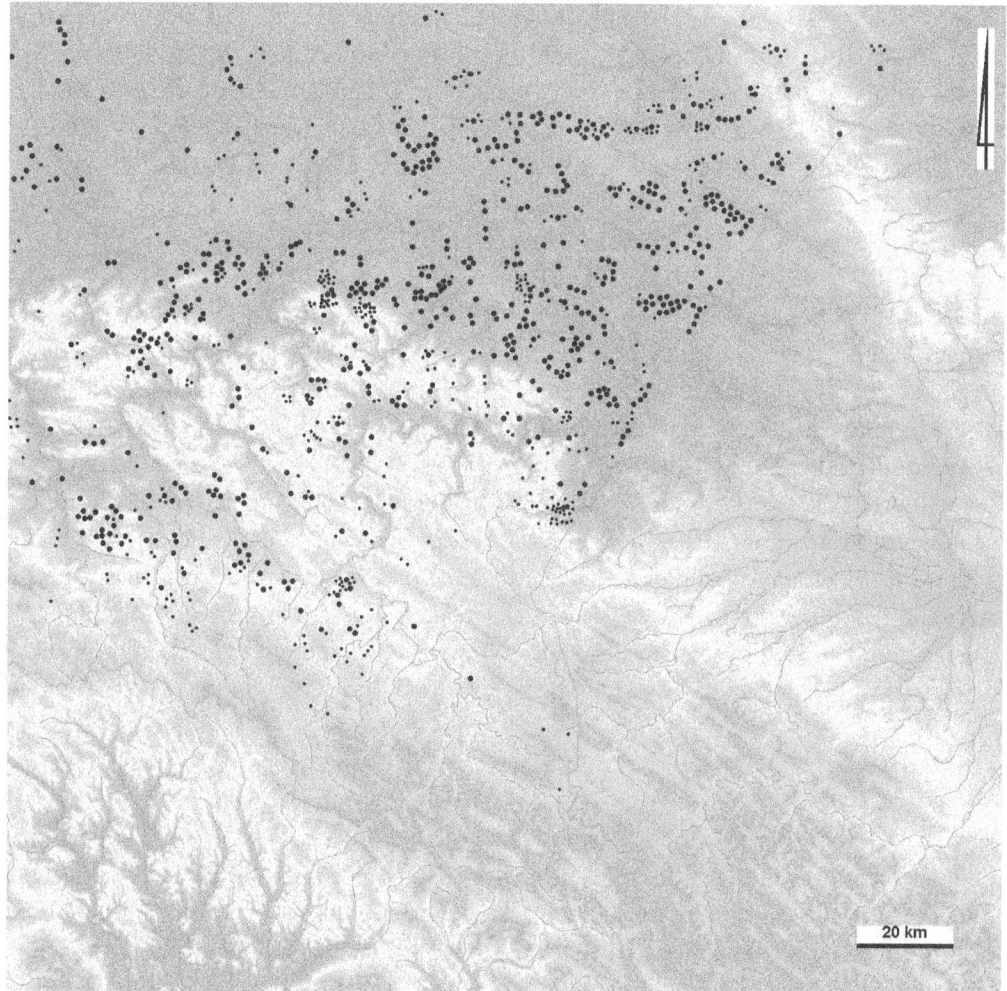

Figure 11.5. The distribution of Early Bronze Age Mierzanowice sites.

present in the lowland, the loess uplands, the foothills, and in the Jasło-Sanok Depression (Czopek 2007; Gancarski 1999; Madej 1998, 1999, 2000, 2001, 2003; Pelisiak and Sosnowska 2005; Przybyła and Blajer 2008; Valde-Nowak 1988; Zych 2002, 2004). Within the foothills zone (up to 400 m asl) the settlement sites are often located in the same place where the sites of the Funnel Beaker culture and/or Corded Ware culture were situated. Settlement sites differ in size and duration. In the zone of lowland settlement, sites are up to one hectare in size, but mostly they are much smaller, such as the Grodzisko Dolne site 22 (Czopek 2007). Concentrations of the Mierzanowice culture sites were recognized in loess uplands.

The biggest excavated site within this region (Sietesz site 5) is probably more than 20 hectares in size (Madej 1998, 1999, 2000, 2001, 2003). Other excavated Mierzanowice culture sites from the loess zone of southeastern Poland are much smaller; for example, Kańczuga site 5 (Przybyła and Blajer 2008) and Kosina site 35 (Przybyła and Blajer 2008). There were fortified settlement sites here, such as Trzcinica site 1, which is located on top of a small hill within the Jasło-Sanok Depression (Gancarski 1999). Numerous but small and short-lived settlement sites were discovered in the foothills zone (up to 400 m asl) of southeastern Poland. Some of them were excavated, including Tarnawka sites 9 and 13 (Zych 2002, 2004). Moreover, numerous single finds of flint artifacts (bifacial knives and bifacial axes) have been registered here (e.g., Pelisiak and Sosnowska 2005).

Neolithic Sites without Cultural Affiliation

About 3,100 Neolithic sites with unclear cultural affiliation were discovered in southeastern Poland (Figure 11.6). These sites are represented by single finds of stone tool fragments, flint blades, blade tools, rectangular axes and their fragments, and small assemblages of flint artifacts. They are located in the lowland, the loess uplands, the Jasło-Sanok Depression, and the foothills.

Relatively numerous finds are located in higher parts of foothills and in the Beskid Mountains, in areas more than 450 m asl. The significant feature of flint artifacts found within this high landscape zone is the lack of tools that might have been connected with cultivation and harvesting of cereals (e.g., sickle blades). Because of the fact that the Danubian culture people did not settle and utilize foothills and mountains, these finds can be connected with populations belonging to the Funnel Beaker and/or Corded Ware cultures.

The Neolithic or Early Bronze Age Sites without Cultural Affiliation

More than 3,000 sites without identified cultural affiliations, dated to the Neolithic period or Early Bronze Age, have been registered in southeastern Poland (Figure 11.7). They are single finds of flint and stone artifacts, fragments thereof, and small assemblages of flint artifacts. The sites are located in the lowland, the loess upland, the Jasło-Sanok Depression, the foothills, and the mountains. Their locations within the foothills and in mountain areas can be connected with the activity of the Funnel Beaker, Corded Ware, and/or Mierzanowice culture sites. Moreover, there are no tools connected with cereal cultivation within the two above-mentioned zones.

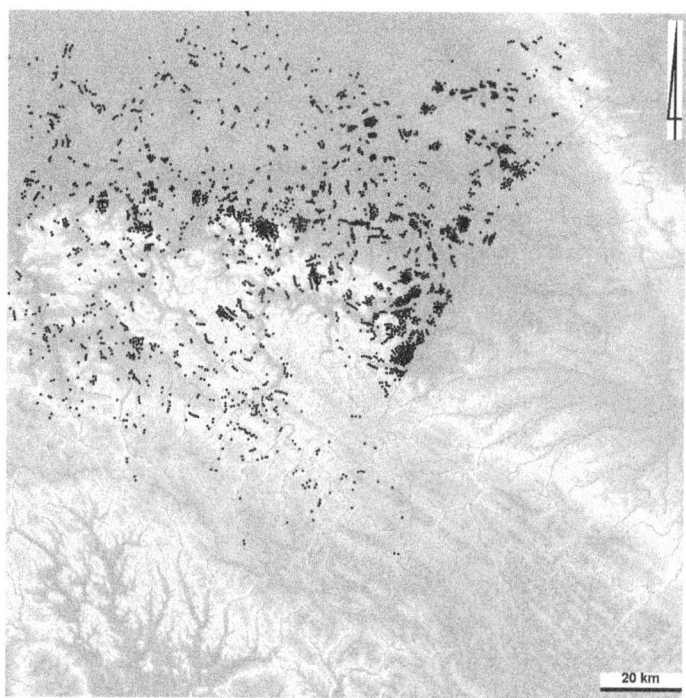

FIGURE 11.6. The distribution of Neolithic sites without cultural affiliation.

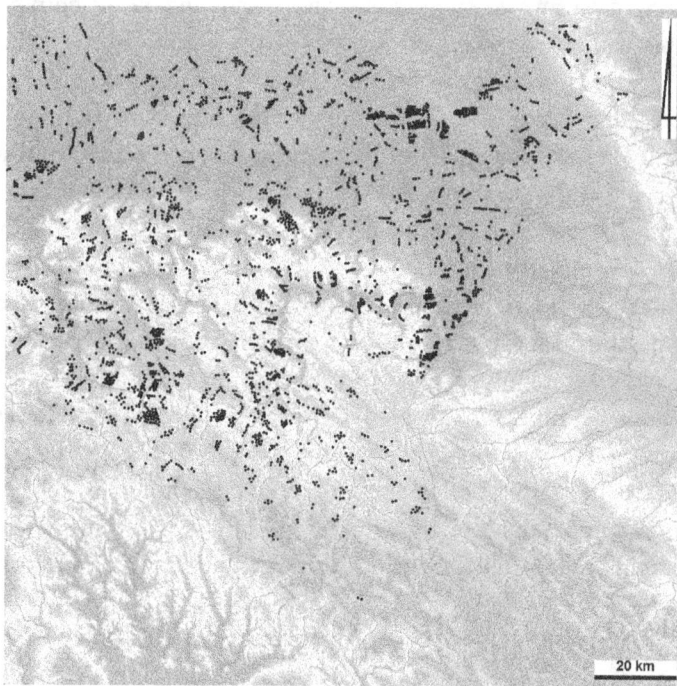

FIGURE 11.7. The distribution of Neolithic or Early Bronze Age sites without cultural affiliation.

This presented review shows that the loess uplands were inhabited by sedentary communities from ca. 5300 B.C. (the Linear Pottery culture). People belonging to the Danubian cultures penetrated the lowlands but they did not settle them. The loess uplands, the lowlands, and lower parts of the Carpathian foothills all were inhabited by the sedentary Funnel Beaker culture and the Early Bronze Age Mierzanowice culture populations. Within these zones, cultivation of plants as well as animal husbandry is confirmed by archaeological and botanical data. Several lines of evidence support the argument that the higher parts of the Carpathian foothills and mountains were utilized only for pastures. First of all, the flint and stone artifacts found there were not used for cultivation of cereals. Second, the poor and thin soils were unsuitable for plant cultivation. Third, there are no permanently inhabited settlements in higher parts of the foothills or in the mountains that could have been the core sites of sedentary communities. Fourth, the palynological data do not confirm plant cultivation. Instead, they give evidence of animal husbandry in these zones.

Natural Vegetation and Needs

Dense forests covered the Carpathian foothills and the Beskidy Mountains during the Atlantic and Subboreal periods. Because of poor undergrowth and deficiency of animal fodder, these primary environments were unsuitable for grazing herds, particularly cattle and sheep.

Such an environment must have been transformed and adapted to the needs of economic activity of the late Neolithic people: plant cultivation and stock breeding. The foothills up to 400 m asl covered by loess-like covers were suitable for both plant cultivation and herding. This is the zone where settlements of the Funnel Beaker culture were located. Subsistence strategies of these communities were based on slash and burn agriculture, which was successfully realized there. People opened new areas by burning off the natural vegetation. In this way, the same part of this area was deforested for cultivated fields and pastures. Both plant cultivation and animal husbandry are confirmed by archaeological materials that were discovered during the excavation of the settlements of the Funnel Beaker culture located within the lower (up to 400 m.asl) parts of the Carpathian foothills (Gancarski et al. 2008). Previously opened areas within these regions were subsequently settled and exploited by semi-nomadic pastoral people of the Corded Ware culture.

The thin and poor soils in higher (between ca. 400 and 1200 m asl) parts of the foothills and mountains were not suitable for cultivation. Both archaeological and palynological data suggest that some parts of these regions could have been utilized for herding by populations of both the Funnel Beaker and Corded Ware cultures, as well as the Mierzanowice culture communities. The primary natural vegetation of these areas, dense forest, must have been transformed by partial clearing in order to prepare pastures for cattle and sheep.

Climate Anomalies and Human Activity

Besides pollen diagrams from southeastern Poland, important climatic data came from Ukraine, eastern and southern Russia, and Moldova (Kremenetski 2003). There were many regional climatic variations, but the main changes were global in range (Europe and West Asia), and there is not complete chronological compatibility of sub-phases (wet/dry, cold/warm periods) in Europe. The climate situation during the Subboreal Period was not stable. There were several oscillations between 3400 and 600 B.C. Between 3400 and 3200 B.C. there was a sharp continental climatic phase, followed by a more moist and warm oscillation beginning in 3300/3200 B.C. A much drier and colder phase then prevailed, but the most continental interval is dated between 2800 and 2000 B.C. In southern Russia and Ukraine from ca. 3200 to 1800 B.C. forest area declined and the role of steppe formations increased. From ca. 1700 to 900 B.C. the climate became moister. A moist climatic phase for the region north of the Caspian Sea between 4000 and 3000 B.C. is confirmed by buried dark chestnut soils (Kremenetski 2003). On the other hand, humidity of climate before ca. 3300/3200 was confirmed by recent palynological research from western Ukraine (Harmata et al. 2006). This can be associated with floods and channel avulsions recorded in the upper Vistula Basin (Starkel 1994).

The beginning of the Neolithic occupation and utilization of the Carpathian foothills and mountains is dated to ca. 3300/3200 B.C. and is connected with the Funnel Beaker culture communities. This process can be chronologically correlated with climatic oscillations reflected by more or less sharp sub-phases characterized by an increase or decrease of temperature and humidity. Climatic oscillations may have resulted in many weather anomalies, such as intense wind, icing, intense rains, and landslips. In this respect, there are spectacular modern observations that can illustrate such anomalies and their results in the natural environment, such as those recorded in the woodlands of southern and central Poland during the spring of 2010. During March and April of that year there was a short (lasting only several days) but sharp decrease in temperature (with nighttime lows of –6 to –9° C) accompanied by heavy rains in several places. Suddenly, the tree crowns were covered by a thick and heavy icy coat (Figure 11.8). The tree crowns were

FIGURE 11.8. Contemporary forest Southeast Poland in winter showing tree crowns covered by a thick, heavy icy coat.

too heavy to survive and wide areas of forest broke down or collapsed at the root. As a result of this icing, many hectares of woodland were destroyed or almost deforested by natural forces (Figure 11.9). The incidents described above took place mainly in areas where soil cover was thin, the ground was composed of gravel and/or rock, and on the slopes of the hills and hillocks; they were frequently observed in the Polish Jura and on the Carpathian foothills, mainly on the places located higher than 300 m asl.

Contemporary observations such as these support the hypothesis that the natural destruction of forests might have significantly supported human activity, enabling the clearance of natural vegetation and preparing land suitable for stockherding. It is proposed that the broken and dried tree crowns and trunks were burned by the Late Neolithic people, and that this combination of natural clearance and anthropogenic burning was a very successful approach for creating open space or thinned, park-type tree stands (Figure 11.10).

Climate and Economic and Social Changes

An increasingly mobile way of life gave opportunities for new subsistence strategies, but also allowed people to create new social customs. For instance, we may point to the increasing social importance of adult men, who had become pastoralists as well as guardians who protected livestock and people. They became warriors, eventually leading

FIGURE 11.9. Contemporary forest in Southeast Poland after a fierce winter when the tree crowns became too heavy, resulting in the collapse of wide areas of forest. Natural forces such as icing may result in the destruction and near deforestation of many hectares of woodland.

FIGURE 11.10. Contemporary park-type landscape in Southeast Poland. Late Neolithic peoples may have created very similar open space by subjecting broken and dried tree crowns and trunks to fire.

to the rise of social elites. The complexity of social relations increased between individual persons as well as within groups of people: families and the larger communities (Anthony and Brown 2003). Social power increased and power was a matter of controlling "us" and "others" and their resources. Ancestors, goods, and land were involved within a symbolic system and the barrows became a reflection and manifestation of new ideas, projected onto the landscape.

The emergence of the Corded Ware culture was a result of gradual transformations of communities associated with the Funnel Beaker culture and a switch to a new socioeconomic system within a modified region. These processes took place on the marginal zones of the area settled by the Funnel Beaker culture communities, in regions of extreme natural conditions ill-suited to previous Neolithic activity (foothills and mountains).

An important part of the Late Neolithic economy in the Carpathian foothills and mountains could have been some forms of transhumance. The archaeological remains (flint and stone artifacts) are unclear in respect to their cultural affiliation (the Funnel Beaker culture or/and Corded Ware culture) because they are connected with specific economic activity (reflecting the pastoralism of the Late Neolithic people in these regions), but cultural identification of these finds has proved to be not very important. It is important instead that they indicate and confirm a socioeconomic transformation and the beginning of a new way of life, reflected in the archaeological materials by the appearance of the Corded Ware culture.

This model of transformations is confirmed by radiocarbon determinations of the oldest Corded Ware culture sites located in southeastern Poland and northwestern Ukraine. Two sites are especially significant: the first one, a barrow at Wola Węgierska in the Dynów foothills, is dated to ca. 2900–2700 B.C. (Machnik and Sosnowska 1998); and the second one, at Side (a camp site) in western Ukraine, to ca. 2800 B.C. (Machnik et al. 1997). Unfortunately, there is a notable lack of radiocarbon dates from the late Funnel Beaker culture site in the Carpathian foothills, but the chronology of some sites can be estimated within the last two centuries of the fourth millennium and the first centuries of the third millennium B.C.

Conclusions

Significant changes of human activity (settlement patterns, economy, social organization) in southeastern Poland took place between ca. 3200 and 2500 B.C. A sedentary way of life was gradually replaced by mobile husbandry and pastoralism. Moreover, for the first time during the Neolithic period, high zones of landscape (between 400 and 800 m asl, and as high as 1200 m asl) were used. Exploitation of mountains is also confirmed by palynological data from this area, and there is no evidence of earlier human activity in the mountains. The most important questions are: Why did the Neolithic people not exploit the mountains and higher parts of the Carpathian foothills until 3300/3200 B.C.?; and Where, when, and to what extent could the climate change the human way of life?

The processes described above took place in the marginal zone of the area inhabited by the sedentary and/or semisedentary Late Neolithic communities of the Funnel

Beaker culture (the loess belt of southeastern Poland and lower parts of the Carpathian foothills). Higher foothills and mountains spread south of this area, which was unsuitable for cultivation of cereals, though it might have been adopted for the husbandry probable within transhumance in the form of stock breeding. The presented example concerns a relatively small area but it probably can be the illustration of processes that took place at the end of the fourth and the beginning of the third millennium B.C. in the wide area of Europe and resulted in the origin of a new socioeconomic system—pastoralism. This system in Central Europe is reflected archaeologically by the Corded Ware culture.

There is no doubt that the processes of the creation of pastoralism might have been different in different parts of marginal zones occupied by the sedentary Late Neolithic and Eneolithic communities, and that they depended on the local (regional) cultural traditions, environment, or political situation.

Processes of gradual turning to a mobile way of life began probably in the middle of the fourth millennium B.C. and occurred among the people of different Neolithic and Eneolithic cultures from the Atlantic Ocean to the steppe and forest-steppe zone of eastern Europe (Anthony and Brown 2003; Chapman 2002; Marciniak 2004). In Central Europe, the Funnel Beaker, Baden, and Tripolye cultures were the main cultural groups involved in these transformations. Moreover, there is chronological coincidence between these processes and climatic changes that began ca. 3600 B.C. in the whole area. It should be stressed that the climate change was only one of many factors involved in the processes of socioeconomic transformations during the second part of the fourth and first part of the third millennium B.C. It is difficult to unequivocally answer the question: How far and to what extend was it influential in these processes? I may provoke different and controversial opinions because there is little reliable archaeological and paleogeographical evidence of the relations between climate change and socioeconomic transformations during the Late Neolithic period. Probably in some regions it could have played a significant role, but everywhere in this period it was connected with both internal and external social, economic, and political events.

References Cited

Aksamit, T. 1968 Prace wykopaliskowe na osadzie neolitycznej we Fredropolu, pow. Przemyśl. In *Materiały i Sprawozdania Rzeszowskiego Ośrodka Archeologicznego za rok 1966*, pp. 116–123. Rzeszów.

Anthony, D. W., and D. R. Brown 2003 Eneolithic Horse Rituals in the Steppes: New Evidence. In *Prehistoric Steppe Adaptation and the Horse*, edited by M. Levine, C. Renfrew, and K. Boyle, pp. 55–68. McDonald Institute for Archaeological Research, Cambridge.

Bałaga, K. 1998 Post-Glacial Vegetational Changes in the Middle Roztocze (E. Poland). *Acta Palaeobotanica* 38(1):175–192.

Bałaga, K. 2002 Przemiany szaty roślinnej Roztocza Tomaszowskiego w ostatnich 12000 lat. In *pradziejów Roztocza. Na ziemi zamojskiej*, edited by B. Balcer, J. Machnik, and J. Sitek, pp. 223–230. Instytut Archeologii i Etnologii PAN, Kraków.

Chapman, J. 2002 Domesticating the Exotic: The Context of Cucuteni-Tripolye Exchange with Steppe and Forest-Steppe Communities. In *Ancient Interactions. East and West in Eurasia,* edited by K. Boyle, C. Renfrew, and M. Levine, pp. 75–91. McDonald Institute for Archaeological Research, University of Cambridge, Cambridge.

Czopek, S. 1997 Nowe odkrycia kurhanów na Pogórzu Dynowskim. *Rocznik Przemyski. Archeologia* 33(5):53–63.

Czopek, S. 2007 Grodzisko Dolne stanowisko 22—wielokulturowe stanowisko nad dolnym i Wisłokiem. Część I. Od epoki kamienia do wczesnej epoki żelaza. Instytut Archeologii Uniwesytetu Rzeszowskiego. Rzeszów.

Czopek, S., and S. Kadrow 1988 Osada kultury pucharów lejkowatych w Białobrzegach, stan. 5, woj. Rzeszów. *Sprawozdania Archeologiczne* 39:73–88.

Czopek, S., and A. Pelisiak 2007 Pierwsze szerokopłaszczyznowe badania wykopaliskowe na odcinku autostrady A4 w obrębie województwa podkarpackiego. *Materiały i Sprawozdania Rzeszowskiego Ośrodka Archeologicznego* 27:253–265.

Czopek, S., and A. Pelisiak 2008 Szerokopłaszczyznowe badania ratownicze prowadzone przez Fundację Rzeszowskiego Ośrodka Archeologicznego w 2006 i 2007 roku w Terliczce i Stobiernej. *Materiały i Sprawozdania Rzeszowskiego Ośrodka Archeologicznego* 28:135–139.

Dębiec, M. 2006 Osada kultury ceramiki wstęgowej rytej w Łańcucie stanowisko 3 (badania w roku 1987 i 1989). Część pierwsza—materiały. *Materiały i Sprawozdania Rzeszowskiego Ośrodka Archeologicznego* 27:27–63.

Dzieduszycka-Machnikowa, A. 1960 Stanowisko kultury ceramiki wstęgowej rytej w Boguchwale. *Materiały Archeologiczne* 2:11–21.

Gancarski, J. 1999 Chronologia grupy pleszowskiej kultury mierzanowickiej i kultury Otomani-Füzesabony w Polsce na podstawie wyników badań wykopaliskowych osad w Trzcinicy i Jaśle. In *Kultura Otomani-Füzesabony—rozwój, chronologia, gospodarka*, edited by J. Gancarski, pp. 145–175. Muzeum Podkarpackie, Krosno.

Gancarski, J., A. Machnikowa, and J. Machnik 1986 Wyniki badań kurhanu A kultury ceramiki sznurowej we wsi Bierówka, gm. Jasło, woj. krośnieńskie. *Acta Archaeologica Carpathica* 25:57–87.

Gancarski, J., A. 1991 Kurhan B kultury ceramiki sznurowej w Bierówce, gm. Jasło w świetle badań wykopaliskowych. *Acta Archaeologica Carpathica* 29:99–126.

Gancarski, J., W. Pasterkiewicz, and A. Pelisiak 2008 Osada kultury pucharów lejkowatych w Przybówce, gm. Wojaszówka, stanowisko 1. In *Archeologia i środowisko naturalne Beskidu Niskiego w Karpatach*. Part II, *Krimská Brázda*. edited by J. Machnik, pp. 347–378. Prace Komisji Prehistorii Karpat 4, Kraków.

Gedl, M. 1997 Starodawne kopce we wschodniej części Pogórza Dynowskiego. *Rocznik Przemyski* 33(5):39–51.

Gerlach, T., L. Koszarski L.,W. Koperowa, and W. Koster 1972 Sediments lacustres postglaciaires dans la depression de Jasło-Sanok. *Studia Geomorphologica Carpatho-Balcanica* 6:37–61.

Gerlach, T., M. Krysowska-Iwaszkiewicz, K. Szczepanek, and S. W. Alexandrowicz 1991 Karpacka odmiana lessów w Humniskach koło Brzozowa na Pogórzu Dynowskim w polskich Karpatach fliszowych. *Zeszyty Naukowe AGH, Geologia* 17(1–2), z. 1–2. Kraków.

Gruszczyńska, A. 1991 Prace wykopaliskowe na osadzie neolitycznej w Łańcucie w latach 1982–1984. *Materiały i Sprawozdania Rzeszowskiego Ośrodka archeologicznego za lata 1980–1984*:149–155.

Harmata, K. 1987a Late-Glacial and Holocene History of Vegetation at Roztoki and Tarnawiec near Jasło (Jasło-Sanok Depression), SE Poland. *Vegetation History and Archaeobotany* 4:235–243.

Harmata, K. 1987b Late Glacial and Holocene History at Roztoki and Tarnowiec, Near Jasło (Jasło-Sanok Depression). *Acta Palaeobotanica* 27:43–65.

Harmata, K., N. Kalinovyč, A. Budek, L. Starkel, and A. Jacyšyn 2006 Environmental Changes during the Holocene. In *Environment and Man at the Carpathian Foreland in the Upper Dnister Catchment from Neolithic to Early Medieval Period*, edited by K. Harmata, J. Machnik, and L. Starkel, pp. 66–91. Prace Komisji Prehistorii Karpat 3, Kraków.

Jarosz, P. 2002 Kurhan kultury ceramiki sznurowej w Średniej, st.3/2, pow. Przemyśl. Wyniki badań wykopaliskowych przeprowadzonych w 2001 r. *Rocznik Przemyski* 38(2):3–21.

Kadrow, S. 1990 Osada neolityczna na stan. nr 16 w Rzeszowie na osiedlu Piastów. *Sprawozdania Archeologiczne* 41:9–76.

Kadrow, S. 1992 Osada kultury lubelsko-wołyńskiej ze stan. 35 w Kosinie, gm. loco, woj. rzeszowskie. *Materiały i Sprawozdania Rzeszowskiego Ośrodka Archeologicznego za lata 1985–1990*:141–150.

Kadrow, S. 1997 Osada kultury ceramiki wstęgowej rytej na stanowisku 3 w Rzeszowie-Staromieściu. *Materiały i Sprawozdania Rzeszowskiego Ośrodka Archeologicznego* 18:5–27.

Koperowa, W. 1970 Późnoglacjalna i holoceńska historia roślinności wschodniej części Dołów Jasielsko-Sanockich. *Acta Palaeobotanica* 11(2):21–42.

Kremenetski, K. V. 2003 Steppe and Forest-Steppe Belt of Eurasia: Holocene Environmental History. In *Prehistoric Steppe Adaptation and the Horse*, edited by M. Levine, C. Renfrew, and K. Boyle, pp. 11–27. McDonald Institute for Archaeological Research. Cambridge.

Lubelczyk, A. 2001 Nowe odkrycia kurhanów oraz wapiennika na Pogórzu Strzyżowskim. *Rocznik Przemyski* 37(1):71–74.

Łanczont, M., K. Klimek, M. Komar, J. Nogaj-Chachaj, A. Poręba, and B. Żegała 2003 Nieinwazyjne badania kurhanów. Na przykładzie kurhanu nr. 2 na st. 37 w Cieszacinie Wielkim. *Rocznik Przemyski* 39(2):5–19.

Łanczont, M., J. Nogaj-Chachaj, and K. Klimek 2001 Kolejny nowy kurhan w okolicach Cieszcina Wielkiego (w pobliżu Jarosławia). *Rocznik Przemyski* 37(1):65–69.

Łanczont, M., J. Nogaj-Chachaj, K. Klimek, A. Poręba, B. Żogała, M. Komar, and W. Zuberek 2004 Wybrane problemy badań starożytnych kurhanów z okolic Jarosławia (Wysoczyzna Kańczucka). *Rocznik Przemyski* 40(2):7–15.

Łanczont, M., J. Nogaj-Chachaj, M. Racka, M. Fabiańska, and G. Bzowska 2007 Geochemiczne badania kurhanu I w Cieszacinie Wielkim. *Rocznik Przemyski* 43(2):11–20.

Machnik, J. 1989 Wyniki najnowszych badań archeologicznych w Karpatach polskich oraz ich znaczenie dla innych dyscyplin naukowych. *Rocznik Oddziału PAN Kraków za rok 1988*:83–104.

Machnik, J. 1990 The Kurgan-Culture and Its Substratum in the Carpathian Zone (Including Settlement and Economic Aspects). *The Journal of Indo-European Studies* 18:1–14.

Machnik, J. 1992 Aus den Forschungen über die Schnurkeramikkultur auf den nördlichen Vorfield des Niederen Beskid. *Acta Archaeologica Carpathica* 31:67–88.

Machnik, J. 1995 Zapomniany kurhan kultury ceramiki sznurowej w Morawsku koło Jarosławia. *Rocznik Przemyski* 31(1):3–22.

Machnik, J. 1998 Uwagi o najstarszym osadnictwie pasterskiej ludności kultury ceramiki sznurowej (III tysiąclecie przed, Chr.) w strefie karpackiej. In *Pradzieje Podkarpacia*. vol. 2, edited by J. Gancarski, pp. 99–120. Museum Podkarpackie, Krosno.

Machnik, J. 2001 *Archaeology and Natural Background of the Lower Beskid Mountains, Carpathians*. Part 1, Polska Akademia Umiejętności, Kraków.

Machnik, J., and E. Sosnowska 1996 Starożytna mogiła z początku III tysiąclecia przed Chrystusem, ludności kultury ceramiki sznurowej w Średniej, gm. Krzywcza. *Rocznik Przemyski* 32(3):3–28.

Machnik, J., and E. Sosnowska 1998 Kurhan ludności kultury ceramiki sznurowej z przełomu III i II tysiąclecia przed Chrystusem w Woli Węgierskiej, gm. Roźwienica, woj. przemyskie. *Rocznik Przemyski* 34(3):3–20.

Machnik, J., and E. Sosnowska 1999 Badania archeologiczne na kurhanie 2/98 w Średniej, gm. Krzywcza. *Rocznik Przemyski* 35(2):19–40.

Machnik, J., E. Sosnowska, and W. Cyhyłyk 1997 Osada ludności kultury ceramiki sznurowej z początku III tysiąclecia przed Chr. w Side koło Sambora. *Rocznik Przemyski* 33(5):3–28.

Madej, P. 1998 Grupy episznurowe w Karpatach Polskich. In *Pradzieje Podkarpacia*. vol. 2, edited by J. Gancarski, pp. 177–199, Muzeum Podkarpackie, Krosno.

Madej, P. 1999 Wyniki badań osady kultury mierzanowickiej na stanowisku 5 w Sieteszy, gm. Kańczuga, woj. podkarpackie. *Rocznik Przemyski* 35(2):41–58.

Madej, P. 2000 Sprawozdanie z badań wykopaliskowych przeprowadzonych w 1999 r na stanowisku 5 w Sieteszy, pow. Przeworsk. *Rocznik Przemyski* 36(1):11–31.

Madej, P. 2001 Uwagi o kulturze mierzanowickiej w dorzeczu Sanu. In *Neolit i początki epoki brązu w Karpatach polskich. Materiały z sesji naukowej. Krosno, 14–15 grudnia 2000 r*, edited by J. Gancarski, pp. 295–303, Muzeum Podkarpackie, Krosno.

Madej, P. 2003 Sprawozdanie z badań wykopaliskowych w 2000 roku na stanowisku 5 w Sieteszy, pow. Przeworsk (AZP 104-80). *Rocznik Przemyski* 39(2):47–57.

Madeja, J. 2001 Historia lokalnej szaty roślinnej w okolicy Wolicy Ługowej koło Sędziszowa Małopolskiego. In *Neolit i początki epoki brązu w Karpatach polskich. Materiały z sesji naukowej. Krosno, 14–15 grudnia 2000 r.*, edited by J. Gancarski, pp. 201–205, Muzeum Podkarpackie, Krosno.

Mamakowa, K. 1962 Roślinność Kotliny Sandomierskiej w późnym glacjale i holocenie. *Acta Palaeobotanica* 3(1):1–57.

Marciniak, A. 2004 Mikrospołeczny wymiar pasterstwa i nomadyzmu i tafonomiczne podstawy ich identyfikacji w materiałach faunistycznych. In *Nomadyzm i pastoralizm w międzyrzeczu Wisły i Dniepru (neolit, eneolit, epoka brązu)*, edited by A. Kośko, and M. Szmyt, pp. 35–43, Wydawnictwo Poznańskie, Poznań.

Mitura, P. 2007 Workshop of Flint Processing and Reparation of Rectangular Axes on Site 34 in Niedźwiada, Ropczyce Commune, Podkarpackie Voivodship. *Sprawozdania Archeologiczne* 59:305–324.

Mitura, P., and R. Zych 1999 Sprawozdanie z badan stanowiska 20 w Olchowej, gm. Iwierzyce, woj. podkarpackie w 1999 roku. *Materiały i Sprawozdania Rzeszowskiego Ośrodka Archeologicznego* 20:261–276.

Moskwa, K. 1963 Badania wykopaliskowe w Albigowej powiat Łańcut (neolit i kultura łużycka). *Sprawozdanie Rzeszowskiego Ośrodka Archeologicznego za rok 1963*:14–15.

Nowak, M. 1998 Karpackie osadnictwo kultury pucharów lejkowatych. In *Pradzieje Podkarpacia*. vol. 2, edited by J. Gancarski, pp. 89–98. Muzeum Podkarpackie, Krosno.

Pelisiak, A. 2003 Ze studiów nad osadnictwem w późnym neolicie i w początkach epoki brązu w Karpatach wschodnich i Kotlinie Sandomierskiej-rejon Husowa. In *Epoka Brązu i wczesna epoka żelaza w Karpatach polskich*, edited by J. Gancarski, pp. 13–41. Muzeum Podkarpackie, Krosno.

Pelisiak, A. 2004 Ogólne wzorce osadnictwa w neolicie i początkach epoki brązu w strefie brzeżnej Pogórza Dynowskiego (Husów). *Zeszyty Naukowe UR 23, Archeologia* 1:71–120.

Pelisiak, A. 2005 Osadnictwo i gospodarka w neolicie we wschodniej części Karpat polskich. Konfrontacja informacji archeologicznych i palinologicznych. In *Roślinne ślady człowieka*, edited by K. Wasylikowa, M. Lityńska-Zając, and A. Bieniek, pp. 29–52. Botanical Guidebooks 28, Kraków.

Pelisiak, A., and E. Sosnowska 2005 Nowe zabytki krzemienne z Pogórza Dynowskiego, Średnia gm. Krzywcze, stan. 56 na obszarze 107-82 AZP. *Rocznik Przemyski* 41(2):165–169.

Przybyła, M. 2004 Osada z okresu neolitu i epoki brązu na stanowisku 17 w Husowie, pow. Łańcut. *Rocznik Przemyski* 40(2):53–72.

Przybyła, M., and W. Blajer 2008 Struktury osadnicze w epoce brązu i wczesnej epoce żelaza na obszarze podkarpackiej wysoczyzny lessowej między Wisłokiem a Sanem. Uniwersytet Jagielloński, Kraków.

Ralska-Jasiewiczowa, M. 1980 Late Glacial and Holocene Vegetation of the Bieszczady Mts. (Polish Eastern Carpathians). Państwowe Wydawnictwo Naukowe, Warszawa.

Szczepanek, K. 1987 Late-Glacial and Holocene Pollen Diagrams from Jasiel in the Low Beskid Mts. (The Carpathians). *Acta Palaeobotanica* 27(1):9–26.

Szczepanek, K. 2001 Anthropogenic Vegetation Changes in the Region of the Dukla Pass, the Lower Beskid Mountains. In *Archaeology and Natural Background of the Lower Beskid Mountains, Carpathians*, Part I, edited by J. Machnik, pp. 171–182. Prace Komisji Prehistorii Karpat, vol. II, Polska Akademia Umiejętności, Kraków.

Starkel, L. 1994 Frequency of Floods during the Holocene in the Upper Vistula Basin. *Studia Geomorphologica Carpatho-Balkanica* 27–28:3–13.

Valde-Nowak, P. 1988 *Etapy i strefy zasiedlenia Parpat polskich w neolicie i na początku epoki brązu.* Ossolineum, Wrocław.

Valde-Nowak, P. 1998 Z badań najstarszego osadnictwa w Karpatach Polskich. In *Dzieje Podkarpacia*, vol. 2, edited by J. Gancarski, pp. 39–54. Muzeum Podkarpackie.

Valde-Nowak, P. 2001 Etapy i strefy w badaniach nad neolitem w polskich Karpatach. In *Neolit i początki epoki brązu w Karpatach polskich. Materiały z sesji naukowej. Krosno, 14–15 grudnia 2000 r.*, edited by J. Gancarski, pp. 89–106. Muzeum Podkarpackie, Krosno.

Wacnik, A. 1995 The Vegetation History of Local Flora and Evidences of Human Activities Recorded in the Pollen Diagrams from Site Regetovka, NE Slovakia. *Acta Palaeobotanica* 35(2):253–274.

Wacnik, A. 1999 Antropogeniczne przekształcenia lokalnej szaty roślinnej w okolicy Regetovki (północno-wschodnia Słowacja), w świetle analizy pyłkowej. In *Rośliny w dawnej gospodarce człowieka. Warsztaty archeobotaniczne'97*, edited by K. Wasylikowa, pp. 127–137. Polish Botanical Studies Guidebook Series 23. Instytut Botaniki, Kraków.

Wacnik, A. 2001 Late-Holocene History of the Vegetation Changes Based on the Pollen Analysis on the Deposits at Kružlova, Slovakia. In *Archaeology and Natural Background of the Lower Beskid Mountains, Carpathians*, Part I, edited by J. Machnik, pp. 127–135. Prace Komisji Prahistorii Karpat, vol. II, Polska Akademia Umiejętności, Kraków.

Wacnik, A., K. Szczepanek, and K. Harmata 2001 Ślady działalności człowieka neolitu i brązu obserwowane w diagramach pyłkowych z okolic Przełęczy Dukielskiej i terenów przyległych. In *Neolit i początki epoki brązu w Karpatach polskich. Materiały z sesji naukowej. Krosno, 14–15 grudnia 2000 r.*, edited by J. Gancarski, pp. 207–221. Muzeum Podkarpackie, Krosno.

Zych, R. 2002 Sprawozdanie z badań na stan. 13 w Tarnawce, gm. Markowa, woj. podkarpackie w 2000 roku. *Materiały i Sprawozdania Rzeszowskiego Ośrodka Archeologicznego* 23:199–205.

Zych, R. 2003 Sprawozdanie z badań wykopaliskowych na stan. 7 w Manasterzu, gm. Jawornik Polski, woj. podkarpackie. *Materiały i Sprawozdania Rzeszowskiego Ośrodka Archeologicznego* 24:329–335.

Zych, R. 2004 Stanowisko KPL w Tarnawce 9, gm. Markowa, woj. podkarpackie. *Materiały i Sprawozdania Rzeszowskiego Ośrodka Archeologicznego* 25:281–302.

Zych, R. 2008 Kultura pucharów lejkowatych w Polsce południowo-wschodniej. Instytut Archeologii, Rzeszów.

CHAPTER TWELVE

Climate and the Definition of Archaeological Periods in Sweden

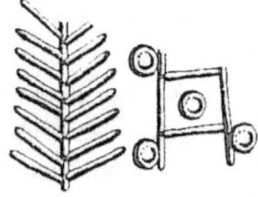

Daniel Löwenborg and Thomas Eriksson

Abstract *This paper discusses to what extent the definition of archaeological periods in Sweden can be linked to climatic events. These climatic changes may have caused some of the cultural changes that are being recognized by archaeologists examining the material remains, and thus inspire the division of prehistory into different chronological periods. Changing climate was to some extent included in the discussion as these periods were first being established, but climate has in later times largely been disregarded in the archaeological academic dialogue. Following the theoretical development of the discipline in the late twentieth century, climate was seen as too crude a way to explain changes in human behavior. With climate returning as a topic in present-day general debate, archaeologists are again turning to climate as a possible explanation to some of the background to what underlies cultural change, as one of many contributing factors. In combination with the increasing resolution of climatic information produced in the natural sciences, and the possibilities to use information technology to make detailed quantitative analyses of the archaeological record, this holds great promise to improve our abilities as archaeologists to understand the dynamics of climatic impact on culture.*

INTRODUCTION

The chronological framework for Northern Europe and the three-period system was invented in Scandinavia during the nineteenth century (Gräslund 1987). The archaeologists collaborated from the beginning with biologists, geologists, and other scientists as well as with archaeologists with a more cultural anthropologic approach. The interdisciplinary approach cross-fertilized all branches and the innovative milieu was the origin of many new ideas and interpretations. Quaternary geology had a strong position around 1900, and traces of the last glacial period are abundant in Scandinavia. Geologists Rutger Sernander and Lennart von Post, who invented the method of pollen analysis and laid the foundations for modern palynology, and Gerhard de Geer, the discoverer of the varves in glacial clay, worked together with archaeologists. In the case of Sernander, he also made excavations on his own. He further combined fluctuations in pollen strata with climatic changes and temperature (De Geer and Sernander 1909; Sernander 1897). This was the origin of the first interpretation of the *Fimbulvetr or Fimbulwinter* (the great winter). With the description of the *Fimbulwinter* in the old Icelandic poems of the *Eddas,* written down in the twelfth century A.D., Sernander connected the transition between the Bronze and Iron Age to the onset of a more humid and colder climate around 500 B.C. (Sernander 1912). It was described as a long winter during three years without summers that preceded the *Ragnarökr*—the end of the world. Sernander connected this with the transition from the Subboreal Period to the colder and more humid Subatlantic Period.

The concept of the *Fimbulwinter* at the transition between the Late Bronze Age and the Early Iron Age was soon accepted by archaeologists (Bergeron et al. 1956; Lindqvist 1920). The Late Bronze Age (1100–500 B.C.) seemed to be a prosperous period, with rich hoards of bronze artifacts, an abundance of varieties of decorations and rock carvings, and in Southern Scandinavia many graves. The following period, the Pre-Roman Iron Age (500 B.C.–A.D. 1), especially the early part (500–150 B.C.) was considered to be a period without dateable graves and artifacts in Central Sweden. Proposed explanations for this focused on emigration, climatic deterioration, or a Celtic trade blockade (Lindqvist 1920).

The study of Early Iron Age settlements was dominated by research on the large islands of Öland and Gotland in the Baltic Sea during the first half of the twentieth century. This was primarily due to the well-preserved remains of the houses there (Stenberger 1933; Stenberger and Klindt-Jensen 1955). The abandonment of this type of settlement and stone-walled houses during the sixth century was soon obvious. The phenomenon was recorded in many different regions in Scandinavia and Finland with disappearing settlements, graves, and frequent depositions of gold around the middle of the sixth century (Näsman and Lund 1988).

The focus on climate and adaption to the natural sources and circumstances had a revival during the processualist era in 1970s and 1980s archaeological discourse. A close collaboration between quaternary geology and archaeology made possible large-scale investigations in different regions in Sweden (e.g., Larsson et al. 1992; Welinder 1974). The discussion, and the discourse, was a mirror of contemporary political debate. A major switch can be seen during the end of the twentieth century, as the political tide had turned. Climate and environmental settings were no longer in focus as a cause of changes in human behavior. Instead, the free will of the individual was emphasized. This was clearly inspired by new neoliberal tendencies in society and widely accepted among archaeologists, even though they usually were not neoliberals themselves.

A new paradigm can be seen in the debate around 2010. Climate is again on the political agenda, and climate has gained a revival among archaeologists as an explanation of cultural change.

Investigating the Impacts of Climate

Climate has always had an impact on society and human possibilities. The idea that shifts in population are connected to climate has its roots in the Malthusian paradigm. A quantitative method for studying the intensity of human activities is to gather all carbon dates and make a combined calibration. The distribution of the samples must in some way be connected to the intensity and volume of the population. There are of course also cultural causes behind the quantity of dates recovered from archaeological excavations. The pattern of how activities were carried out in the landscape, and whether the activities left visible remains with dateable material, partly depends on cultural behavior, proprietorship, and settlement patterns. Their discovery may be influenced by the antiquarian knowledge of where to look for ancient remains. Nevertheless, it is perhaps one of the best ways to study human impact on the landscape. Rescue excavations have made immense discoveries in Central Sweden during the last decades through large-scale excavations of settlements and grave fields covering the full span of time of the last several millennia up to the present. The excavations have made the northern part of the Lake Mälaren basin to one of the most excavated areas in Europe (Figure 12.1).

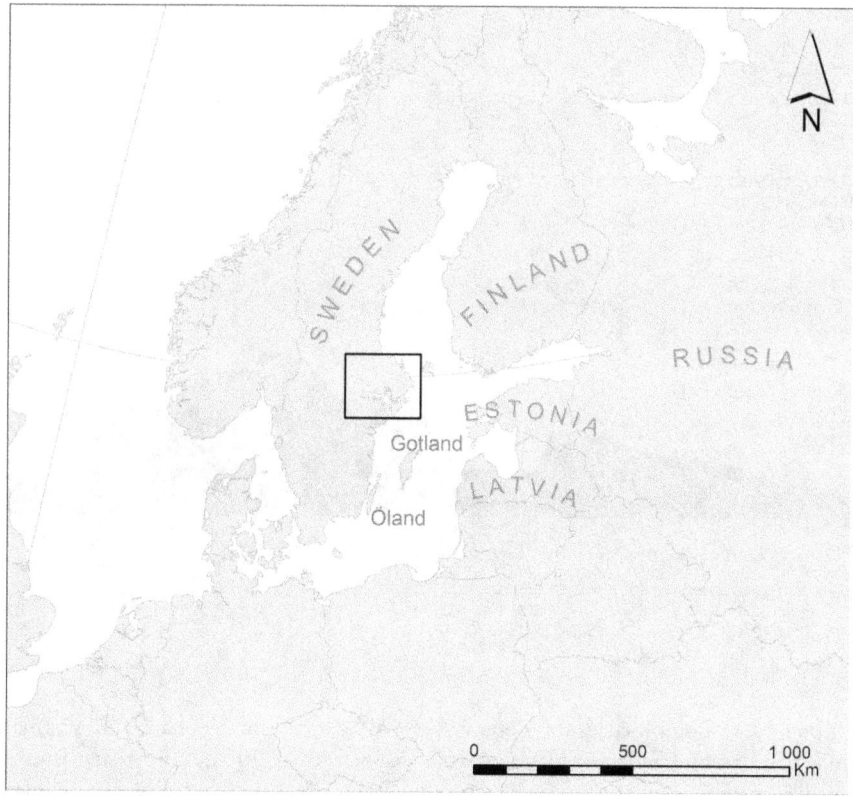

FIGURE 12.1. Map of northern Europe showing the location of the main research area in central Sweden, the Mälaren Basin (square) and the location of sites mentioned in the text.

Almost two thousand absolute dates have been collected from all kinds of features: wells, postholes, hearths, charcoal from funeral pyres, artifacts, and other features such as occasional hunting grounds and pastoral and agrarian remains. The material has been treated with two different methods. The first method is a combined calibration in IntCal04.14C. The benefit of this is that it uses the calibrated dates that are adjusted to our culturally defined periods (Figure 12.2). The disadvantage is that the dates have a tendency to cluster and reflect the calibration curve. Therefore, a second method has been used, a method where the uncalibrated time spans with one sigma have been used and the number of dates has been counted by every 50 years (Figure 12.3). This is done in an effort to dissolve the problem of clusters in dates due to the calibration curve.

Another way of achieving the volume of information necessary for making quantitative analyses of large-scale social changes is to create syntheses of the results of archaeological excavation projects for whole areas. In Sweden and elsewhere, more effort is now being put into the creation of archives for digital information, and this should ideally

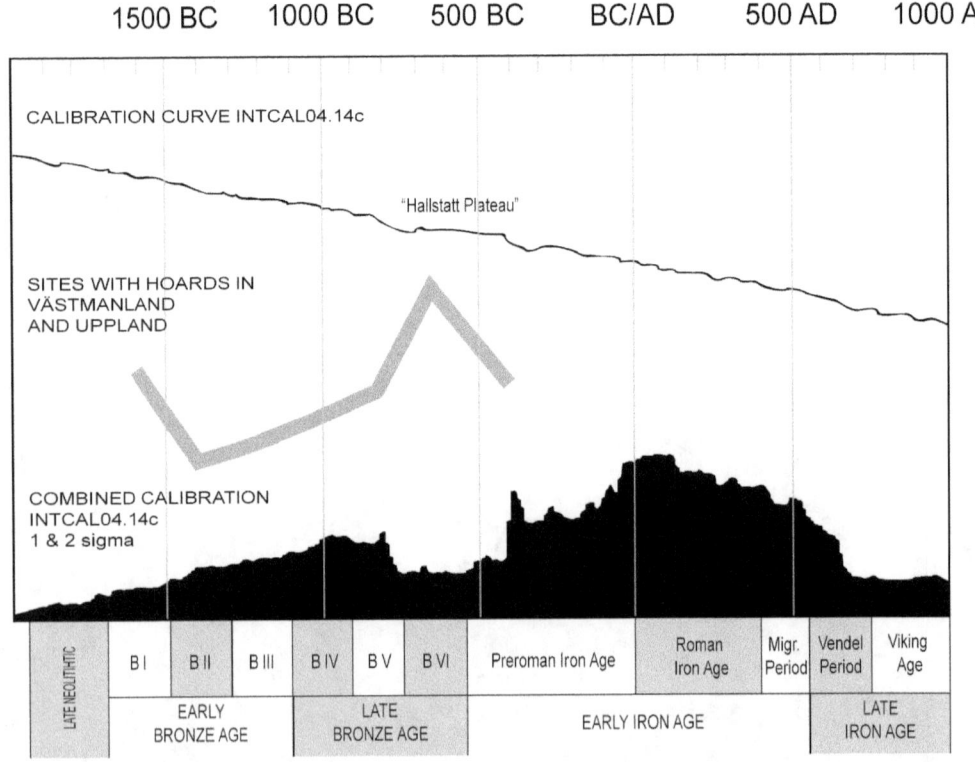

FIGURE 12.2. Combined calibrations of the 14C dates from Uppland and Västmanland. Calibration made by IntCal04.14C (Reimer et al 2004). The result is compared to the number of sites with bronze hoards in the same area as well as climatic proxy-data from lakes in Northern Sweden (Grudd et al 2002; Gunnarsson 2008). There are clear overlaps with periods with cold and humid climate and low numbers of 14C-datings.

Climate and the Definition of Archaeological Periods in Sweden 261

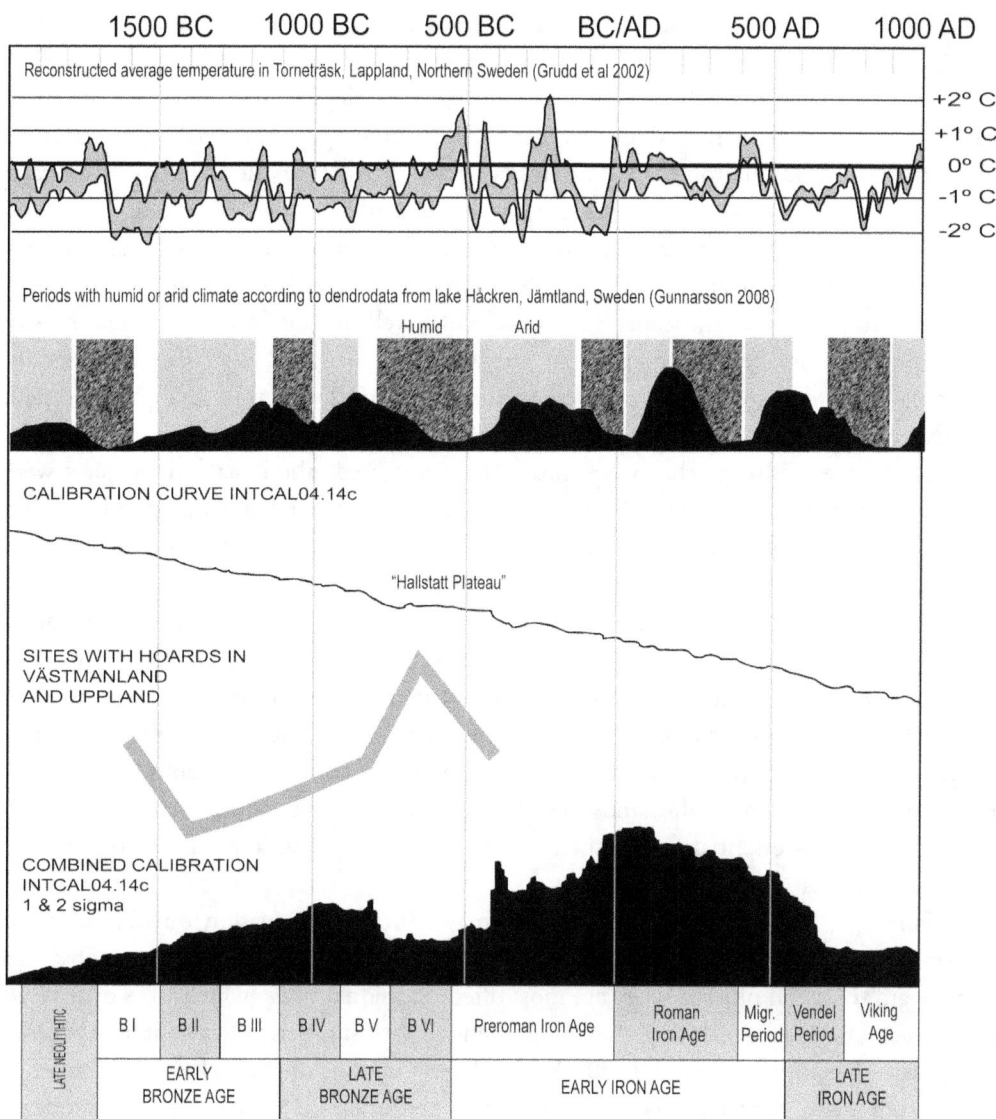

FIGURE 12.3. Number of uncalibrated time spans of 14C-datings from the northern part of Lake Mälaren Basin (staples). The diminishing numbers around 800 B.C. (ca 2650 B.P.), 400 B.C. (ca 2450 B.P.) and after A.D. 500 (ca 1500 B.P.) are clearly visible. The correspondence between the peak of sites with hoards, numbers of bronzes in hoards, and the end of the Bronze Age is also obvious.

result in a situation where it is possible to collect GIS (Geographical Information Systems) data from excavations in a structured manner. This will enable further research beyond the excavation report, and facilitate detailed, high-resolution analyses of the archaeological material of larger areas than would otherwise be possible. Ideally, all excavations should be gathered and correlated so that the information can be combined in analyses that take into account all archaeological material at once as a base for archaeological knowledge and interpretation.

This becomes especially interesting as the archaeological data can be combined with detailed models of the landscape, using GIS technology. This can be used both to reconstruct historic environments and to calculate variables to describe the landscape to be used for further analyses. An example of this could be the case of a GIS analysis of the representativity of burial grounds in central Sweden, based on the results from a set of large archaeological excavation projects. In combining these and using them as a statistical sample of the landscape, they could ideally tell us more about the total, mostly unexcavated, landscape. The method of *Geographically Weighted Regression* was used to extrapolate the results of the excavations to the entire landscape, thus giving an estimate of how many burial grounds there were in total in the region (Löwenborg 2010). In the parts of the landscape that were most intensively used, about half of the sites were missing from the record, primarily due to intensive agricultural use that has destroyed many sites. With the original frequency of sites estimated in the computer simulation, it is easier to appreciate the role of the burial grounds in the society. Similarly, it is possible to describe and compare how different sites are located in the landscape through statistics of variables from models of the landscape in a GIS. This would thus provide a better way of understanding how the sites are related to the landscape, to improve the interpretation of their function. As sites often carry great symbolic meaning they may mirror important aspects of the landscape, and some of that meaning could be interpreted from their location in the landscape. Proper utilization of digital archives and GIS technology is thus essential for synthesizing what is known and key for interpreting the social dimensions within the landscape.

Finally, specific types of find contexts may contain information on the effects of climate change. In Sweden, one of the first types of categories of finds that was observed were hoards of metal objects. They are most often occasional finds, which were discovered by farmers and workers during the most intensive farming period without mechanized equipment, between 1850 and 1920. Archaeologists seldom uncover them, but hopefully many of them were reported and brought to the museum collections. The modern, mechanized work in the fields and peat bogs leaves us few new finds of this kind today.

The discussion about the causes behind the phenomena of deposits is as old as it is vast. The majority of the hoards in Central Sweden during the Bronze Age come from the later part, ca. 800–500 B.C. (Larsson 1986; Levy 1982). The discussion about the hoards during this period has mostly been focused on ritual behavior with a theological rather than a political and societal context. The richness of hoards and the complexity of artifacts within them have often been seen as a proof of wealth during the Late Bronze Age (Figures 12.2–12.4).

This kind of consumption of metals, both in wetlands and in dry contexts, almost disappears during the Early Iron Age in Central Sweden. There are stray finds of gold and silver objects during the Roman Iron Age (0–A.D. 400) but this cannot be compared to the richness of, for instance, gold treasures on the islands of Gotland and Öland in the Baltic Sea.

Climate and the Definition of Archaeological Periods in Sweden 263

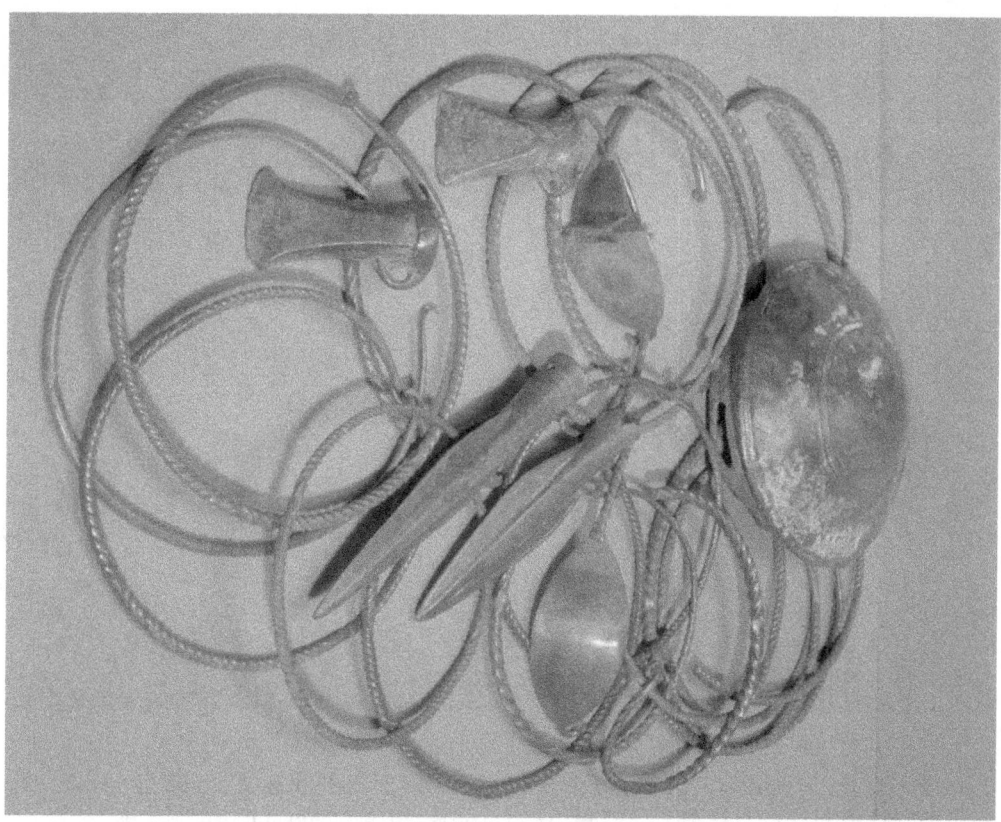

FIGURE 12.4. The hoard from Spelvik, Spelvik parish in Södermanland (SHM inv. 813) The hoard consists of eleven intact neck-rings and fragments of more rings, two celts, two spears, and a hanging-bowl. It can be dated ca. 900–700 B.C. Photo Statens Historiska Museum © Historiska museet.

The hoards of Roman coinage and indigenous gold artifacts on these two islands during the Roman Iron Age and Migration Period (A.D. 400–550) were observed early by archaeologists. Especially the finds from the latter period were interpreted as signs of a disordered period or a sign of some kind of climatic disaster (Stenberger 1933). The high frequency of deposited bracteates in the sixth century has been connected to the A.D. 536 event and the climatic crisis at that time (Axboe 2004:266–272). Deposits of gold and bracteates peak in the first half of the sixth century, and after that they virtually disappear from production. Axboe suggests that they were deposited in order to appease the Gods during this troublesome time (Figure 12.5).

THEORETICAL FRAMEWORK

As theories and concepts in archaeology are constantly developing and becoming more diversified, it is increasingly possible to speak in terms of how the environment may

FIGURE 12.5. Some of the gold bracteates from the Söderby hoard, Danmark parish in Uppland. The hoard consists of ten bracteats, gold bullion for sword-beads, and other, small gold objects (SHM inv. 5802 & 33022). It was deposited south of Uppsala during the first half of the sixth century A.D. The décor on the bracteates consists of geometrical symbols and a male figure that can be interpreted as the god Odin and his two ravens. (Lamm et al 1999). Photo Ulf Bruxe; Statens Historiska Museum © Historiska museet.

influence social change. Acknowledging the post-processual critique regarding the danger of viewing society as determined by external factors, the human-environment relationship is now usually seen as much more complex and nonlinear. As human society is not seen as separate, but part of the environment, it is evident that they are intertwined in a mutual dynamic dependence, where "the social informs the natural and the natural informs the social" (McGlade 1995:114). Putting man back inside his environment thus reflect how both are part of a whole, and cannot be understood separately.

Climate change has often been considered as a long-term gradual increase or decrease of temperature that society might have time to adapt to, since most ecosystems are flexible and have certain resilience. These adaptations might be so gradual that they are embedded within an existing system and are hardly noticeable. However, everything has a limit, and if a crisis occurs which surpasses an ecological threshold that cannot be adequately compensated for within the system, then some kind of cultural response would have to be the result (Crumley 1994:10). Those responses are, however, active choices;

the chosen action will determine the outcome of any environmental challenge, both in history and today. Here, archaeology, more than other disciplines, might be informative by bringing a long temporal perspective capable of providing examples of how societies have responded to crises and change, and some of the long-term effects of those changes.

During the 1980s, the reaction against adopting climatic changes and landscape settings as the major explanation for changes in culture was focused on the sometimes-simplified discussion and the eco-deterministic perspective. Humans and societies adopted their behavior and social orders unintentionally and without any opportunities to express choices. The climate made changes inevitable. The pendulum in the discussion shifted to a perspective that held that climate and ecosystems did not have any cultural effects. Instead, cultural and economic changes depended on ideology. Today we must combine those two perspectives. Climate has and has always had an impact on all living organisms; what makes humans unique is our possibility to adopt changes in different ways. We have a free will even though it is defined by the limits of social and cultural behavior and natural conditions. Therefore, we must remember the different possibilities within human culture to adopt changes, and look at humans as actors within a system (Adger et al. 2009; Folke 2006; Hodder 2008; Lahsen 2010).

Case Studies

In order to illustrate some aspects of the argument outlined above, two case studies of how climatic change might have had an impact on prehistoric societies will be presented. In both cases, it is argued that the radical changes in material culture that have been observed and that have inspired archaeologists to make distinctions among different chronological periods has a background in climatically induced cultural change. Scandinavia would be well suited for investigating climatic impact on human culture, as its location in the northern part of Europe makes it a border zone between a temperate and an arctic climate. Warm air transported northward by the Gulf Stream provides warm summers and a climate that is milder than would be expected from the location, which borders the northern polar circle. Winters are often cold and with much snow, and only parts of Scandinavia are suitable for cultivating crops in any volume, as the annual growth period is short. As Scandinavia in many ways might be described as a border zone of cultivation, any fluctuation in temperature, and especially decline, potentially has dramatic consequences. Failing crops caused by bad weather is common, but richer crops nearby can usually compensate for these local phenomena. More extensive failures, both geographically and chronologically, could, however, be devastating.

Scandinavia 800–400 b.c.

During the middle of the twentieth century, the eastern part of Central Sweden, in the Lake Mälaren Basin, was considered to have had a blooming period between 800 and 400 B.C., Montelius period (IV) V–VI, as an independent and prosperous region. The rich hoards, gold finds, and rich finds from burials were considered as proof. Another strong indication was the so-called Mälardal-celts, a type of celt that was seen as indigenous to the region, due to the numerous finds here and in Central Russia, in the Khazar-province. The chorology of the celts was interpreted as indicating a long-distance trade from Sweden eastward. The end of this expansion was believed to have come at around 500 B.C. (Baudou 1950, 1953, 1960; Nerman 1957). Recent studies have shown that the so-called Mälardal-celts were cast in copper from the Ural region and are therefore a sign of contact between east and west, and they are now known under the Russian name *Kelty Akozino-Melarskie* (Kresten 2005; Kuz'minych 1996). Molds for the celts have been found in Russia, Estonia, Finland, and Latvia but not in the Lake Mälaren basin (Eriksson 2009:247ff.). Thus, the celts of this kind cannot be seen as a sign of an indigenous and rich culture but rather of new influences from the east during the Late Bronze Age. These influences can also be traced in the pottery after 900 B.C. (Eriksson 2009:129–136, 248). During the entire Bronze Age (1800–500 B.C.), the majority of the metal as well as the influences in pottery style came from Central and Southern Europe.

Instead of seeing the Late Bronze Age society in the Lake Mälaren Basin as a prosperous period we should see it as a system under pressure. The climate grew colder and more humid and therefore the conditions for an agrarian economy deteriorated, something that can be seen in Southern Scandinavia as well (Kristiansen 1998). A shift from different kinds of wheat and nude barley to more sustainable hulled barley can be seen. During this period we can also see traces of new iron production in which iron was made from local ore and gradually replaced the imported copper and tin (Hjärthner-Holdar 2008; Hjärthner-Holdar and Risberg 2003). The more humid climate would also have improved the quality of the limonite ores that were the raw material for iron production. The interregional network of contacts for acquiring bronzes was disrupted, something that was especially troublesome for the elite that had maintained the system, as their position came into question. Their responses were to pursue an augmentation of rituals, feasting, and sacrifices, in order to maintain their hegemony and common ideology during a troublesome period.

The number of ^{14}C dates in Uppland and Västmanland peaked at about 1000–800 B.C. and experienced a marked dropoff around 800 B.C. (Figures 12.2 and 12.3). Simultaneously with the drop in datings, there was an increase in the number of depositions, as weapons, mainly swords and lances, were laid down primarily in bogs and lakes (Figure 12.4). They can be viewed as signs of a violent period. During the succeeding periods, jewelry, drinking bowls, and dress garments rather than weapons were deposited in the hoards. The richness of hoards could well indicate a system under pressure rather than a prosperous society. The elite and the ritual leaders tried to maintain *l'ancienne régime* by sacrifices to the deities. Drinking rituals become more important, something that could be seen in pottery. The region did not become desolated, but the population did probably diminish. The isostatic uplift continuously made more arable land available for pasture, especially during this phase. This process have been a counterforce to the deteriorating climate and partly counterbalanced its effects, at least in a long-term perspective.

The Bronze Age system was maintained until ca. 500–400 B.C. At that time, the rituals with bronze hoards almost disappeared, and while some neck rings were probably sacrificed in wetlands, the tradition was becoming rare. This coincides with a shorter climatic crisis that can be seen in different quaternary studies (Figure 12.2). The bronze casting technique was also abandoned and was replaced by locally produced iron products. The centuries from 500 to 150 B.C. comprise one of the most materially barren eras in the prehistory of the region, as viewed through the criteria of datable metal artifacts and prestige items. In contrast, however, judging from ^{14}C dates and the number of settlements it was socially one of the most outstanding periods in prehistory. Even the graves seem to be numerous, although hard to date due to their lack of artifacts. The society seems to have changed to a more equal one, with few visible traces of hierarchies.

This period coincided with the so-called Hallstatt Plateau in the radiocarbon calibration curve. It is characterized by one of the most distinguished plateaus with a constant level of ^{14}C, with remarkable fluctuations in the levels at both beginning and end of the period (Mauquoy et al. 2004; Speranza et al. 2003). The events can also be seen in geomagnetic shifts and in paleomagnetic changes in the layers in bogs in Central Sweden. The levels of ^{14}C-amounts in the atmosphere as well as the climatic and paleomagnetic changes all depended on alterations in solar activity (Stanton 2011).

The process of these adaptations took almost four centuries. At the first stage, around 800 B.C., the climate change necessitated a new agricultural system relying on hulled barley and manuring. The latter also transformed the pattern of distributing waste at the settlements, a pattern that made them cleaner, as organic decay products became valuable as fertilizers. The new crop also made brewing beer easier. A relocation of settlements can also be seen at this stage, where new settlement grounds were founded and old ones abandoned. The isostatic land rise also continuously made more arable land available, a fact that compensated for the climatic deterioration and ensured good opportunities for food production. The old system, and the elite who controlled it, managed to survive due to their role as ritual leaders and suppliers of copper, tin, and flint from Southern Scandinavia and Central Europe. Bronze casting was common handicraft at almost all settlements at this stage. Ideology was maintained by ritual feasting and the consumption of metal objects as sacrifices. Clay products such as ceramic vessels, molds, and crucibles played an important role in the society, something that could be seen in pottery and other artifacts (Eriksson 2008). The economic system and political system were threatened not only by the more humid and colder climate, but also by an ever-growing production of iron.

Around 500–400 B.C., the system collapsed. The artifacts of bronze and gold disappeared, the contacts with Central Europe ceased, and the tradition of producing pottery vessels and feasting terminated. This is the stage that Sernander connected to the so-called *Fimbulwinter*, but the climatic deterioration happened about 300 years earlier than he had believed. The event around 800 B.C. was a trigger that started the process that changed the system 300 years later, but the system had the resilience to survive. But climate also changed around 500/400 B.C. due to the fluctuations in the ^{14}C levels and in the paleomagnetic directions. The breakdown came when the economic base was changed, and that is what we can see in the material remains, with a new tradition of metal working, making of pottery, and using vessels. The new Pre-Roman society was characterized by immense land reclamation, new settlements, and a more egalitarian society.

Scandinavia in the Sixth Century A.D.

A severe climatic event occurred in A.D. 536. This was probably due to a massive volcanic eruption somewhere in the Pacific, which caused a cloud of sulfur and volcanic ash high up in the atmosphere for 18 months, and had a cooling effect on the climate for about ten years. The veil of ash covered most parts of the Northern Hemisphere to varying degrees, although it can be assumed that the consequences of this would be very different in different parts of the world. The event is well documented in written sources, and Roman historians described the event and told how the sun did not warm the skin, and even at mid-day did not leave a shadow on the ground beneath you. From Iran, there are descriptions of harvests failing due to the lack of sunlight, and the same is the case in China, where this led to extensive famine (Axboe 2004; Charpentier Ljungqvist 2009:95–103; Gräslund 2007).

To what extent this would have affected countries in northern Europe is more difficult to estimate, but recently Gräslund has suggested that the experience of this event might be the background to the legend of the *Fimbulwinter* in the Old Norse literature, rather than the previous theories that associated the phenomena with the events around 500 B.C. (Gräslund 2007:105). That brings the question whether the transition in material and social culture from Early to Late Iron Age, in Sweden usually said to have occurred around A.D. 550, might be understood in terms of responses to this climate crisis. The climatic event in the sixth century can clearly be seen in different scientific proxies. Ice cores from Greenland show a distinct layer of sulfate from this time, and a combination of different types of information on climate in Europe shows a distinct drop in temperature (Büntgen et al. 2011; Larsen et al. 2008) On top of this, and possibly caused by the event, large parts of Europe and the Middle East were further struck by the Justinian plague in the decades directly following.

The decline, or crisis, in the sixth century is something that has been noted on many occasions and has been much debated. It is sometimes referred to as "the Migration Period crisis" and considered in relation to the turmoil on continental Europe following the decline of the Roman Empire, or at other times interpreted as an agricultural crisis with internal causes. There are indications of whole regions being more or less deserted sometime around the middle of the first century A.D. Still, there have been many theories about what happened, and at least since the 1980s many scholars have suggested continuity rather than any breaks in continuity, and even if it has often included a shift in location, have seen this as contradicting previous theories of upheaval (Carlsson 1979:163).

However, when large volumes of results of excavation projects are compared, it is clear that there is a marked lack of continuity, as a majority of sites were abandoned around the middle of the first millennium (Figure 12.6). Further, there were considerably

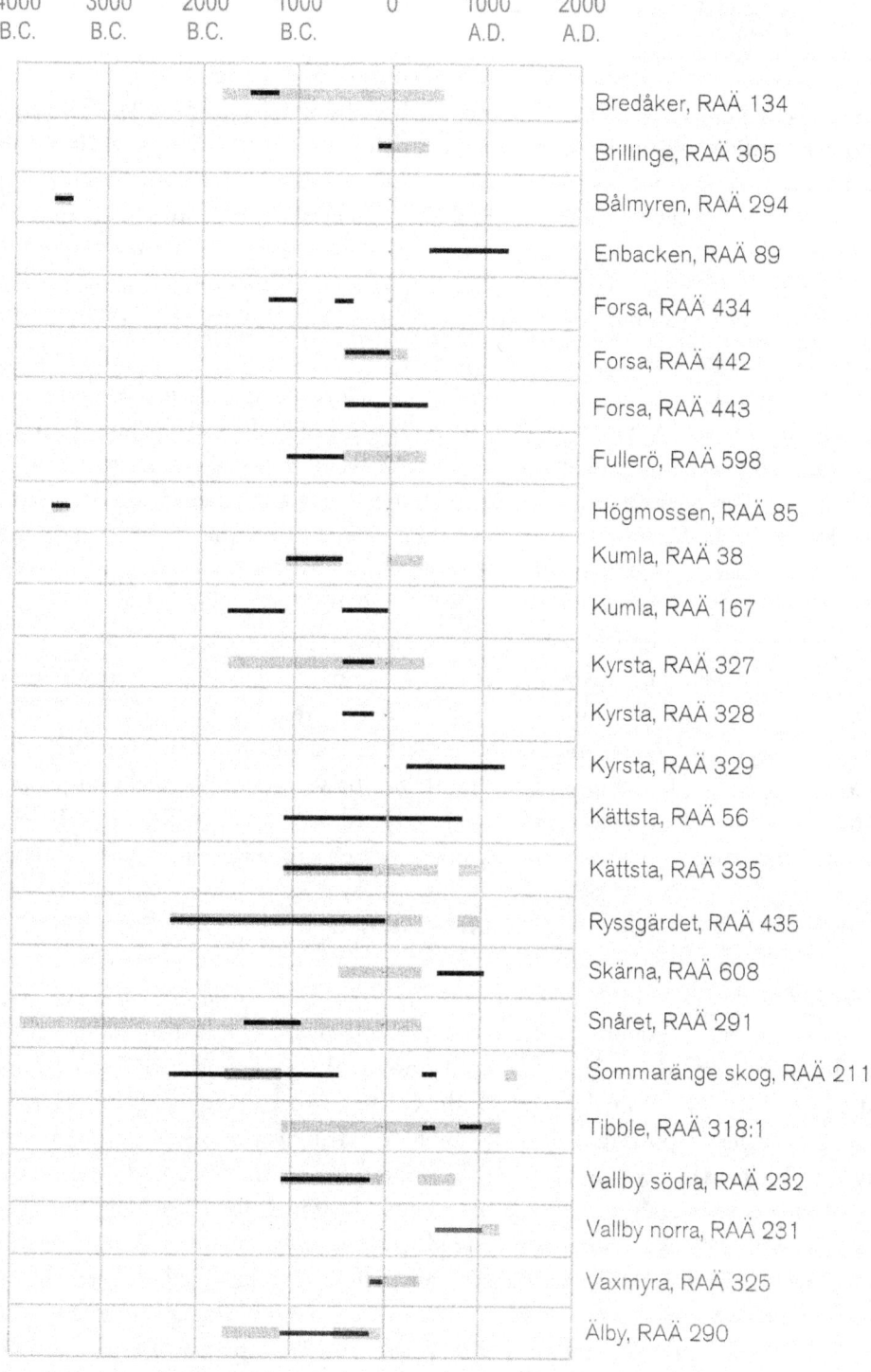

FIGURE 12.6. Illustration of the chronology of sites from a recent large archaeological project in central Sweden. Lines represent burial grounds (black) and nearby settlements (grey) (after Wikborg 2007:179).

fewer settlements after this period, compared to the rather fully settled Early Iron Age (Wikborg 2007).

The decline is also apparent from a marked drop in the number of ^{14}C dates during the sixth century A.D. (Figures 12.2 and 12.3; Eriksson 2009:265–271). This suggests a dramatic decline in population, which in turn would explain why there are far fewer indicators of cultivation in parts of the land, and instead an increasing amount of forested areas in the Mälaren basin and throughout large parts of Europe (Berglund 2003; Welinder 1974). The decline in population would have resulted in large areas of land being unused and available to be claimed by those who survived, thus making it possible to accumulate large landholdings. This is reflected in the social and economic situation in the Swedish Late Iron Age, where society was dominated by wealthy elites that controlled large domains of land, and displayed their status by exclusive and expensive, often imported, objects and weaponry and richly furnished graves. Part of this shift in property rights might be seen in the changing locations of burial grounds around the same time. While burials during the Early Iron Age primarily were positioned on higher ground in areas that, presumably, constituted a border zone between settlements, the Late Iron Age burial grounds are instead more often situated on clay on level ground, much similar to the locations chosen for settlements (Löwenborg 2010b). A plausible explanation for this relocation of burial grounds would be that abandoned land was claimed by new owners through the placement of graves in close relation to the land, in order to mark the renegotiated property rights (Löwenborg 2010c). Indeed, graves from this time are often found directly on top of settlements and houses that had been abandoned shortly before (Eriksson 1994; Renck 2008). The climatic event must thus be seen as having had a major impact on the population, and society was largely reshaped as a response to the crisis, even if the material culture at first glance rather might suggest a period of prosperity and wealth.

Conclusions

To be able to handle and analyze the complexity of these social events, it is suggested that it is essential to examine large volumes of information with quantitative methods, in order to understand the impact of the events on the society as a whole. A more limited and narrower scope might overlook the dynamics that shape the whole society, and there is a danger in putting too much emphasis on individual sites that, interesting as they may be in their own right, would be limited for explanations regarding long-term social change. In this respect, the advanced use of information technology for creating synthesis of archaeological information holds great promise.

If we accept that climate and environment are important for the underlying structure and framework of human societies, than climate should once again be used to evaluate how material culture has been reshaped with the development of social institutions and traditions. The cultural periods in Swedish archaeology were established in accord with typological and stylistic considerations. Some of the transitions between phases were so marked in their expression of material culture that archaeologists coined new periods to describe the differences. Two of these were the transitions between Bronze Age and

Iron Age, and another is the transition between Early and Late Iron Age. The oldest definitions were based mainly on typology but with an awareness that climate could influence culture, as a reason for change. Some of the explanations were rather simplistic and eco-deterministic. This was most explicitly expressed during the processual era of archaeology during the 1970s. The reaction against this eco-determinism arose during the 1990s in reaction to the Swedish debate with the postmodern position, and even today the most established views for explaining human behavior and culture reflect a post-processual orientation, in which culture, ritual, and ideology are emphasized, and where humans always can make their own free choices. What is acceptable in archaeological discourse is determined by the general ideology and debate in contemporary society. Clearly the archaeological explanations presented are influenced by the surrounding worldview, and to some extent archaeological theory might influence ideas in contemporary society. It is thus no coincidence that climate is now back in the archaeological debate when it is on the forefront of the general political and scientific agenda.

While it is important to avoid oversimplified and deterministic explanations, it is also important to recognize that human societies are part of the environment. Human behavior and our domesticated animals affect the environment in a mutual way. But the environment always sets the boundaries for our behavior in a long-term perspective. Climatic changes affect humans, especially those who live near the ecological borders. Scandinavia is one of those milieus, close to the North Pole and depending on the Gulf Stream for our climate. This makes peoples living in the north more vulnerable to fluctuations, where small differences might overwhelm the system. Therefore, it is important to renew the awareness of climate in archaeological theories and explanations.

Our two examples have in common a climatic deterioration and subsequent cultural change. The older example from 800–400 B.C. seems to reflect a prolonged process where the climate affected a vulnerable societal structure, featuring a society that depended on long-distance trade and external contacts. These were probably associated with the higher levels in the social hierarchy, where rituals, feasting, and a redistributive system were important. The system could be maintained during ca. 300 years due to the appearance of new settlements and grazing ground created by land-rise, and for as long as the production of iron did not compete with the importation of copper and tin. The moment the locally produced iron replaced bronze as the most important material, the base for this social structure broke down. The system had been under stress for a long time, and the wetland hoards can be seen as one sign of this. The climatic changes also created new conditions for agriculture, and naked barley and different kinds of wheat were replaced with the more resistant hulled barley ca. 800–400 B.C. This affected the cuisine, and the new barley was also better for making beer, which was fundamental for feasting during the Late Bronze Age. The collapse of the Bronze Age Society did not mean a decolonization of Central Sweden during the Pre-Roman Iron Age. Instead it was followed by a period during the Early Pre-Roman Iron Age of more intensive agricultural activity and new settlement, but with a more equal social organization.

The A.D. 536 event instead represents a rather rapid crisis, where a radical drop in temperature within ten years had a dramatic impact on the means of subsistence in present-day Scandinavia. The resulting catastrophic decline in population seems to have had a vast impact on the society as it was reformed after the crisis. The material culture was expressed in new ways, and the use of land and the economic system underwent fundamental changes. This was reflected in a whole new system of landownership, land use, settlement structures, burial customs, and language. According to the collected ^{14}C samples, it might appear that the landscape was almost deserted after the crisis. However, if more archaeological categories such as burials are included we can see that this was not the case, but rather a result of changing settlement patterns and the use of the landscape. As somewhat of a paradox, the society we see once it had recovered was one of prosperity and wealth, as a result of resources being concentrated in the hands of a limited number of people. These tendencies can be seen in the construction of some of the most prominent burial grounds in all of Sweden, in Old Uppsala, immediately in the aftermath of the crisis (Ljungkvist 2006). Therefore, it has for a long time been difficult for archaeologists to accept the idea of a crisis in the sixth century A.D., and the discontinuation of sites has often been reasoned away in favor of promoting the notion of a gradual and uniform growth and expansion of settlement and population (Ambrosiani 1964; Hyenstrand 1974).

Climate change would not have been the sole reason for cultural changes, and humans could always use different strategies of response to meet these challenges. Consequences for historical development would always be complex and diverse. Every historical situation needs to be understood in its specific context, where a multitude of factors are involved. It seems clear, however, that the transitions between some of the main periods used in Swedish archaeology closely coincide with major climatic deviations, suggesting the possibility of a causal link that will be worth further investigations.

References Cited

Adger, W. N., S. Dessai, M. Goulden, M. Hulme, I. Lorenzoni, D. R. Nelson, L. O. Naess, J. Wolf, and A. Wreford 2009 Are There Social Limits to Adaptation to Climate Change? *Climatic Change* 2009(93):335–354.

Ambrosiani, B. 1964 *Fornlämningar och bebyggelse: studier i Attundalands och Södertörns förhistoria*. Stockholms universitet, Uppsala.

Axboe, M. 2004 *Die Goldbrakteaten der Völkerwanderungszeit: Herstellungsprobleme und Chronologie*. Walter de Gruyter. Berlin.

Baudou, E. 1950 Regionala grupper i norden under yngre bronsåldern. *Fornvännen* 1950:19–36.

Baudou, E. 1953 De svenska holkyxorna under bronsåldern. *Fornvännen* 1953:241–261.

Baudou, E. 1960 *Die regionale und chronologische Einteilung der jüngeren Bronzezeit im Nordischen Kreis*. Studies in North-European Archaeology 1. Almqvist and Wiksell, Stockholm.

Bergeron, T., M. Fries, C. A. Moberg, and F. Ström 1956 Fimbulvinter. *Fornvännen* 1956:118.

Büntgen, U., W. Tegel, K. Nicolussi, M. McCormick, D. Frank, V. Trouet, J. O. Kaplan, F. Herzig, K.-U. Heussner, H. Wanner, J. Luterbacher, and J. Esper 2011 2500 Years of European Climate Variability and Human Susceptibility. *Science* 331(6017):578–582.

Carlsson, D. 1979 *Kulturlandskapets utveckling på Gotland: en studie av jordbruks- och bebyggelseförändringar under järnåldern* [*The Development of the Cultural Landscape on Got-*

land: *A Study of Changes in Agriculture and Settlement during the Iron Age*. Visby: Press. Visby.

Charpentier Ljungqvist, F. 2009 *Global nedkylning: klimatet och människan under 10 000 år*. Norstedt, Stockholm.

Crumley, C. 1994 *Historical Ecology: Cultural Knowledge and Changing Landscapes*. School of American Research Press, Albuquerque.

Eriksson, T. 1994 Hus och gravar i Norrtälje. In *Arkeologi i Sverige 3*, pp. 225–40. Fornminnesavdelningen, Riksantikvarieämbetet, Stockholm.

Eriksson, T. 2008 Pottery and Feasting in Central Sweden. In *Breaking the Mould: Challenging the Past through Pottery*, edited by I. Berg, pp. 47–55. BAR International Series 1861. Archaeopress, Oxford.

Eriksson, T. 2009 *Kärl och social gestik: keramik i Mälardalen 1500 B.C.–400 AD*. Aun 41. Riksantikvarieämbetet Arkeologiska Skrifter No 76. Uppsala universitet, Uppsala.

Folke, C. 2006 Resilience: The Emergence of a Perspective for Social-Ecological Systems Analyses. *Global Environmental Change* 16(3):253–267.

De Geer, G., and R. Sernander 1909 *On the Evidences of Late Quaternary Changes of Climate in Scandinavia*. P.A. Norstedt and söner. Stockholm.

Gräslund, B. 1987 *The Birth of Prehistoric Chronology: Dating Methods and Dating Systems in Nineteenth-Century Scandinavian Archaeology. New Studies in Archaeology*. Cambridge University Press, Cambridge.

Gräslund, B. 2007 Fimbulvintern, Ragnarök och klimatkrisen år 536–537 e. Kr. *Saga & Sed, Kungl. Gustav Adolfs akademiens årsbok*, 2007:93–123.

Grudd, H., K. R. Briffa, W. Karlén, T. S. Bartholin, P. D. Jones, and B. Kromer 2002 A 7400-Year Tree-Ring Chronology in Northern Swedish Lapland: Natural Climatic Variability Expressed on Annual to Millennial Timescales. *The Holocene* 12(6):657–665.

Gunnarson, B. E. 2008 Temporal Distribution Pattern of Subfossil Pines in Central Sweden: Perspective on Holocene Humidity Fluctuations. *The Holocene* 18(4):569–577.

Hjärthner-Holdar, E. 2008 Iron Production in Bronze Age Sweden. In *The Introduction of Iron in Eurasia: Papers Presented at the Uppsala Conference on October 4–8, 2001*, edited by S. Forenius, E. Hjärthner-Holdar, and C. Risberg, pp. 9–15. UV GAL, Riksantikvarieämbetet. Uppsala.

Hjärthner-Holdar, E., and C. Risberg 2003 The Introduction of Iron in Sweden and Greece: Theories and Methods. In *Prehistoric and Medieval Direct Iron Smelting in Scandinavia and Europe: Aspects of Technology and Society Proceedings of the Sandbjerg Conference 16th to 20th September 1999*, edited by C. L. Nørbach, pp. 83–86. Aarhus University Press. Aarhus.

Hodder, I. 2008 The "Social" in Archaeological Theory: An Historical and Contemporary Perspective. In *A Companion to Social Archaeology*, edited by L. Meskell and R.W. Preucel, pp. 23–42. Blackwell, Malden.

Hyenstrand, Å. 1974 *Centralbygd-randbygd. Strukturella, ekonomiska och administrativa huvudlinjer i mellansvensk yngre järnålder. [Central Settlement, Outlying Districts, Principal Structural, Economic, and Administrative Features in the Late Iron Age in Middle Sweden]*. Studies in North-European Archaeology, 5. Almqvist and Wiksell, Stockholm.

Kresten, P. 2005 Analysis of LBA Celts from the Collections of the Museum of Nordic Antiquities, Uppsala University. *Activity Report 2000–2001*, Geoarchaeological Laboratory, Department of Archaeological excavations, UV GAL. National Heritage Board. Uppsala.

Kristiansen, K. 1998 *Europe Before History*. New Studies in Archaeology. Cambridge University Press, Cambridge.

Kuz'minych, S. V. 1996 Osteuropäische und Fennoskandische Tüllenbeile des Mälartyps: Ein rätsel der Archäologie. In *Fennoscandia Archaeologica XIII. Suomen arkeologinen seura. Arkeologiska sällskapet i Finland. The archaeological society of Finland*, XIII:3–27.

Lahsen, M. 2010 The Social Status of Climate Change Knowledge: An Editorial Essay. *Wiley Interdisciplinary Reviews: Climate Change* 1(2):162–171.

Lamm, J-P., H. Hydman, and M. Axboe 1999 "Århundradets brakteat": kring fyndet av en unik tionde brakteat från Söderby i Danmarks socken, Uppland. I: *Fornvännen* 94:4:225–243.

Larsen, L. B., B. M. Vinther, K. R. Briffa. T. M. Melvin, H. B. Clausen, Jones, P. D. Jones, M.-L. Siggaard-Andersen C. U. Hammer, M. Eronen, H. Grudd, B. E. Gunnarson, R. M. Hantemirov, M. M. Naurzbaev, and K. Nicolussi 2008 New Ice Core Evidence for a Volcanic Cause of the A.D. 536 Dust Veil. *Geophysical Research Letters* 35:L04708.

Larsson, L., J. Callmer, and B. Stjernquist 1992 *The Archaeology of the Cultural Landscape: Field Work and Research in a South Swedish Rural Region*. Acta archaeologica Lundensia. Series in 4°. 19. Almqvist and Wiksell International, Stockholm.

Larsson, T. B. 1986 *The Bronze Age Metalwork in Southern Sweden: Aspects of Social and Spatial Organization 1800–500 B.C. Archaeology and Environment; 6*. Umeå: Dept. of Archaeology, University of Umeå.

Levy, J. E. 1982 *Social and Religious Organization in Bronze Age Denmark: An Analysis of Ritual Hoard Finds*. British Archaeological Reports. International series; 124. Oxford.

Lindqvist, S. 1920 Den keltiska Hansan eller huvudorsaken till kulturnedgången i Norden vid järnålderns början. *Fornvännen, 1920, häfte 3*, 113–135.

Ljungkvist, J. 2006 *En hiar atti rikR: om elit, struktur och ekonomi kring Uppsala och Mälaren under yngre järnålder*. Aun, 0284-1347; 34. Uppsala: Institutionen för arkeologi och antik historia, Uppsala universitet.

Löwenborg, D. 2010a Using Geographically Weighted Regression to Predict Site Representativity. *Making History Interactive: CAA 2009. Computer Applications and Quantitative Methods in Archaeology. Proceedings of the 34th Conference, Williamsburg, United States*, April 2009. (ed. B Frischer).

Löwenborg, D. 2010b An Analysis of the Location of Burial Grounds in Västmanland, Sweden. *Acta Archaeologica* 81:124.

Löwenborg, D. 2010c *Excavating the Digital Landscape: GIS Analyses of Social Relations in Central Sweden in the 1st Millennium AD*. Aun 42.

McGlade, J. 1995 Archaeology and Ecodynamics of Human-Modified Landscapes. *Antiquity* 69:113–132.

Mauquoy, D., B. van Geel, M. Blaauw, A. Speranza, and J. van der Plicht 2004 Changes in Solar Activity and Holocene Climatic Shifts Derived from ^{14}C Wiggle-Match Dated Peat Deposits. *The Holocene* 14(1):45–52.

Nerman, B. 1957 Yngre bronsåldern—en första svensk vikingatid. *Fornvännen* 1957:257–285.

Näsman, U., and J. Lund 1988 *Folkevandringstiden i Norden: en krisetid mellem ældre og yngre jernalder: rapport fra et bebyggelsearkæologisk forskersymposium i Degerhamn, Öland, d. 2–4 oktober 1985*. Universitetsforl, Aarhus.

Plicht, J. van der 2004 Radiocarbon, the Calibration Curve, and Scythian Chronology. In *Impact of the Environment on Human Migration in Eurasia*, edited by E. M. Scott, A. Y. Alekseev, and G. I. Zaitseva, pp. 45–61. NATO science series (Series IV, Earth and environmental sciences), Vol. 42. Kluwer, Dordrecht.

Reimer, P. J., M. G. L. Baillie, E. Bard, A. Bayliss, J. W. Beck, C. Bertrand, P. G. Blackwell, C. E. Buck, G. Burr, K. B. Cutler, P. E. Damon, R. L. Edwards, R. G. Fairbanks, M. Friedrich, T. P. Guilderson, K. A. Hughen, B. Kromer, F. G. McCormac, S. Manning, C. Bronk Ramsey, R. W. Reimer, S. Remmele, J. R. Southon, M. Stuiver, S. Talamo, F. W. Taylor, J. van der Plicht, and C. E. Weyhenmeyer 2004. IntCal04 terrestrial radiocarbon age calibration, 0–26 cal kyr BP. *Radiocarbon* 46:1029–1058.

Renck, A. M. 2008 Erövrad mark-erövrat landskap. In *Hem till Jarlabanke: jord, makt och evigt liv i östra Mälardalen under järnålder och medeltid*, edited by M. Olaussonpp, pp. 91–111. Historiska Media, Lund.

Sernander, R. 1897 *Våra torfmossar: deras sammansättning och utvecklingshistoria samt deras betydelse för kännedomen om Nordens fornvärld. Studentföreningen Verdandis småskrifter, 64*. Stockholm.

Sernander, R. 1912 Die geologische Entwicklung des Nordens nach der Eiszeit in ihrem Verhältnis zu den archäologischen Perioden. (G. Kossina, Red.) *Mannus*, Der erste Baltische Archäologen-Kongress zu Stockholm 13. bis 17. August 1912, 4. Würzburg.

Speranza, A., B. van Geel, and J. van der Plicht 2003 Evidence for Solar Forcing of Climate Change at ca. 850 Cal B.C. from a Czech Peat Sequence. *Global and Planetary Change* 35(1–2):51–65.

Stanton, T. 2011 *High Temporal Resolution Reconstructions of Holocene Paleomagnetic Directions and Intensity: An Assessment of Geochronology, Feature Reliability, and Environmental Bias.* Lundqua Thesis 63. Department of Quaternary Geology, Lund University, Lund.

Stenberger, M. 1933 *Öland under äldre järnåldern: en bebyggelsehistorisk undersökning*. Kungl. vitterhets historie och antikvitets akademien, Stockholm.

Stenberger, M., and O. Klindt-Jensen (editors) 1955 *Vallhagar: A Migration Period Settlement on Gotland, Sweden*. 2 vols. E. Munksgaard, Copenhagen.

Welinder, S. 1974 *Kulturlandskapet i Mälaromrädet. 4, Sammanfattande del*. Report submitted to the University of Lund, Department of Quaternary Geology. Lund.

Wikborg, J. 2007 De levande och de döda: gravfältens kontinuitet och relation till bebyggelsen. In *Att nå den andra sidan: om begravning och ritual i Uppland*, edited by M. Notelid, pp. 173–200. Arkeologi E4 Uppland Vol. 2. Riksantikvarieämbetet UV GAL, Uppsala.

PART III

Commentary

CHAPTER THIRTEEN

Epilogue to a Prologue

The Changing Climate of the Past, Present, and Future

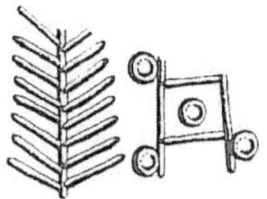

Ezra B. W. Zubrow

PROLOGUE TO AN EPILOGUE

Mark Twain famously stated, "Everybody talks about the weather but nobody does anything about it." But like many things that refer to the weather, there always is controversy. No one is actually sure who first wrote these words. Perhaps it was Mark Twain (Samuel Clemens), or it may have been his collaborator on *The Gilded Age*, Charles Dudley Warner. Or it may have been someone whose name has not come down to us. All of us hear weather talk. "It's been the warmest summer in New York State" or "One doesn't know what to wear anymore . . . we used to be able to tell, when I was a child." Or put in a nutshell in the words of a famous BBC weatherman regarding rain for the morrow. "Massive unpredictability is absolutely certain, maybe."

This chapter will be divided into three parts.

First, there will be an overview of general issues around climate change. Second, there will a discussion of what is known about modern and prehistoric climate change and how long-term climate change studies are relevant to a modern understanding of climate change. Third, there will be a summary discussion of the chapters and how they fit into the broad general scheme of climate change in the twenty-first century.

OVERVIEW OF GENERAL ISSUES

There has been considerable progress in understanding climate change. Partially it is the result of the large number of very good scientists who have been working on climate change. Partially it is the result of new methodologies and new data. Partially it is the result of the development and testing of the world climate model. Underlying all of the above is the increase in public concern and interest.

Some of the discoveries are relevant to both the past and the present. I will focus in this epilogue on the former. However, what is very clear is that we have a far more sophisticated understanding of climate change than we did even a decade ago.

Today, rising temperatures and changing rainfall are affecting human society and have placed climate change in the forefront of human consciousness. One needs only to note the melting of the polar ice cap and the recently announced melting of the Greenland ice sheet. Incidentally, the systematic archaeological examination of the edge of the ice results in many new finds of perishable materials. It has created a bonanza of "ice men" and newly exposed information (Lorenzi 2012; McKie 2010). Similarly, fluctuations in temperature and rainfall have affected past cultures in a variety of ways. Whether these trends are recent or not is a matter of controversy. One knows they happened before, indeed millions of years ago. Although there is some evidence that the recent "anthropogenic" era of climate change started with the Industrial Revolution, there are also proponents of the view it started far earlier. For example, William Ruddiman (2003) and his followers suggest that the beginning of global warming was approximately 8,000 years ago when people began changing landscapes during the Bronze and Iron ages by clearing the forests for agriculture. The opponents generally view Holocene CO_2 trends by looking at the natural terrestrial biomass loss. One of these is Pongratz et al. (2009) whose simulations indicate that land cover clearance is not sufficient in size or scale to explain atmospheric CO_2 increase during the Holocene.

What Do We Know about Climate Change?

In short, the major points are:

1. Climate is changing and appears to be changing at an increasing rate.
2. Climate change is scalar.
3. It is highly variable across time and space.
4. There are global changes with local impacts and local changes with global impacts.
5. Climate change has cascading effects.
6. Human actions sometimes are causal sometimes not.
7. Earth is getting warmer.
8. The ocean is acidifying.
9. The ice sheets and glaciers are retreating.
10. The sea level is rising.
11. Weather patterns are changing, becoming less predictable and more extreme.

Changing climate patterns transformed and continues to change ecosystems on an extraordinary scale and at an extraordinary pace. Although climate change is taking place at a global level, the ecological impacts are local and are place variable. Conversely, local changes, because of complicated interaction patterns, have global impacts.

At a macro scale hundreds of thousands of square kilometers change their vegetative, faunal, and cultural cover. Ice moves forward and retreats. Areas become wetlands. Other areas become deserts. These are neither independent phenomena, nor are their impacts on life independent. As each species responds to the changing environment, its interactions with the physical world and the organisms around it change too. What is true at the macro level may be scaled down to the most micro level where individual members of a species react.

Climate change creates a cascade of impacts throughout the entire natural and cultural ecosystem. Among the impacts are invasion of species into new ranges, intermingling of formerly geographically distinct species, changing of phenology, including reproductive timing, and even extinction. Examples of these types of impacts have been observed in many species, in many regions, and over long periods of time. About 40 percent of wild plants and animals are relocating to stay within their tolerable climate ranges. Zone maps used by gardeners to determine which areas are suitable for certain plants are being redrawn every year. Some organisms—those that cannot move fast enough or those whose ranges are actually shrinking—are being left with no place to go. For example, in the Arctic areas that I study, as the sea ice contracts, the animals that depend on it will literally reach the end of the Earth—an imaginary last Harp seal sitting in regal isolation on the North Pole.

Seasonal behaviors for many species now happen 15–20 days earlier than a half-century ago. Plants flower earlier, migrant birds travel north earlier, and larvae emerge sooner. This need not be a problem if all the species in an ecosystem shifted their seasonal behavior in exactly the same way. But when a species depends upon another for survival and only one changes its timing, these shifts can disrupt important ecological interactions, such as those between predators and their prey.

What Is Known about Prehistoric Climate Change and How Long-Term Climate Change Studies Are Relevant to a Modern Understanding of Climate Change

Most scientists see human actions as the cause of the observed climate changes. However, that is not always the case prehistorically and historically. One knows that climate change happened hundreds of thousands and millions of years ago without human causation.

However, whether human-caused or not, the three major environmental changes that have impacted *Homo sapiens sapiens* and have climate change implications are the Wisconsin/Wurm ice age (110,000–10,000 years ago) and its Holocene termination approximately 12,000–10,000 years ago, the deforestation of the world beginning about 10,000 B.C., and the increasing desertification of the Middle East and other parts of the world.

During the Wurm/Wisconsin, from a climate very similar to that of today, mid-latitude temperatures declined an average of 7° to 10°C. High latitude temperatures were at least 15°C cooler and huge ice sheets covered northern North America, Western Europe, the British Isles, and ice caps covered most mountain ranges. By the Holocene, temperatures were modulated by the ice sheets, the mid-latitudes were drier and tropics wetter than today. The increase in the average global temperature was not a result of human causation. (Sherilyn C. Fritz, Sarah E. Metcalfe et al. 2001). This is not true for the other two.

Desertification, frequently a byproduct of overgrazing, also begins about 10,000 years ago and diminishes biological diversity. This in turn cascades into changing soil conservation, changing surface water regulation, and changing local climate as well as changing the amount of carbon released into the atmosphere, carbon impacting global climate change (Whitford and Wade 2002:275–304).

The human deforestation of the world begins with the beginning of agriculture as people use fire as an easy method to clear the land for crops. Beginning around 9000 B.C. it increases so rapidly that, by 1000 B.C., in many areas 50 to 70 percent of land is deforested. Slash and burn agriculture was not only efficient in producing new land for agriculture and returning minerals to the soil; it was, for many people, a type of ownership strategy and an investment (Angelsen 1999; Kaplan et al. 2009; Krumhardt et al. 2009).

In this broad, global, and long-term temporal context, the 9.3 ka, 8.2 ka, 6.2 ka, and 536 AD events are small perturbations.

In short, one may say prehistoric cultures, through activities such as agriculture, water management, transportation, fishing, biological conservation, and many others, impact ecosystems that in turn influence the ways and the extent to which climate change will alter the natural world. Furthermore, humans also change their behavior in ways that can not only reduce the rate of past and future climate change but have helped wild and domestic species adapt to climate change that cannot be avoided.

Before turning to this book's contributions, I would like to point out a couple of other contributions that long-term climate change studies are making. As part of a study showing how settlements had changed with climate change at approximately the same time as the 8.2 event, but in areas far away from the Middle Eastern and Eurasian focus of the book, the International Circumpolar Archaeological Project (ICAP) looked at similarities of climate and the adaptation by prehistoric societies around the circumpolar area. ICCAP, NSF 038633251, was an international team of scientists, students, and laboratories from the United States, Finland, Canada, Russia, England, and Germany that I directed. Its purpose was to finish creating climate and archaeological datasets that would be comparable from the Russian Far East, northern Finland, and northern Canada. The objectives were, first, to determine the similarities in environmental change experienced by prehistoric inhabitants of:

- the Ust-Kamchatsk area of the Kamchatka peninsula, on the shores of Nerpichye and Kultuchnoye lakes, which are part of the Kamchatka River estuarine and paleo-estuarine system (KRE);

- Old Factory Lake near Wemidji, Quebec, on the James Bay and part of the Wemidji river estuarine and paleo-estuarine system (OFL);

- and the Yli-ii area of northern Finland near Oulu and part of the "Ii" river and estuarine and paleo-estuarine system (YLI).

The second objective was to determine the nature of the resultant social change.

ICAP identifies interaction between humans and the environment between 6,000 and 4,000 years ago and seeks information regarding how prehistoric groups created resilient adaptive systems in response to environmental challenges while developing historically unique sets of life-ways. Data was gathered on geological, climatic, and anthropogenic determinants and consequences of environmental change, and human responses.

Figure 13.1 shows the three areas of scientific investigation. Climate was reconstructed by a variety of proxies, including using the World Climate Model and pollen and bog data. There are many conclusions that one may make from these studies. I just want to make two superficial points without going into great detail about the climate proxies and settlement data. Figure 13.2 shows simulated climate data for these three areas. The data are the January Minimum Temperatures for 7000 to 3900 B.P. Other proxies show the same patterns.

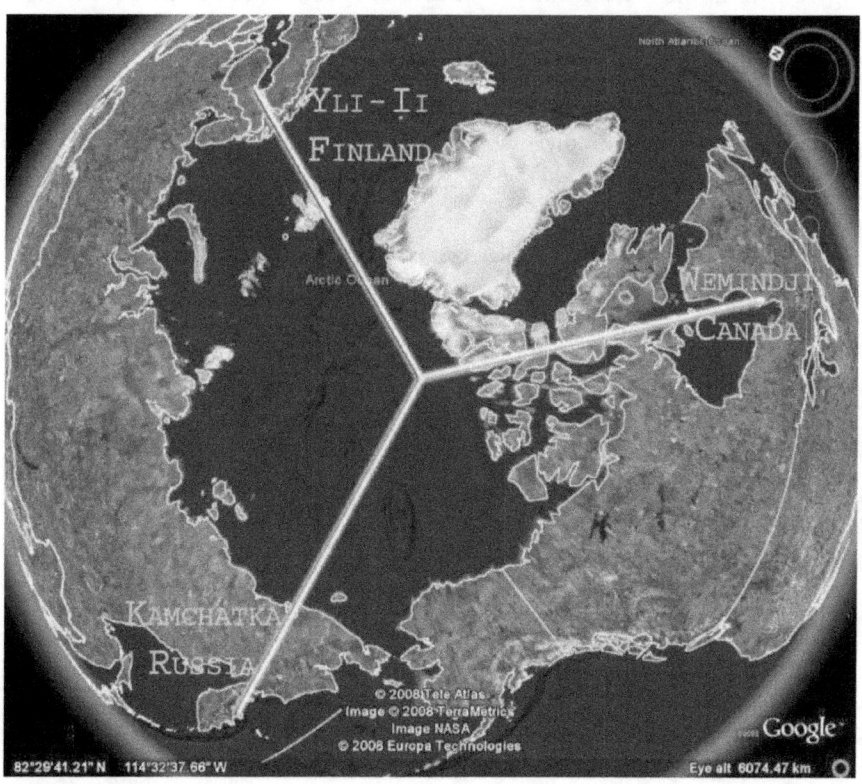

FIGURE 13.1. The general locations of ICAP field sites (Kamchatka Russia, Old Factory Lake Canada, Yli-ii Finland).

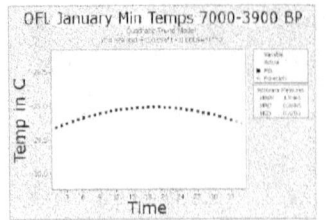 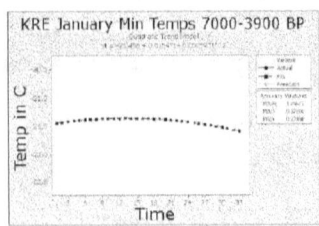 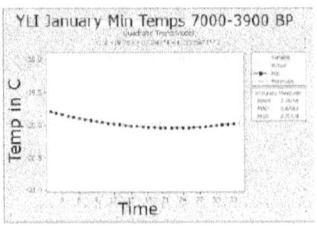

FIGURE 13.2. Simulated Temperatures in Centigrade from 7000 to 3900 BP from Finland, Canada, and Kamchatka at the locations of ICAP field sites.

The first point to note is that if there were correlations between settlement pattern (expansion, contraction, density, nearest neighbors, etc.) and climate measured through the variety of proxies, then one would expect the same correlations with the "cold, warm, cold" sequences in Old Factory Lake (Canada) and Ust-Kamchask, Kamchatka. In contrast, there should be a different correlation order corresponding to "warm, cold, warm" in Yli-ii Finland. This does not occur. Climate is not consistently impacting human adaptation in different areas of circumpolar north the same way.

One may use long-term climate data derived through archaeological research to predict the future.

Figure 13.3 shows the distribution of *Betula, Pinus,* and *Picea* over time. Figure 13.4 shows a more derived proxy—the trend analyses on the above pollen data. It is *Betula* (birch), *Pinus* (pine), *Picea* (spruce) from northern Finland surrounding the YLI simulated data above. The data are from sites in Finland between 64 and 67 degrees north.

Birch is of course a short-lived strongly pioneer species widespread in the northern areas. They are well known for open area colonizing following fire and are early members

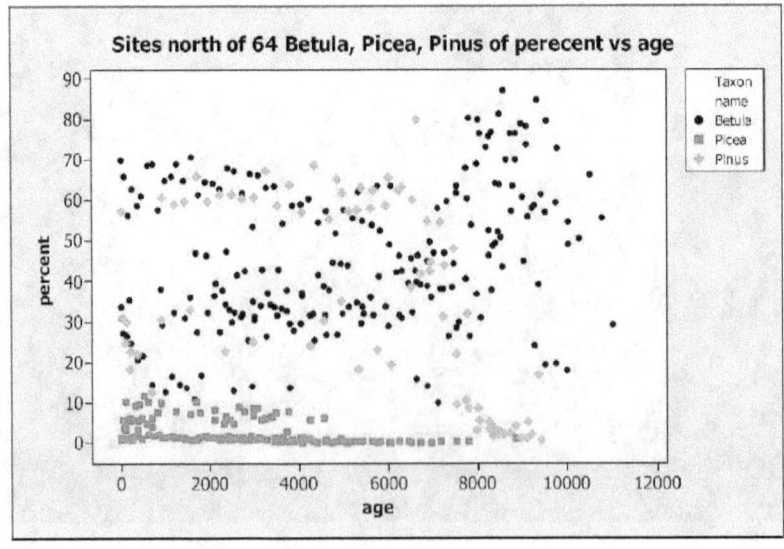

FIGURE 13.3. The distribution of Betula, Pincea, and Picea over time from Finnish sites north of 64 degrees.

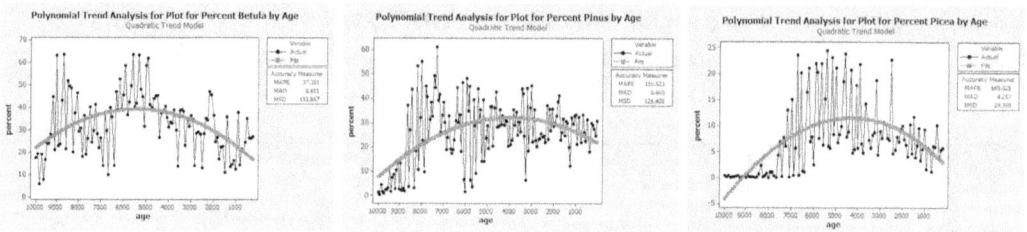

FIGURE 13.4. Trend analyses of Betula (birch), Pincea (pine), and Picea (spruce) pollen from northern Finland sites between 64 degrees to 67 degrees North.

of primary successions. Pine is an evergreen, coniferous, resinous tree that grows well in acidic soils. Their distribution goes as far as 67 degrees north. Spruce is a conifer found in northern temperate and taiga areas. It is a climax species for many successions. *Betula* and *Pinus* arrive first and both increase together. *Pinus*'s growth is uninterrupted until 8000 while Betula has a small regression. Once *Picea* arrives in significant numbers around 8000, then it with *Betula* grow until both reach their climax about 5000. By this time *Pinus* has stabilized and both *Betula* and *Picea* decrease as we approach modern environments.

Polynomial trend analyses were fitted to each taxon; Figure 13.4 above. It is a form of regression in which the relationships between the independent variable and the dependent variable are modeled as an nth order polynomial and is used to get the equation for nonlinear relationships.

The polynomial trend analysis procedure calculates the best equation for the data and draws a graph containing the observations, predicted values (the fitted trend equation), and forecasts, versus time. The predicted or forecast value at time t is obtained by simple calculations inputting the desired t and using the fitted equation to calculate the best y.

Table 13.1 shows the predicted values for the percentage of pollen 50 years, 100 years, and 150 years in the future. There are several important inferences to be drawn from this table. First, the ratio of data to prediction is more than 100:1. In other words, each predicted data point is based on 117, 159, and 118 known data points. One may remove the noise in the data by running various filters and running averages. For the noise-removed data, the number of points is about one-fifth less—97, 139, and 98—but still a very high number from which to make predictions. Second, the predictions in all cases are below the mean (and, incidentally, the median). Third, if one compares the predictions to the means and the standard deviations, one gets an impression of how far away from the average environment of the last 10,000 years these predictions are. As a rough "rule of thumb," if the prediction is within one standard deviation, it is considered quite close to the mean. If the prediction is between one standard deviation and two standard deviations then it is approximately part of the one-third of the values that are the farthest distance from the mean. If the value is beyond the second deviation then it is part of the 5 percent of values that are the farthest from the mean.[1]

For *Betula* with noise, all of the predictions (50, 100, and 150 years) are significant between one and two standard deviations. *Betula* without noise are well beyond

TABLE 13.1. PREDICTED VALUES FOR THE PERCENTAGE OF POLLEN BY TAXA WITH AND WITHOUT NOISE FOR 50 YEARS, 100 YEARS, AND 150 YEARS IN THE FUTURE

Expected Percentages of Pollen by Taxa with and without Noise	50 yrs	100 yrs	150 yrs	N	Mean	Std Dev
Betula	16.45	16.02	15.58	117	32.58	13.7
Betula without noise	13.18	12.71	12.22	97	34.375	6.44
Pinus	21.6	21.33	21.06	159	25.82	13.03
Pinus without noise	21.44	21.21	20.98	139	27.317	5.825
Picea	2.77	2.56	2.34	118	7.258	6.837
Picea without noise	7.27	6.96	6.65	98	8.142	4.071

two standard deviations. For *Pinus* with noise the predictions are within one standard deviation. *Pinus* without noise for all predictions are between one and two standard deviations. For *Picea* with noise and without noise are all within one standard deviation.

Clearly, even if one is being conservative, then one needs to be concerned about the future. In particular one needs to be aware of the changes that are occurring to *Betula*. The predicted decrease is very significant and well beyond what would be expected according to the normal variation of the changing environment. For the others, the decrease is real but within expected variation for the last ten thousand years. So when one begins to look at climate change in the far north, it is clear that at least some of the environmental impacts are within the tolerance limits (pine and spruce) but it is well beyond these limits for birch and probably other early entrants to succession. Northern societies in Finland can expect "unexpected environmental changes" in the "fast growing disturbed environments," while in the more stable pine and spruce more consistent future change may occur.

We know that we live in a global world today. Societies need to react and do react to changes around the globe more and more rapidly. However, prehistoric societies frequently needed to react to more global events than one would have expected. Changes in the climate in one area of the world impact changes in the climates of others.

In the Arctic, the changes have been four degrees for every degree in the temperate world. Since it is a global system, there is a form of telelinkage between different areas of the world based upon the connections through ocean currents (Gulf Stream, Labrador Current, North and South Equatorial and counter currents) and the atmospheric circulation system (trade winds, westerlies, and easterlies). This telelinkage also operated for past periods.

Indeed, another two studies that we are doing show this—one in Sri Lanka and one in the Yucatan. What we have shown in the Yucatan is that NAO (the North Atlantic Oscillation) not only impacts the weather in the Yucatan but that the classic sites such as Uxmal, or the site on which we are working, Xcoch, is not only impacted but adapts to the change (see figure 13.5, pg. 290, based on Knudson et al. 2011, Fritz et al. 2001, and Zubrow 2012).

We are in the process of finding evidence that shows that (1) as the amount of Arctic-induced precipitation decreases, the size, number, and complexity of cultural hydrological features increase, (2) changes in artifact patterns and distributions should reflect the inverse correlation between decreased Arctic-induced precipitation and increased frequency of water storage vessels, and (3) as the amount of Arctic-induced precipitation decreases, the location of domestic units will expand from site centers to the site peripheries to increase the construction of water storage facilities.

Similarly, the second study is showing how a change in the Arctic changed the monsoon calendar, which in turn impacted settlements from Southern Sri Lanka along the Bay of Bengal coast.

A Summary Discussion of the Chapters and How They Fit into the Broad General Scheme of Climate Change in the Twenty-First century

The chapters cover a broad Old World spread including Eurasia, Anatolia, the Middle East, Central Europe, and Scandinavia. In particular, Turkey, Syria, Iraq, Greece, Cyprus, Serbia, Rumania, Poland, and Sweden are emphasized and if you include this epilogue, Finland and Russia as well. Some sites are Palmyra, Tell Sabi Abyad, Catal Huyuk, Khirokitia, Franchthi, as well as many others. The cultural range is from the Upper Paleolithic, through the Mesolithic, Neolithic, Eneolithic, and Chacolithic, to the Bronze and Iron ages. Particular cultures that are highlighted are the Mousterian, Natufian, Pre-pottery Neolithic B, Halafian, Linearbandkeramik, Funnel, and Trypolie, cultures as well as others.

Most of the chapters focus on the 8.2 ka event. Eight focus on it directly and two focus on it as part of longer temporal issues. Several chapters consider other events including the 9.3 ka, 7.5 ka, 6.2 ka, and A.D. 536 events. However some go as far back as 35,000 years.

The methodologies of the chapters include archaeological surveys, site excavations, comparative analysis, demographic analysis, settlement dynamics, correlations of sites, dates, and climate proxies, and a range of laboratory analyses including paleobotanical and, phytoliths,

Did the 8.2 ka event cause changes in culture, cultural adaptation, or impact local sites and regional cultures? The answer varies depending upon the author, the area, and the specific analysis. Some believe it did. Others believe it did, but was not necessarily mono-causal or necessarily direct. Cremaschi and Zerboni, Mottram, Biehl, and Rosenstock, Ryan and Rosen, Bonsall et al., and Gronenborn are positive of the effects of the 8.2 event. Pelisiak, Löwenborg, and Eriksson also are positive, but point out that climate change is just one of the multicausal factors. Niewenhuyse et al. suggest that the archaeological record is too poor for meaningful understandings of the causal relationships. Perlès sees no unambiguous relationship, and Düring argues that the found synchronicity is not proof of causation.

The following table, which categorizes each chapter by title, author, issue, time, area, culture, and methodology, provides a more detailed summary (Table 13.2). One of the most striking patterns in these chapters is that the 8.2 ka event is not fixed in time and/or space. It does not occur around the Mediterranean at exactly 8,200 years ago. It moves in time and in space around the Mediterranean and is not synchronic throughout

TABLE 13.2. EACH CHAPTER OF THIS VOLUME CLASSIFIED BY TITLE, AUTHOR, ISSUE, TIME, AREA, CULTURE AND METHODOLOGY

Title	Author	Issue	Time	Area	Culture	Methodology
Oasis in Palymyra…	Cremaschi and Zerboni	Since 8.2 event, the contraction of oasis resulting from lack of rainfall is synchronic with sedentism and reduction of site size	Pleistocene/ Holocene transition and 8.2 event	Central Syria (Sites at Oasis of Palymyra and surrounding Sabhkat al Mouh	Mousterian, Natufian, PPNB, Late Chacolithic	Multidisciplinary, Archaeological survey, excavation, geomorphological and laboratory analysis
When the Going Gets Tough	Mottram	The emergence of the Halafian painted pottery were climate driven symbolic representations of risk management social processes	8.2 event, 6300–5950 BC	Northern Mesopotamia (Iraq, Syria, Turkey)	Halafian	Comparative Analysis
8.2 Event in Upper Mesopotamia	Niuwenhuyse, Akkermans, Plicht, Russell, Kaneda	Climate and culture debate has been based upon a poor archaeological record	8.2 event 7000–5500 BC focus 6200–6000BC	Northern Syria (Tell Sabi Abyad)	Early Pottery Neolithic, Pre Halafian, Transitional, Early Halafian	Settlement dynamics, demographic chance
Catal Huyuk East to West	Biehl and Rosenstock	Climate change creates climate refugees moving East to West (the Mounds at Catal Huyuk are a microcosm of the process)	8.2 event	Turkey (Catal Huyuk)	Late Neolithic	Excavation
Managing Risk through Diversification	Ryan and Rosen	Some societies are more resilient than others and thus continuous occupation replaces a previous 8.2 abandonment interpretation	8.2 event 7000–5000 BC	Turkey (Catal Huyuk)	Neolithic Chalcolithic	Phytoliths

Topic	Author	Finding	Date/Period	Region	Archaeological Period	Method
8.2 Event and the Neolithic Expansion	Düring	Synchronicity is not proof of causation when other explanations are available	8.2 event	Western Anatolia	Neolithic	Correlation of dates and site distributions
Singing in the Rain	Daunne-LeBrun and LeBrun	Shift of the built area and erratic but significant flooding	End of 7th millennium BC	Cyprus (Khirokitia)	Aceramic Neolithic	Stratigraphy and Excavation
Early Holocene Climatic Fluctuations and Human Responses	Perles	No unambiguous relationship between climatic and socio-economic change	Late Pleistocene/ Holocene, Mid Holocene, 8.2 Event	Greece (Franchthi)	Mesolithic, Early and Middle Neolithic	Fine grained and broad grained analysis of disruptions in correlations between climatic events and settlement patterns, site use, and subsistence
Rapid Climate Change and Radiocarbon Discontinuities	Bonsall, Macklin, Boroneant, Pickard, Bartosiewicz, Cook, and Higham	C14 discontinuities at 9.3, 8.2, and 7.5 k years correspond to Rapid Climate Change Events in Iceland Core	15000–5000 BP	Serbia and Romania (Iron Gates)	Mesolithic and Early Neolithic	Discontinuities in cumulative calibrated data probabilities of radiocarbon dates correlated to RCC events
Climate Fluctuations, Human Migrations and Spread of Farming	Gronenborn	Population shifts link to climate stress caused by increase in anomalies	10000–4000 (9.3, 8.2, 6.2 events)	Western Eurasia	Mesolithic and Neolithic (LBK)	Correlations among settlement patterns and changing climate
Economic and Social Changes	Pelisiak	Climate change is one factor creating a "mobile," "pastoral" way of life	3200–2500 BC	South East Poland	Late Neolithic Eneolithic (Funnel Beaker, Baden, Trypolie)	Paleobotanical analysis
Climate and Definition of Archaeological Periods in Sweden	Lowenborg and Eriksson	Climate deterioration is followed by subsequent cultural change but is not the sole cause	800–400 BC and 536 AD Event	Sweden	Bronze Age, Early to Late Iron Age	Calibrated and calibrated date curves correlated to humid and arid periods and hoards

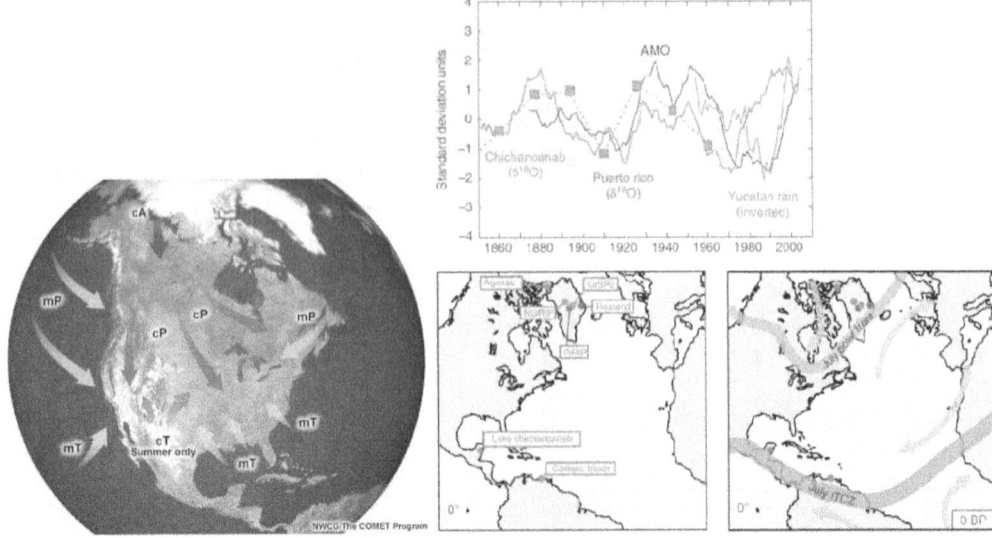

FIGURE 13.5. General North American Air Masses Patterns (mPMaritime Polar air masses, cPContinental Polar air masses). Linking Holocene climate records to the AMO index (top right). The AMO index (black) and the instrumental precipitation record from the Yucatan peninsula (green) is shown together with the coral based δ18O record from Puerto Rico14 (red) and the δ18O record from lake Chichancanab29 (dashed line). (Bottom left) Location of the climate proxy records (dots) and schematic overview of the major atmospheric systems (bottom right). Modern atmospheric systems of the North Atlantic region. This scenario represents average conditions for the last 3,000 years BP, where an overall 'neoglacial' regime with more frequent meridional atmospheric circulation patterns and an ITCZ located close to the Cariaco site was prevalent during Northern Hemisphere summer. The arrows indicate the dominant wind directions.

the area. In some areas it appears earlier and in others later. Thus, in two contingent areas at exactly the same time, one might see a positive correlation of culture change to the 8.2 event in one area but see a negative correlation in another. In fact, the 8.2 event might not be occurring in the second area at all at this time. This does not preclude finding the universal correlation of the 8.2 event to particular cultural changes but it makes it far more difficult, for one has to readjust, recalibrate, and lag the correlations.

What causes this conundrum? One cause might be what is called the Mediterranean seesaw effect. In 2004, in a short report entitled "Abrupt Temperature Changes in the Western Mediterranean over the Past 250,000 Years," Martrat et al. (2004) pointed out that abrupt changes were more common during warming periods than cooling periods. In addition, numerous studies have shown what is sometimes called the Mediterranean oscillation or, more familiarly, the Mediterranean seesaw effect (Moreno et al. 2012). The North Atlantic Oscillation (NAO) has strongly influenced inter-annual precipitation variations in the western Mediterranean, while some eastern parts of the basin have shown an anti-phase relationship in precipitation and atmospheric pressure (Roberts et al. 2012). It not only applies to the large temporal scales of Martrat but the shorter

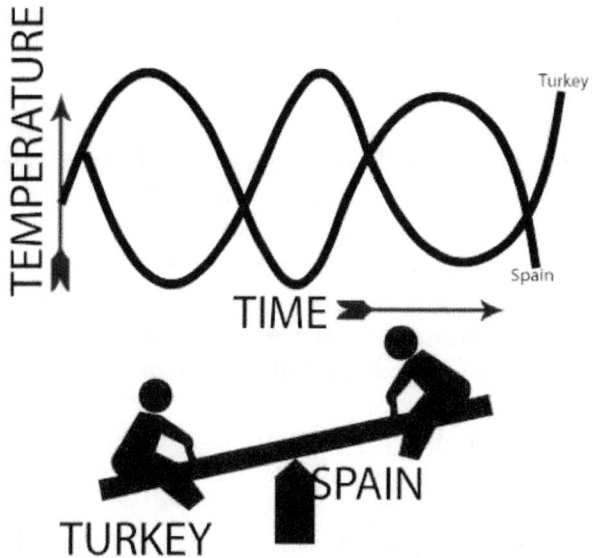

FIGURE 13.6. The Mediterranean seesaw.

time scales of the 9.3 ka, 8.2 ka, 6.2 ka, A.D. 536, Little Ice Age (LIA), and Medieval Climate Anomaly (MCA) events. It is supported by high-resolution paleolimnological, geochemical, sedimentological, isotopic, and paleoecological analyses and shows the oscillation at decadal and centennial levels of resolution.

Figure 13.6 illustrates this Mediterranean seesaw. Climate change has not been and is not simple. What we know is that previous societies similar to modern societies needed to consider greater degrees of uncertainty as the seesaw goes up and down nonsynchronically and noncontingently in the Old World. So perhaps it is appropriate to provide a level of childhood support with a modern nursery rhyme for our children and grandchildren.

> See-saw, Marjorie Daw,
> Turkey descending
> Spain ascending
> Environment needed mending.
> Seesaw, Marjorie Daw
> Temperature ascending
> Once ice now a thaw
> Environment needs mending.

Note

1. This is a rough "rule of thumb" because it applies exactly as stated only to normal distributions. The data above are not exactly normal. However, the detailed calculations not presented here show similar results for nonnormal distributions.

References Cited

Angelsen, A. 1999 Agricultural Expansion and Deforestation: Modelling the Impact of Population, Market Forces, and Property Rights. *Journal of Development Economics* 58(1):185–218.

Fritz, S. C., S. E. Metcalfe, and W. Dean 2001 Holocene Climate Patterns in the Americas Inferred from Paleolimnological Records. In *Interhemispheric Climate Linkages*, edited by V. Markgraf, pp. 241–263. Academic Press, New York.

Kaplan, J. O., K. M. Krumhardt, and N. Zummermann 2009 The Prehistoric and Preindustrial Deforestation of Europe. *Quaternary Science Reviews* 28:3016–3034.

Knudsen, M. F., M.-S. Seidenkrantz, B. H. Jacobsen, and A. Kuijpers 2011 Tracking the Atlantic Multidecadal Oscillation through the Last 8,000 Years. *Nature Communications* 2(178), doi:10.1038/ncomms1186.

Lorenzi, R. 2012 Ancient Weapons Emerge from Melting Arctic Ice. Electronic document, http://news.discovery.com/tech/gear-and-gadgets/arctic-weapons-ice-melt.htm, accessed January 29, 2015.

Martrat, B., J. O. Grimalt, C. Lopez-Martinez, I. Cacho, F. J. Sierro, J. A. Flores, R. Zahn, M. Canals, J. H. Curtis, and D. A. Hodell 2004 Abrupt Temperature Changes in the Western Mediterranean over the Past 250,000 Years. *Science* 306(5702):1762–1765.

McKie, R. 2010 How Global Warming Is Aiding—and Frustrating—Archaeologists. Electronic document, http://www.guardian.co.uk/science/2010/sep/26/global-warming-ancient-artefacts, accessed January 29, 2015.

Moreno, A., A. Pérez, J. Frigola, V. Nieto-Moreno, M. Rodrigo-Gámiz, B. Martrat, P. González-Sampériza, M. Morellón, C. Martín-Puertas, J. P. Corella, Á. Belmonte, C. Sancho, I. Cacho, G. Herrera, M. Canals, J. O. Grumalt, F. Jiménez-Espejo, F. Martínez-Ruiz, T. Vegas-Vilarrúbia, and B. L. Valero-Garcés 2012 The Medieval Climate Anomaly in the Iberian Peninsula Reconstructed from Marine and Lake Records. *Quaternary Science Reviews* 43:16–32.

Pongratz, J., C. H. Reick, T. Raddatz, and M. Claussen 2009 Effects of Anthropogenic Land Cover Change on the Carbon Cycle of the Last Millennium. *Global Biogeochemical Cycles* 23, GB4001, doi:10.1029/2009GB003488.

Roberts, N., A. Moreno, B. L. Valero-Garcés, J. P. Corella, M. Jones, S. Allcock, J. Woodbridge, M. Morellón, J. Luterbacher, E. Xoplaki, and M. Türkeş 2012 Palaeolimnological Evidence for an East-West Climate See-Saw in the Mediterranean since AD 900. *Global and Planetary Change* 84–85:23–34.

Ruddiman, W. F. 2003 The Anthropogenic Greenhouse Era Began Thousands of Years Ago. *Climatic Change* 61(3):261–293.

Whitford, W., and E. L. Wade 2002 *Ecology of Desert Systems*. Academic Press, London.

Zubrow, E. B. W. 2012 Under the Global Weather: Teleconnections of Arctic Climate Change and Human Adaptation in the Puuc Region of Yucatan, Mexico. NSF Proposal number 1226819.

Contributors

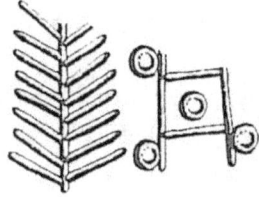

Peter Akkermans, Professor of Near Eastern Archaeology, Leiden University

László Bartosiewicz, Professor, Osteoarchaeological Research Laboratory, Stockholm University

Peter F. Biehl, Department Chair and Professor, Department of Anthropology, State University of New York at Buffalo

Clive Bonsall, Professor of Early Prehistory, School of History, Classics and Archaeology, University of Edinburgh

Adina Boroneanţ, Senior Researcher, Romanian Academy 'Vasile Pârvan' Institute of Archaeology, Bucharest

Gordon Cook, Professor of Environmental Geochemistry, Scottish Universities Environmental Research Centre Glasgow

Mauro Cremaschi, Professor, Dipartimento di Scienze della Terra "A. Desio," Università degli Studi di Milano

Odile Daune-Le Brun, CNRS Senior Researcher, PMO, Archéologies et Sciences de l'Antiquité (ARSCAN), Maison Archéologie et Ethnologie—René-Ginouvès, Nanterre

Bleda S. Düring, Associate Professor, Faculty of Archaeology, Leiden University

Thomas Eriksson, Researcher, The Laboratory for Ceramic Research, Department of Geology, Lund University

Ingmar Franz, Institut für Ur- und Frühgeschichte, Christian-Albrechts-Universität zu Kiel

Anna Fryer (nee Russell), Researcher, Faculty of Archaeology, Leiden University

Detlef Gronenborn, Senior Research Curator/Adjunct Professor, Römisch-Germanisches Zentralmuseum—Leibniz Research Institute for Archaeology / Johannes-Gutenberg University, Mainz, Germany

Thomas Higham, Professor and Deputy Director, Radiocarbon Accelerator Unit, Research Laboratory for Archaeology and the History of Art, University of Oxford
Ceren Kabukcu, Researcher, Department of Archaeology, Classics and Egyptology, University of Liverpool
Akemi Kaneda, Researcher, Archaeology Department, Leiden University
Alain Le Brun, CNRS Senior Researcher, PMO, Archéologies et Sciences de l'Antiquité (ARSCAN), Maison Archéologie et Ethnologie—René-Ginouvès, Nanterre
Daniel Löwenborg, Researcher, Department of Archaeology and Ancient History, Uppsala University
Mark Macklin, Professor, Head of School and Chair of Physical Geography, Director of the Lincoln Centre for Water and Planetary Health, University of Lincoln, Brayford Pool, Lincoln
Mandy Mottram, Researcher, School of Archaeology and Anthropology, Research School of Humanities and the Arts, The Australian National University
Olivier P. Nieuwenhuyse, Researcher, Faculty of Archaeology, Leiden University
David Orton, Lecturer, Department of Archaeology, University of York
Andrzej Pelisiak, Professor, Instytut Archeologii, Uniwersytet Rzeszowski
Catherine Perlès, Professeur émérite Université Paris Ouest Nanterre la Défense
Catriona Pickard, Lecturer in Archaeological Science, School of History, Classics and Archaeology, University of Edinburgh
Johannes van der Plicht, Professor, Faculty of Mathematics and Natural Sciences, University of Groningen
Jana Rogasch, Archaeology Department, Flinders University
Arlene Rosen, Professor, Department of Anthropology, University of Texas at Austin
Eva Rosenstock, Assistant Professor, Institut für Prähistorische Archäologie, Freie Universität Berlin
Philippa Ryan, Researcher, Department of Conservation and Scientific Research, The British Museum
Elizabeth Stroud, Institute of Archaeology, University of Oxford
Patrick T. Willett, Department of Anthropology, State University of New York at Buffalo
Andrea Zerboni, Researcher, Dipartimento di Scienze della Terra "A. Desio," Università degli Studi di Milano
Ezra Zubrow, Professor, Department of Anthropology, State University of New York at Buffalo

Index

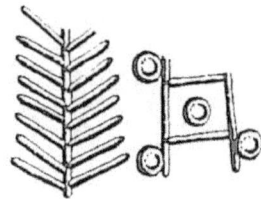

536 (event), 268, 271,
8.2 event (8.2 B.P. event, 8.2ka event, mega 8.2 event), 1, 6–8, 29, 37, 40, 41, 54, 67–81, 95, 118, 139, 142, 143, 145, 146, 169–73, 182–84, 204, 287, 290

Aceramic Neolithic, 153
adaptation, 2, 4, 6–8
adaptive cycles, *217,* 218
Anatolia, 95
anthracology, 104, 108
anthropogenic, 5–6
Arctic, 281, 286, 287

Bademağacı, 140, 142–44
Barcın Höyük, 141
barley, 122
Big Men, 52, 53
Bronze Age, 266, 270, 271
burials, 178, 179
buttress, 100, 109

Carpathians, 237, 238, 240, 246–48, 250, 251
Catalhoyuk, 95, 97, 100, 119
ceremonies. *See* feasting
Chalcolithic, 100

climate event, 1, 4, 6–8, 281
climate fluctuations, 237, 258, 265, 271
collapse, 6–8
constructions, 164
cooking pottery, 46, 47
coping strategies, 37, 41, 58
craft activities, 164
culture, 1–4, 6–8
culture change, 1–3, 6–8, 282, 287, 290
Cyprus, 153, 166

Danube catchment, 195, 202, 203–06
desert-kites, 31
determinism, 5, 7
diversification, 117
    of economic, subsistence activities 42

Early Neolithic, 211, 224
economy, 237–51
enclosure wall, 154, 159, 162, 164
environment, 161, 162, 164
expansion of farming, 135, 136, 142

farmers, farming economy, 170, 171, 180, 185, 186
faunal resources, 161

feasting, 40, 44–48, 58, 179. *See also* cooking pottery
Fikirtepe, 144
Fimbulwinter, 258, 267
fishing, 177, 179
floods, 195, 196, 202–06
funerary practices, 164, 166

Geometric Kebaran, 20
gift exchange (exchange, gifts), 45, 46, 53
global warming, 1–2
gold bracteates, 264
granaries *49*, 50, *51*, 52

Halaf (tradition, sites, pottery), 37–58
heirlooms, 53, 54
Holocene, 1, 5, 7, 104, 281, 282
humidity. *See* precipitations
hunter-gatherers, 170, 171, 174, 180, 187

International Circumpolar Archaeological Project (ICAP), 282, 283
Iron Age, 268, 270, 271
Iron Gates of Danube, 195–205

Jerf el-Ajla cave, 27

Khirokitia Culture, 153, 164

Last Glacial Maximum, 18
Late Neolithic, (period, sites, societies, pottery) 37–58, 239, 246, 248, *249*, 250, 251
life conditions, 166

Mediterranean, 290, 291
Mesolithic, 195–200, 202, 204, 206
Middle Paleolithic, 15
mobility, 43, 44, 54, 55, 57
model of culture, 137

Natufian, 21
Neolithic, 97, 117, 195–200, 202, 206
Neolithization, 107

oasis, 14

Old Norse literature, 268
ornaments, 177–79

painted pottery, 106, 108
paleoclimate, 103
paleoclimate archives, 195, 199, 203–06
Palmyra, 13
phragmites, 125
phytolith, 120
pollen, 104, 106
polygyny (polygynous societies), 57
polynomial trend analysis, *285*
population increase, 223–25
pottery, 266, 267, 268
PPNB-collapse, 136
precipitations, 171, 182, 189
Pre-Pottery Neolithic, 21

radiocarbon discontinuities, 195, 198, 202, 206
ranking, 52, 53
RCC Event, 195, 202, 203, 205
resilience, 108
  to climatic change, 169, 171, 174
resource sharing, 45
risk, 37
  management, 40, 122
  minimization/reduction, 41, 58
river, 155–57

Sabkhat al Mouh, 14
sciences and humanities, 136
seasonality, 119
secondary products, 107
seesaw 290 *291*
settlement dispersal, 44. *See also* mobility
settlement patterns, 170, 183, 185, 187, 188
social changes, 237–51
Southeastern Poland, 237–51
stability, 3
storage, 117. *See also* granaries
  social, 45
  food/wealth/buildings/jars 43, 46–48, *49,* 50, 52, 55, 57, 58
subsistence, 103, 107
  subsistence economy/subsistence activities, 170, 177, 185, 187

summed radiocarbon probability distributions, 195–98, 200, 201, 206
survival, 3
Sweden, 265, 266, 267
synchronicity, 1–7
system, 3, 7

Tell Page, 25
temperate Europe, 212, 219, 225
temperatures, 171, 179–82, 185, 189
Tenaghi Philippon, 138

transformation, 1

Ulucak, 140–45

vegetation, 161, 169, 171, 173, 174, 181, 182
volcanic eruption, 268

Wadi Aid, 17
wetland plants, 125
White Ware, 26
Wisconsin/Wurm ice age, 281, 282

www.ingramcontent.com/pod-product-compliance
Lightning Source LLC
LaVergne TN
LVHW060411280426
837443LV00039B/680